New Tools or Trends for Large-Scale Mapping and 3D Modelling

New Tools or Trends for Large-Scale Mapping and 3D Modelling

Editors

Tarig Ali
Jorge Delgado García
Fayez Tarsha Kurdi

MDPI

Basel • Beijing • Wuhan • Barcelona • Belgrade • Novi Sad • Cluj • Manchester

Editors

Tarig Ali
American University of Sharjah
Sharjah
United Arab Emirates

Jorge Delgado García
University of Jaén
Jaén
Spain

Fayez Tarsha Kurdi
Griffith University
Nathan
Australia

Editorial Office
MDPI
St. Alban-Anlage 66
4052 Basel, Switzerland

This is a reprint of articles from the Special Issue published online in the open access journal *Remote Sensing* (ISSN 2072-4292) (available at: https://www.mdpi.com/journal/remotesensing/special_issues/New_Tools_Trends_Mapping_and_3D_Modelling).

For citation purposes, cite each article independently as indicated on the article page online and as indicated below:

Lastname, A.A.; Lastname, B.B. Article Title. *Journal Name* **Year**, *Volume Number*, Page Range.

ISBN 978-3-7258-0651-5 (Hbk)
ISBN 978-3-7258-0652-2 (PDF)
doi.org/10.3390/books978-3-7258-0652-2

Contents

About the Editors

Tarig Ali

Dr. Tarig A Ali is a Professor of Civil Engineering at the American University of Sharjah, UAE. He holds a BSc degree (honours) in Civil Engineering and an MS and Ph.D. in Geoinformation and Geodetic Engineering from the Department of Civil, Environmental, and Geodetic Engineering at the Ohio State University, USA, which he received in 2003. He was an Assistant Professor at the East Tennessee State University and the University of Central Florida, USA, between 2003–2009 before joining the American University of Sharjah, UAE.

Dr. Ali's research interests are in geo-information science (GIS) and digital mapping, geospatial informatics, GIS applications in smart cities, and geospatial artificial intelligence (GeoAI). Dr. Ali is the recipient of many awards including the 2023 Earle J. Fennell Award from the United States National Society of Professional Surveyors (NSPS), the Ohio State University's Duane C. Brown Jr. Award in 2003, the American Shore and Beach Preservation Association's Educational Award in 2004, the ETSU Faculty Excellence Award in 2005, and the Esri Award for Best Scientific Paper in Geographic Information Systems in 2006.

Jorge Delgado García

Prof. Dr. Jorge Delgado has a Ph.D. in Geological Sciences (University of Granada, Spain, 1993) and has been a Professor of Cartographic Engineering, Geodesy, and Photogrammetry at the University of Jaén (Spain) since 1994.

The main field of his work is the procedures of the capture and analysis of geographic information, having participated in several research projects and having been the author of a large number of publications in journals and communications in specialized congresses. For several international projects, he has participated as a cartography specialist . He is a member of the Copernicus Academy and member of the Board of Delegates in the EuroSDR.

Fayez Tarsha Kurdi

Dr. Fayez Tarsha Kurdi has received various positive appraisals. Kurdi has ten years of industrial experience in surveying and laser scanning, and holds a Ph.D. in geomatics and the automatic processing of Lidar data from the Graduate School of Science and Technology of Strasbourg (INSA of Strasbourg), ULP University (Strasbourg I), since 2008. He was a fellow researcher at the Institute for Integrated and Intelligent Systems at Griffith University, Australia, in April 2019. He has been an adjunct lecturer and research assistant at the School of Civil Engineering and Surveying, University of Southern Queensland, Australia, since May 2021, with his current focus being on machine learning techniques and 3D modeling being. Keeping his position in UniSQ, he worked in the East Coast Surveys Ltd firm as a Survey Systems and Research Manager. Dr. Tarsha Kurdi serves on the editorial boards of several remote sensing journals.

remote sensing

MDPI

Article

Modeling Multi-Rotunda Buildings at LoD3 Level from LiDAR Data

Fayez Tarsha Kurdi [1,*], Elżbieta Lewandowicz [2], Zahra Gharineiat [1] and Jie Shan [3]

[1] School of Surveying and Built Environment, Faculty of Health, Engineering and Sciences, University of Southern Queensland, Springfield Campus, Springfield, QLD 4300, Australia; zahra.gharineiat@usq.edu.au
[2] Faculty of Geoengineering, Institute of Geodesy and Civil Engineering, Department of Geoinformation and Cartography, University Warmia and Mazury in Olsztyn, 10-719 Olsztyn, Poland; leela@uwm.edu.pl
[3] School of Civil Engineering, Purdue University, West Lafayette, IN 47907, USA; jshan@purdue.edu
* Correspondence: fayez.tarshakurdi@usq.edu.au

Abstract: The development of autonomous navigation systems requires digital building models at the LoD3 level. Buildings with atypically shaped features, such as turrets, domes, and chimneys, should be selected as landmark objects in these systems. The aim of this study was to develop a method that automatically transforms segmented LiDAR (Light Detection And Ranging) point cloud to create such landmark building models. A detailed solution was developed for selected buildings that are solids of revolution. The algorithm relies on new methods for determining building axes and cross-sections. To handle the gaps in vertical cross-sections due to the absence of continuous measurement data, a new strategy for filling these gaps was proposed based on their automatic interpretation. In addition, potential points associated with building ornaments were used to improve the model. The results were presented in different stages of the modeling process in graphic models and in a matrix recording. Our work demonstrates that complicated buildings can be represented with a light and regular data structure. Further investigations are needed to estimate the constructed building model with vectorial models.

Keywords: LiDAR; LoD2; LoD3; 3D buildings; automatic building modeling

Citation: Tarsha Kurdi, F.; Lewandowicz, E.; Gharineiat, Z.; Shan, J. Modeling Multi-Rotunda Buildings at LoD3 Level from LiDAR Data. *Remote Sens.* **2023**, *15*, 3324. https://doi.org/10.3390/rs15133324

Academic Editor: Enrico Corrado Borgogno Mondino

Received: 11 May 2023
Revised: 15 June 2023
Accepted: 26 June 2023
Published: 29 June 2023

1. Introduction

Three-dimensional urban models represented in the CityGML 3.0 standard have considerable potential for numerous applications, in particular navigation systems. These applications are useful for designing transport systems for autonomous vehicles [1]. To meet such needs, building models must be developed at the LoD3 (level of detail 3). In LoD3, a building is represented as a solid, closed 3D geometry with separate components for the walls, roof, and architectural elements to accurately depict structural details and ornamental features [2–4]. In addition, LoD3 level models are also widely utilized in urban microclimate studies to identify buildings in urban space, generate energy-saving plans, and identify the sources of noise and noise propagation routes. Urban morphology models will play an increasingly important role in the future [5].

Three-dimensional city models are often developed based on light detection and ranging (LiDAR) data, which are collected with the use of aerial and terrestrial remote sensing techniques [6,7]. The process of building modeling at various levels of detail, from LoD0 to LoD2, has been extensively investigated [8–17]. New approaches to modeling buildings are being proposed based on the density of point clouds [18], normal vectors on minimal subsets of neighboring LiDAR points to determine characteristic points in roof creases [15], shape descriptors, and cubes that divide the point cloud into roof surface segments [19]. However, even sophisticated techniques will not be able to handle some intrinsic modeling problems [20,21]. The density of point clouds acquired during airborne scanning of urban areas differs sometimes between roofs and walls, and the presence of outliers and noisy data can lead to errors in the process of generating point clouds and incorporating clouds into the reference system [17,22,23].

Once the LiDAR point cloud is classified into main classes such as terrain, buildings, and vegetation [24], various methodologies have been proposed for automating the generation of mass-building models. Individual buildings must be distinguished and selected [25] from compact dense urban development [26], and then modeled in 3D [27]. An algorithm for identifying flat roofs and modeling individual buildings at the LoD3 level based on planar structures was proposed in [28]. A similar solution [29,30] for modeling buildings based on planar primitives produces structures with more elaborate shapes. Planar primitives are generated from a point cloud and are then reconstructed with the use of characteristic lines identified in the acquired images. In the last step of the process, the generated models are optimized by a polynomial curve fitting (PolyFit). Planar primitives are also used to model buildings based on a dense triangulate irregular network (TIN) mesh [31].

Other algorithms for 3D building modeling integrate various sources of data. The first solutions relied on old maps, plans, and cadastral data [32,33]. At present, LiDAR data are increasingly combined with remote sensing datasets, machine learning methods, and neural networks [10,34–38]. Window and door openings on walls are modeled at the LoD3 level based on terrestrial laser scanning images and segmented 2D images [39]. These methods rely on deep machine learning techniques. In a graph-based model [40], the structural complexity of a building facade can be automatically modeled, and geometric data can be combined with semantic input.

Several automatic solutions have been proposed for generating mass building models, in particular roofs, at the LoD2 level, based on aerial images and high-resolution remote sensing data by artificial intelligence methods [41–43]. Artificial intelligence is also useful for 3D modeling at the LoD3 level based on street view images [44]. These methods produce satisfactory results when the modeled buildings have regular shapes, in particular, when terrestrial laser scanning data are available. In spite of all these efforts, atypical and irregularly shaped buildings with complex ornaments continue to pose a challenge to state-of-the-art solutions. These buildings are particularly difficult to model based solely on aerial images. The presented study in this paper was undertaken to further explore this issue based on the authors' previous findings [45].

2. Research Objectives

Buildings with irregularly shaped features often constitute landmarks in urban spaces. They are important in navigation. One of the first attempts to automatically model atypical buildings composed of rotational surfaces was made by Lewandowicz et al. [45]. This cited study proposed an algorithm for rendering ornamental features in greater detail and capturing these buildings' unique ambiance. The method proposed in [45] was based on modeling the rotunda based on only one point cloud cross-section.

The presented study in this paper intends to improve and extend the algorithm proposed by Lewandowicz et al. [45] to capture and enhance the presentation of unique structural elements of buildings. In this context, the novelties of our work, as well as the objectives, are formulated as follows:

- Improvement of the method for determining the axis of buildings represented by solids of revolution;
- Introduction of a new approach for the automatic generation of building cross-sections and a gap-filling strategy when a complete set of points is not available;
- Evaluation and interpretation of deviated data points (outliers) in the process of incorporating these data into the developed model.

As a result, a matrixial form of modified building models was developed in the last stage of the study. The results were presented and visualized in different stages of the modeling process.

3. Datasets

The presented study uses point clouds acquired with the airborne laser scanning (ALS) methods, representing distinctive buildings in the analyzed cities. These data were obtained from the Polish Spatial Data Infrastructure (SDI). Buildings with atypical shapes, features, and heights often constitute landmarks in urban spaces. They include sports and entertainment arenas, water towers, buildings with domed roofs (such as planetariums), or industrial buildings with tall chimneys (Figure 1).

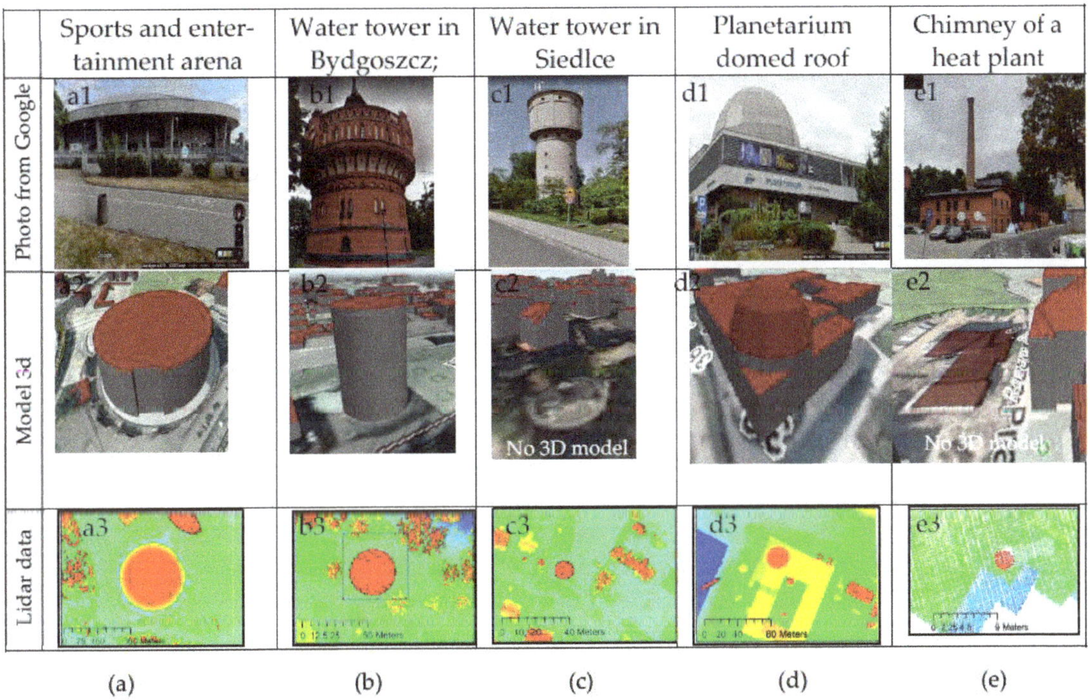

Figure 1. Visualization of buildings (**a1–e1**), their 3d models with a database (**a2–e2**). Vertical cross-sections of a point cloud of buildings selected for the study. The analyzed buildings are visible with a red circle: (**a3**) sports and entertainment arena in Łódź; (**b3**) water tower in Bydgoszcz; (**c3**) water tower in Siedlce; (**d3**) domed roof of the Nicolaus Copernicus Planetarium in Olsztyn; and (**e3**) chimney of a heat plant in the Kortowo campus of the University of Warmia and Mazury in Olsztyn.

These building models are largely simplified at the LoD2 level in the 3D models of Polish cities developed. They are represented by cylinders or are overlooked in models (Figure 1(a1,a2,b1,b2,c1,c2,d1,d2,e1,e2)). Points representing buildings that are rotational surfaces can be extracted from a LiDAR. When classifying points by height and viewing the vertical projections of the LiDAR sets, one can distinguish clusters of points showing the tested objects in the shape of circles (Figure 1). Selected for the study are different types of buildings (Figure 1a–c) and elements of building (Figure 1d,e).

Data files are acquired in LAZ format, while the point coordinates are expressed in the ETRS_1989_Poland_CS92 (EPSG 2180) coordinate system. All points are assigned class ID, signal intensity values, and RGB values from aerial images. The scanning was acquired in 2017–2022 with a resolution of 12 or 4 points per square meter, depending on the year of acquisition.

4. Method

Successive stages of the modeling process are described in the following subsections.

4.1. Improve Vertical Cross-Section Point Cloud

From Figure 2a,b, it can be noted that the point density, as well as the point distribution on the vertical walls of a tower point cloud, are heterogeneous. As an example, the calculation of two vertical cross-sections of the point cloud illustrated in Figure 2b according to two different directions (direction 1-1 and direction 4-4, as shown in Figure 2e) provides two different results shown in Figure 2c,d. In fact, the difference between the two obtained results is due to the irregular distribution of LiDAR points on the building facades. At this stage, the major question that arises is in which direction (according to Figure 2e) the vertical cross-section must be calculated to obtain the best representative result. This paper proposes a new approach to calculate the best cross-section that considers all LiDAR points describing the tower building.

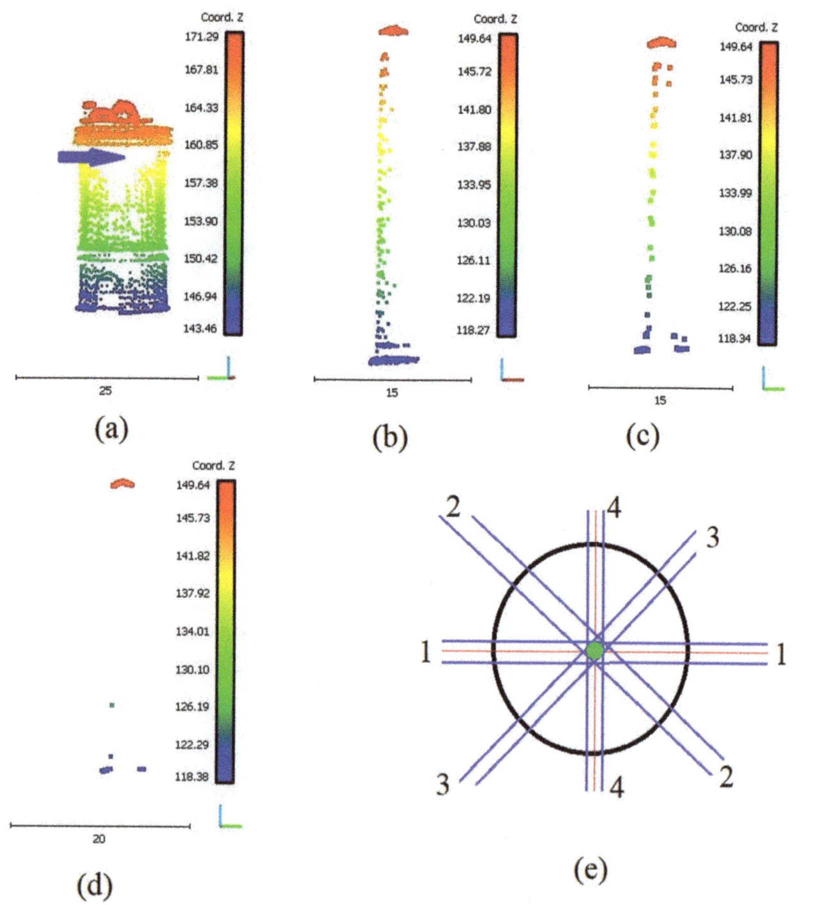

Figure 2. (**a**) Tower_1 point cloud. (**b**) Tower_2 point cloud. (**c**) Point cloud of vertical cross-section according to the direction 1-1 in (**e**). (**d**) Point cloud of vertical cross-section according to the direction 4-4 in (**e**). (**e**) Black circle is the horizontal cross-section of the given rotating tower, green circle is the gravity center of the horizontal cross-section, and the blue lines are directions of suggested vertical cross-sections.

The first step is to project all points according to a circular trajectory and then group them into one half vertical plane located on one side of the tower (Figure 3). To carry out this operation, the cloud coordinates (X, Y, and Z) are transformed into a plane coordinate system (X_1 and Y_1) according to Equation (1).

$$X_1 = Y_g + \sqrt{(X - X_g)^2 + (Y - Y_g)^2} \quad Y_1 = Z \tag{1}$$

where X_g and Y_g are the coordinates of the point cloud gravity center according to Lewandowicz et al. [45] as shown in Equation (2).

$$X_g = \text{minX} + \frac{\text{Xmax} - \text{Xmin}}{2},\ Y_g = \text{minY} + \frac{\text{Ymax} - \text{Ymin}}{2} \tag{2}$$

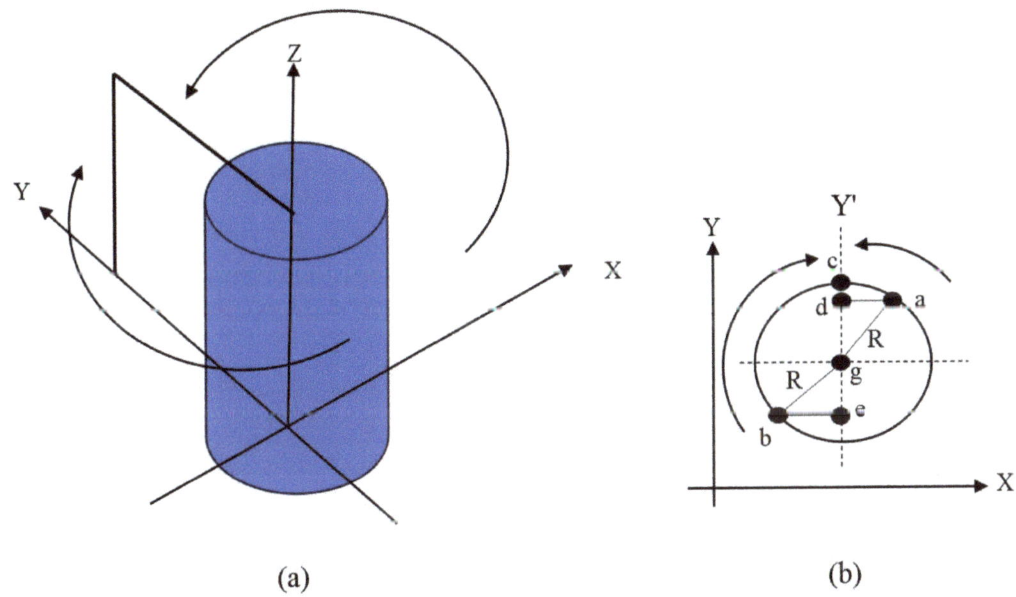

Figure 3. Rotating of points and grouping them to a half vertical plane located on one side of the tower. (**a**,**b**) Consecutively 3D and 2D rotating illustrations.

To clarify this operation, the example illustrated in Figure 3b is detailed. In Figure 3b, the distances are equal between all points of the circle and the gravity center (the circle center) and equal to R (the circle radius). Point 'a' is projected in a circular trajectory on the same circle, the obtained result is Point 'c'. The same operation is applied to Point 'b', and the obtained result will also be Point 'c'. In Equation (1), as Points 'a' and 'b' have the same Z value and their distances to the gravity center are the same, the new coordinate X_1 of the two points will be the same. At this stage, it is important to refer that this operation does not represent a projection on gY′ axis. Indeed, the projection of Points 'a' and 'b' on axis gY′ are consecutively Points 'd' and 'e'.

Thereafter, the new point cloud $\{X_1, Y_1\}$ which represents the vertical cross-section, should be put in descending order regarding the Z coordinate values. At this stage, it can be noted that according to the point density, it is possible to present groups of points having the same Z coordinate value. In fact, there are three considerations to present this kind of point: LiDAR point accuracy, texture smoothing, and building architecture. The basic hypothesis in the suggested approach is that one building surface consists of a main rotating surface and some decoration parts added to this surface. Hence, in the case of several points having the same Z values and different distances from the rotating axis, the nearest point to the rotating axis is located on the main rotating surface, whereas further points are likely located on the decorations. If the building architecture reason is neglected, then the basic frustum of a cone must pass through the central point. Moreover, if the point accuracy and texture smoothing are neglected, then the basic frustum of a cone must pass through the nearest point to the rotating axis. Furthermore, if it is desired to consider all of the three reasons together, the points of the same Z coordinate value must be divided into two groups: one group of points that belong to the main building surface and the other that belongs to the decoration. This analysis needs more experiments to decide if it is efficient or not. Finally, as all available points will be considered in the model equation, the errors will only be located at places where points are missing.

In this paper, a new rule is added, as follows: if a group of points has the same Z coordinate value, only the farthest point from the rotation axis is kept; the other points are temporarily eliminated until the last modeling step.

This procedure allows for a reduction of the number of points of the vertical cross-section. The new cross-section point cloud is noted as a reduced point cloud. In this context, a new list that has the same length as the reduced point cloud is defined and named the point-frequency list. This list represents, for each point of the reduced point cloud, the number of points having the same Z coordinate value in the original point cloud. Another list named dev_list is defined. For the points having point-frequency values greater than one, the value of the corresponding dev-list cell is equal to the subtraction of the nearest and farthest distances from the rotating axis. If the point-frequency value is equal to one, in this case, the corresponding dev-list cell is assigned zero.

Figure 4a visualizes the reduced vertical cross-section of the tower point cloud shown in Figure 2b. Figure 4b uses the histogram illustration to visualize the frequent list of point clouds shown in Figure 4. In this figure, it can be noted that the frequency of the most reduced point clouds is equal to one. The maximum value of the frequency is equal to 10. In fact, the importance of this list as well as the dev_list will be highlighted in the third improvement step. Figure 4c utilizes histogram graphics to visualize the dev_list of the point cloud shown in Figure 4a. In this figure, it can be stated that most reduced cross-section points do not have deviations from the building model. Moreover, points with deviations can be classified into two classes. The first class is the points with small and neglected deviations comparable to the LiDAR point accuracy of 0.4 m or smaller. The second class is the points having a deviation greater than 0.4 m due to the presence of decoration or noise.

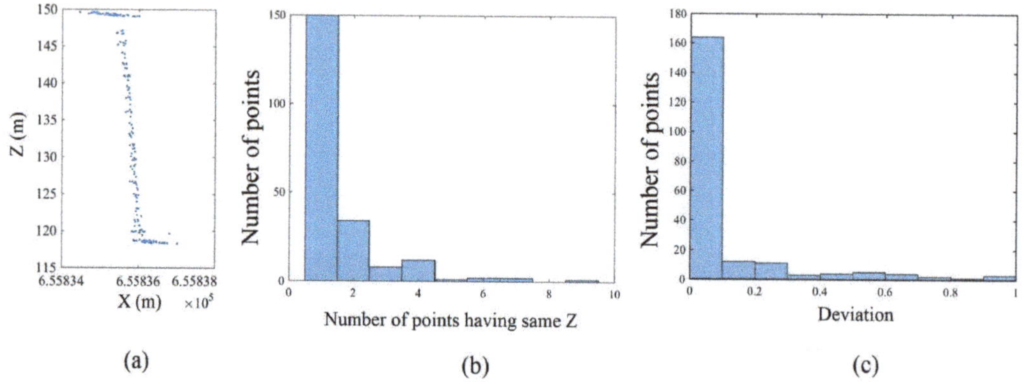

Figure 4. (**a**) Reduced vertical cross-section of the tower point cloud shown in Figure 2b. (**b**) Illustration of the frequency list of the point cloud shown in (**a**). (**c**) Illustration of the dev_list of the point cloud shown in (**a**).

4.2. Gap Analysis and Filling

Once the cross-section point cloud is calculated and reduced, the next step is the vertical cross-section gaps analysis. The mean expected distance between two neighboring points depends on the point density. In the building point cloud, it is common to meet neighboring points separated by distances greater than the mean expected distances; this separation distance is named a gap. However, two kinds of such gaps can be distinguished: horizontal gaps when the greater separation distance is horizontal (see blue arrow in Figure 2a), and vertical gaps when the greater separation distance is vertical. In fact, there are several reasons for the presence of these gaps, such as the obstacles that prevent the laser pulses to arrive at the scanned surface, the geometric form of the reflecting object, the physical nature of the scanned surface (e.g., the surface is made of glass), and the scanning parameters such as the flying height and the building location regarding the sensor location. Though the employment of the vertical cross-section to model the building can cancel the direct influence of the horizontal gaps because it moves all building points through a circular trajectory to group them into a vertical plane. But in the final obtained model, the presence of the horizontal gaps may reduce the building model accuracy in the gap zone due to the lack of information in this area. Concerning the vertical gaps, despite their influence being reduced through using the vertical cross-section described in the last section, sometimes these gaps still appear in the vertical cross-section (Figure 5) because all cloud points are grouped into one vertical plane (Figure 3). Therefore, it is necessary to add a special procedure to process the remaining vertical gaps and reduce the depicted deformation due to these kinds of gaps.

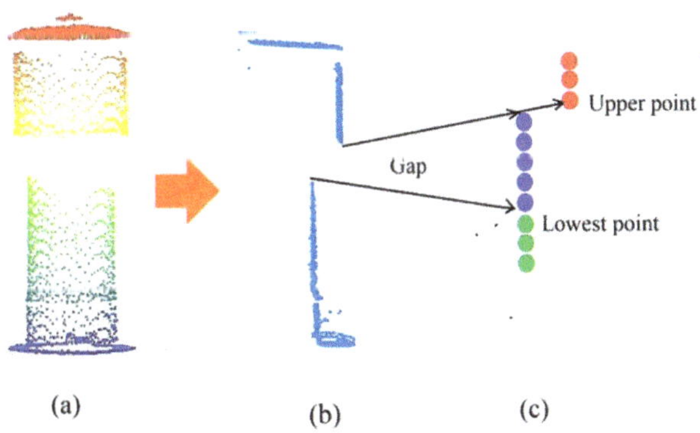

Figure 5. Gap-filling strategy. (**a**) Tower point cloud. (**b**) Vertical cross-section. (**c**) Red points are the gap's upper points, green points are the gap's lowest points, and blue points are points filled inside the gap.

When a line segment is revolved around an axis, it mathematically draws a piece of a cone called the frustum of a cone. This frustum of a cone could be a cylinder when the line segment is parallel to the rotating axis. According to this principle and in the case of vertical gaps (Figure 5a,b), if the upper and the lowest gap points have different distances from the rotating axis (Figure 5b,c), the gap will generate a frustum of a cone connecting the two consecutive frusta of cones or cylinders in the building model (Figure 6b). In fact, this solution does not consider the main reason for the gap presence when the geometric form of the scanned surface prevents the laser pulses to arrive at the scanned object. That is why there is a great deformation in the gap area in the building model presented in Figure 6b. Hence, to improve the calculated building model, this paper proposes a new strategy to fill the gaps in the vertical cross-section as follows.

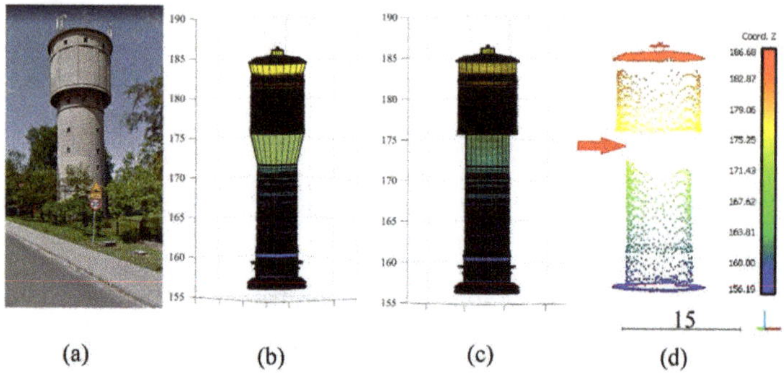

Figure 6. (a) Tower image. (b)Tower model before filling the gaps. (c) Tower model after filling the gaps. (d) Building point cloud; Red arrow shows gap.

In the last section, the reduced point cloud was put in descending order regarding the Z coordinate values, which means that the first point in the list has the greater Z value and the last point in the list has the lowest Z value. At this stage, a new list, named $Z_{spacing}$, is defined. The first cell in this list contains the value zero. Thereafter, the value of each cell is calculated by subtracting the Z coordinates value of the corresponding point in the reduced point cloud from its precedent point.

Figure 7 illustrates the visualization of the distribution of $Z_{spacing}$ cell values. In this histogram, it can be noted that the vertical spacing between most of the reduced cross-section cloud points is around zero. Moreover, the points having vertical spacing smaller than a given threshold (e.g., 0.2 m) can be considered as points having accepted vertical spacing and then having no gaps. Also, points having vertical spacing greater than the same threshold are considered points having gaps. In fact, the employed threshold value ($TH_{spacing}$ = 0.2 m) depends on the point density. Its value can be considered equal to or smaller than the mean expected distance between two neighboring points.

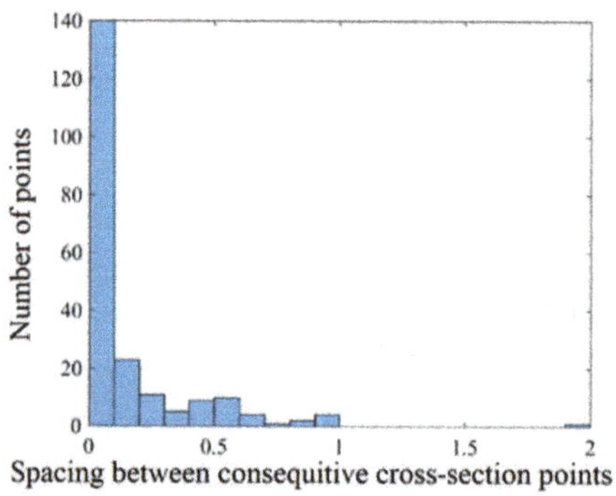

Figure 7. Vertical spacing between consecutive vertical cross-section points shown in Figure 3a.

Once the $Z_{spacing}$ list is calculated and the spacing threshold is determined, the point gaps can be detected by comparing the vertical space values with the spacing threshold. To fill a gap, a list of points is added within the gap. These points have the same abscissa of the gap's lowest point (Figure 5c) and have gradual ordinates starting from the gap's lowest point ordinate added to $TH_{spacing}$ until the gap's upper point ordinate.

Figures 6 and 8 show the modeling results of the building point clouds illustrated in Figures 2b and 5a consecutively in the case of the application of the gap-filling strategy and without applying this strategy. In Figure 8, the gap heights are smaller than those in Figure 6, which is why the influence of the filling gap operation is less notable oppositely to the case of the building presented in Figure 6. In Figure 6b, the building model has a great deformation in the gap area. This deformation disappears in Figure 6c thanks to the gap-filling function. Moreover, the building model becomes more faithful to the original building presented in Figure 6a after applying the gap-filling strategy.

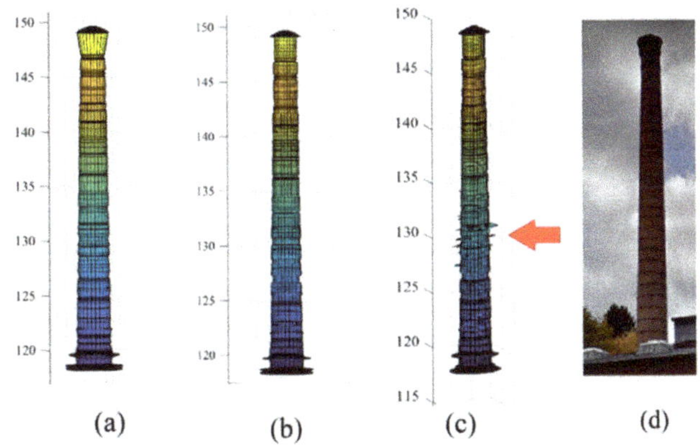

Figure 8. Models calculated from the point cloud shown in Figure 2b. (**a**) Tower model before fillingthe gaps. (**b**) Tower model after filling the gaps. (**c**) Tower model after considering all cloud points. (**d**) Tower image.

4.3. Integrating Deviated Points in the Calculated Model

It can be observed from Figure 4b that most of the building cloud points have distinctive (non-duplicated) Z coordinate values. However, in the presence of several LiDAR points having the same Z coordinate value, the suggested algorithm in the last section considers only the nearest point to the rotating axis and neglects the other points. In this section, the suggested algorithm will be extended to consider all cloud points without neglect. For this purpose, the coordinates of non-considered points are used to modify the rotating surface depicted by Equations (3)–(5) [45].

At this stage, it is important to show how Equations (3)–(5) are deduced. One rotating surface can be divided into n horizontal slices according to the consecutive Z coordinate values of the half -cross-section cloud (see Figure 4a). The points of each slice have the same Z_i coordinate value, which is why the elements of each row in the Z matrix are equal. Each slice represents a circle because it belongs to a rotating surface. This circle can be divided into m angular sectors. One rotating surface is expressed by three matrices X, Y, and Z. This surface is composed of cells. The coordinates of the middle point of each cell will be considered from the three corresponding cells of the last three matrices. The dimensions of one cell can be calculated as a function of the thickness of the horizontal slice, the number of angular sectors, and the cell circle radius value ($R = Y_i - Y_g$). The angle of each angular sector equals $\frac{2\pi j}{m}$, where j is the sector number. In Equation (5), the origin of β is the circle center, but the origin of α is Y_i. The application of basic sine and cosine relationships allows deducing α and β equations where the value $\frac{3\pi}{2}$ is added to the angle value for adapting the signs.

Return to the integration of non-considered points, if one point (X_p, Y_p, Z_p) does not belong to a rotating surface defined by Equations (3)–(5), it is desired to integrate this point within this surface. Hence, this operation can be carried out by calculating the angle θ (see Figure 9) using Equation (6). Thereafter, the angle θ_o measured from the rotating origin R_o (see Figure 9) is calculated according to Equation (7). In the matrices X and Y, the row number of the concerned cell can be calculated depending on the Z coordinate value (Z_p).

Furthermore, the column number of the concerned cell can be calculated depending on θ_o value and the number of columns of the matrix X according to Equation (3).

$$X = \begin{bmatrix} X_g X_g + \beta_{1,1} X_g + 2\beta_{1,2} \ldots \ldots X_g \\ X_g X_g + \beta_{2,1} X_g + 2\beta_{2,2} \ldots \ldots X_g \\ X_g X_g + \beta_{3,1} X_g + 2\beta_{3,2} \ldots \ldots X_g \\ \cdot \quad \cdot \quad \cdot \quad \cdot \\ X_g X_g + \beta_{n,1} X_g + 2\beta_{n,2} \ldots \ldots X_g \end{bmatrix} \tag{3}$$

$$Y = \begin{bmatrix} Y_1 Y_1 + \alpha_{1,1} Y_1 + 2\alpha_{1,2} \ldots \ldots Y_1 \\ Y_2 Y_2 + \alpha_{2,1} Y_2 + 2\alpha_{2,2} \ldots \ldots Y_2 \\ Y_3 Y_3 + \alpha_{3,1} Y_3 + 2\alpha_{3,2} \ldots \ldots Y_3 \\ \cdot \quad \cdot \quad \cdot \quad \cdot \\ Y_n Y_n + \alpha_{n,1} Y_n + 2\alpha_{n,2} \ldots \ldots Y_n \end{bmatrix} \quad Z = \begin{bmatrix} Z_1 Z_1 \ldots \ldots Z_1 \\ Z_2 Z_2 \ldots \ldots Z_2 \\ Z_3 Z_3 \ldots \ldots Z_3 \\ \cdot \quad \cdot \quad \cdot \quad \cdot \\ \cdot \quad \cdot \quad \cdot \quad \cdot \\ Z_n Z_n \ldots \ldots Z_n \end{bmatrix} \tag{4}$$

$$\alpha_{i,j} = (Y_g - Y_i)\sin\left(\frac{2j\pi}{m} + \frac{3\pi}{2}\right),\ \beta_{i,j} = (Y_g - Y_i)\cos\left(\frac{2j\pi}{m} + \frac{3\pi}{2}\right) \tag{5}$$

where X_g and Y_g are the coordinates of the gravity center (Equation (2)); X_i, Y_i, and Z_i ($i = 1$ to n) are the point coordinates of the half cross-section; $j = 1$ to m; n is the number of points in the half cross-section; α_i and β_i are the step values of X and Y, respectively; and m is the number of columns in matrix X.

$$\theta = \arctan\frac{\text{abs}(\Delta Y)}{abs(\Delta X)} = \arctan\frac{abs(Y - Y_g)}{abs(X - X_g)} \tag{6}$$

$$\begin{aligned} &If \Delta X < 0\ and \Delta Y > 0\ then\ \theta_o = \tfrac{3\pi}{2} + \theta \\ &If \Delta X > 0\ and \Delta Y > 0\ then\ \theta_o = \tfrac{\pi}{2} - \theta \\ &If \Delta X < 0\ and \Delta Y < 0\ then\ \theta_o = \pi + \theta \\ &If \Delta X > 0\ and \Delta Y < 0\ then\ \theta_o = \tfrac{\pi}{2} + \theta \end{aligned} \tag{7}$$

$$CN = round\frac{\theta_o}{2\pi} \times m \tag{8}$$

where CN is the column number in matrix X (Equation (3)), m is the number of columns in matrix X, and "round" is a function that provides the round value of a given real number.

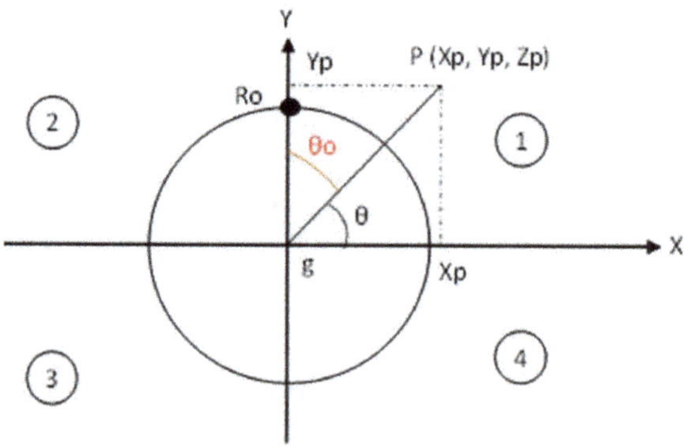

Figure 9. Integration of a new point within the constructed building model, R_o is the rotating origin, g is the gravity center, and P is a point off the rotating surface.

The new value of the corresponding cells in X and Y matrices are calculated using Equation (9).

$$X_n = X_g + (Y_g - Dis_{pg}) \times Cos\theta_o$$
$$Y_n = Y_g + (Y_g - Dis_{pg}) \times Sin\theta_o \quad (9)$$

where X_n and Y_n are the new value of the corresponding cells in X and Y matrices, Dis_{pg} is the distance between the gravity center g and the given point P.

Once the new value of the corresponding cells in the X and Y matrices are calculated, these values can be reassigned to the concerned cell in the matrices X and Y. This operation can be carried out for all deviated points to consider them within the building model.

The red arrow in Figure 8c points to the influence due to the integration of the deviated points within the building model. Unfortunately, the deviated points in the case of the building illustrated in Figure 8 represent noisy points, which is why the constructed model shown in Figure 8c has certain deformations due to the inclusion of noise points. However, the inclusion of the deviated points may sometimes improve the model quality when the deviated point density is high enough, and the deviated points represent building details or decoration. Figure 10a,b show the tower model before and after the inclusion of the deviated points. At the red arrow in Figure 10c, the geometry of the tower part covered by the LiDAR points (see Figure 10c, which shows the superimposition of the point cloud on the building model) was improved due to considering all LiDAR points. Moreover, Figure 10c illustrates that the tower point cloud completely fits the improved constructed model. Nevertheless, more investigations are needed to automatically classify the building point cloud into building points and noise points.

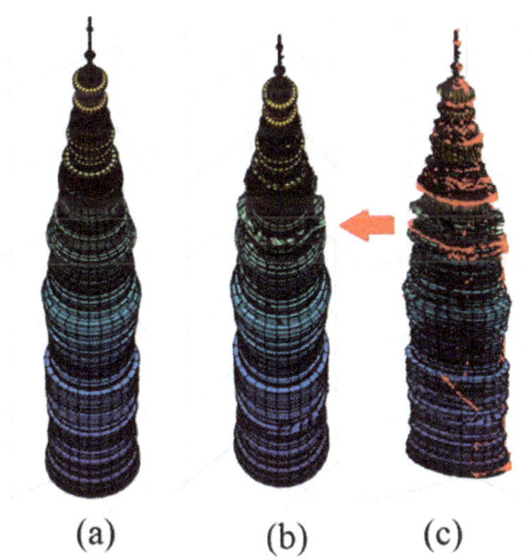

Figure 10. (**a**) Tower model before including deviated points. (**b**) Tower model after including deviated points. (**c**) Laying out point cloud over the tower model after including the deviated points; LiDAR point cloud is in red color.

5. Discussion

In this section, the suggested modeling algorithm will be applied to different samples of the tower point clouds. Then, the modeling accuracy as well as the faithfulness of the obtained models will be discussed.

5.1. Performance of the Method

Figures 6, 8 and 10–12 depict the tower models constructed by the proposed approach. Figures 6 and 8 show the influence of gap-filling operation on the constructed building model, where this influence is humble in the case of the building illustrated in Figure 8 because the geometric form of the tower does not contain a hidden area regarding the airborne scanning, whereas in the case of building illustrated in Figure 6, the gap filling operation is crucial in order to avoid the huge deformation within the hidden area. Nevertheless, the success of the gap-filling procedure needs sufficient points to cover the tower body. This situation can be illustrated in Figure 11i–l. The LiDAR points that cover the building body are concentrated only on the upper part of the tower, in contrast to the other lower parts, where very few points are laid on the building body's outer surfaces. That is why the obtained building model cannot show the steps of building architectural form (Figure 11k,l). In the same context, the gap-filling procedure depends on the vertical spacing threshold value, which is related to the point density as well as the LiDAR point accuracy.

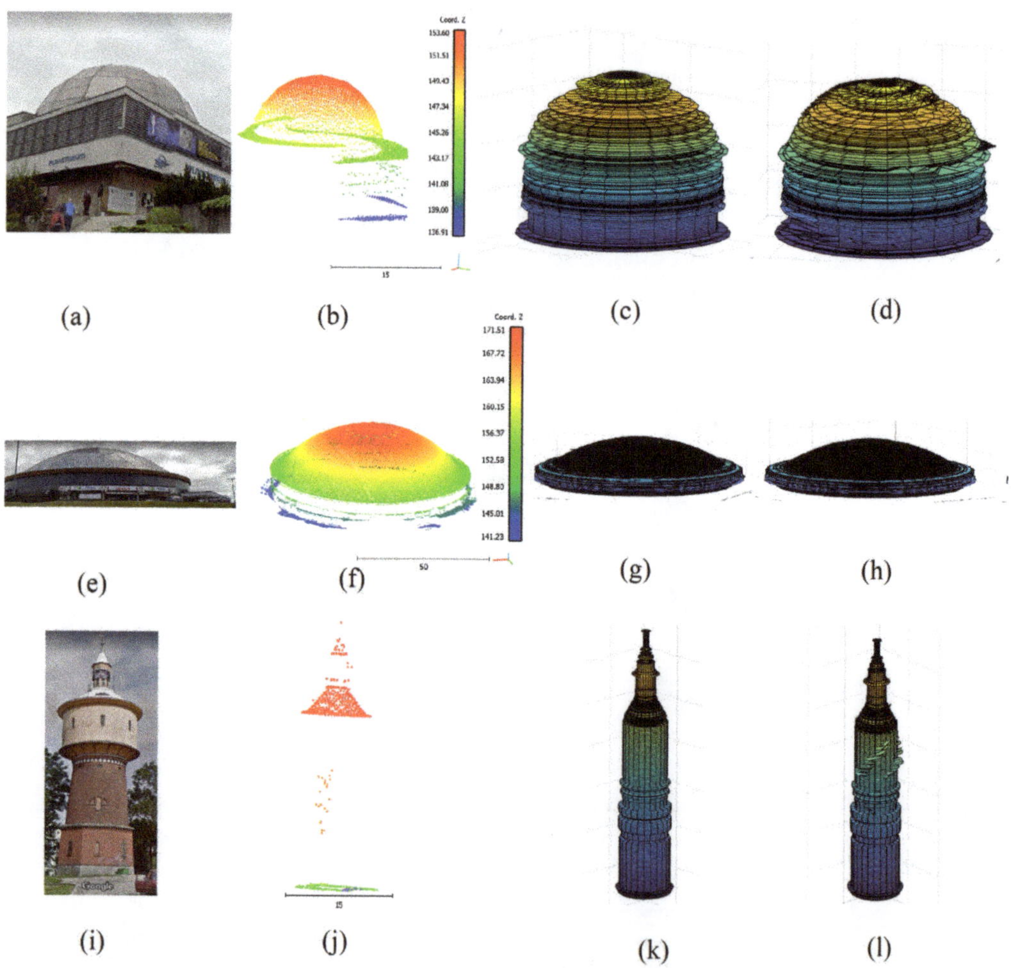

Figure 11. (**a**,**e**,**i**) Tower images. (**b**,**f**,**j**) Tower point clouds. (**c**,**g**,**k**) Tower model before considering all cloud points. (**d**,**h**,**l**) Tower model after considering all cloud points.

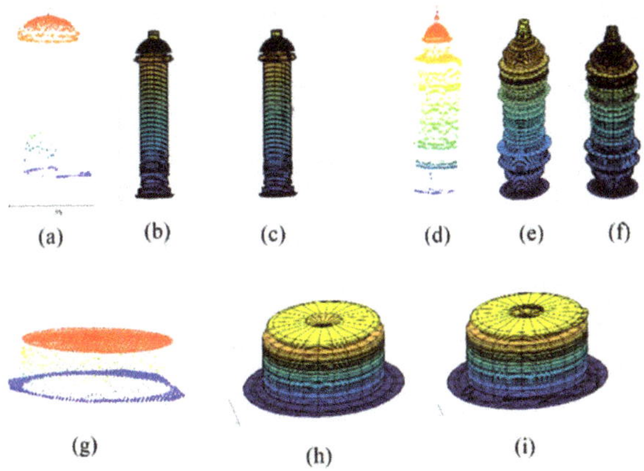

Figure 12. (**a**,**d**,**g**) Tower point clouds. (**b**,**e**,**h**) Tower model before considering all cloud points. (**c**,**f**,**i**) Tower model after considering all cloud points.

Moreover, in the building vertical cross-section, the accepted vertical spacing value between neighboring points is variable regarding the regularity level of point distribution, point accuracy, point density, building architectural form and complexity level, construction material nature, and scanning angle. More investigation is needed to improve the selection and effect of the vertical spacing between neighboring points.

Concerning the integration of deviated points operation, two cases may be envisaged. First, when the deviated points represent noisy points, the integration of these points within the building calculated model will produce undesired deformations (see red arrow in Figure 8c). Second, if the deviated points do not represent noisy points, the integration of these points into the constructed building model may improve its quality if their density value is elevated enough because it will become more faithful to the scanned building (Figure 10b,c, Figure 11, and Figure 12). On the other hand, if the deviated points' density is low, the introduced corrections may make the building model look deformed. Furthermore, the resemblance level between the obtained building model and the original scanned building will be related to the point density and accuracy values. At this stage, more investigations in future research are requested about the effective integration of the deviated points into the constructed tower model.

Though the high efficiency of the suggested approach is demonstrated regarding the architectural complexity of the target buildings, it still suffers from some limitations that deserve future efforts. These limitations can be summarized as follows:

- Undesirable distortions may appear in the constructed model when the input point cloud has inconsistent quality regarding the point density, distribution regularity, and homogeneity. Certain levels of balance may be desired that can comprise the data volume, level of details for presentation, and the accuracy of the model;
- Like many other methods, the developed method can only reconstruct buildings that meet certain assumptions, which in this case are rotating surfaces. Small attachments or decorations of the main surface need to be treated separately. A promising effort is to extend and/or integrate this method with other methods to handle complex and diverse buildings.

5.2. Modeling Accuracy

Concerning the accuracy of the constructed building models, there are two main accuracy estimating approach families [22]. First, the created building model is compared with the reference model constructed manually or semi-automatically using LiDAR data or other data sources such as aerial images [14,20,21]. In the second approach, the LiDAR point cloud is employed as a reference model, where the accuracy can be evaluated by calculating the distances between the constructed building model and the point cloud [20,21,28]. In this paper, the accuracy of modeling will be discussed through three viewpoints. First, despite the undesirable deformations in the constructed building models, the accuracy of the calculated building model is 100% when the building point cloud is considered as a reference model. Indeed, the constructed model fits all building cloud points and, consequently, the building model is completely faithful to the input building point cloud. In this context, Tarsha Kurdi and Awrangjeb [22] compared the building point clouds with the obtained building models. They concluded that the accuracy, regularity, as well as point density of the building point cloud, may affect the faithfulness of the building model to the original building even if the model greatly fits the point cloud.

From a second viewpoint, the cell's dimension of the model can also express the building model accuracy, because the cell size represents the interval where the LiDAR point may be located. From Equations (3)–(5), the building model consists of a matrix of cells connected through robust neighbor relationships. The cell's dimensions are used as an evaluation metric. The width (C_W) and height (C_H) of a cell can be calculated using Equation (10).

$$C_W = \frac{2 \times \pi \times Dis_{pg}}{m}; \qquad C_H = Z_i - Z_{i-1} \neq 0 \tag{10}$$

where Dis_{pg} is the distance between the gravity center g and the given point (Equation (11)), m is the number of columns in matrix X.

$$Dis_{pg} = \sqrt{(X_p - X_g)^2 + (Y_p - Y_g)^2} \tag{11}$$

From Equation (10), it can be noted that the cell's dimensions are related to the number of columns in the building model matrices, the distance from the rotating axis, and the point density. While the number of columns of the model matrices increases, the cell widths will decrease. Also, C_W and C_H values are variable from point to point in the building model. Hence, for each building model, the minimum, maximum, and mean values of these parameters are calculated (Table 1). In this context, the buildings illustrated in Figures 11 and 12 are considered to estimate the modeling accuracy by respecting their order.

Table 1. Accuracy of building models for $m = 61$.

Building Number	Min C_W (m)	Max C_W (m)	Mean C_W (m)	Min C_H (m)	Max C_H (m)	Mean C_H (m)
1	0.01	1.36	0.81	0.01	0.20	0.02
2	0.01	4.55	2.79	0.01	0.20	0.01
3	0.02	0.66	0.40	0.01	0.20	0.07
4	0.01	0.88	0.53	0.01	0.20	0.08
5	0.01	0.49	0.26	0.01	0.20	0.04
6	0.01	1.33	0.75	0.01	0.14	0.02

From Table 1, it can be noted that at least one dimension of the cell is related to the building diameter. That is why it is advised to increase the m value with an increase in the building radius. To conclude, two main factors that influence the dimension of the model cells are the building diameter and the point density.

Finally, in the same context of the building model accuracy, the question of accuracy estimation by comparing the constructed building model with a reference model constructed manually or semi-automatically [14,20,21] will also be discussed. In fact, the target buildings by the proposed modeling approach have complicated architectural forms (see Figures 6, 8 and 10–12), where their geometric forms contain curved surfaces as well as decorations. That is why the construction of accurate models for them to be used as references will need a huge amount of time and extra data and measurements. Moreover, if it is insisted to construct these reference models, a new question will arise concerning the comparison between the reference model and the calculated models. Indeed, the references model is supposed to have vectorial forms, whereas one calculated building model is composed of three matrices (X, Y, and Z). Also, the visualization of the calculated models is carried out using 3D pixel form because the model represents a novel modeling strategy in the world of LiDAR data and is based on the concept of the rotating surface. Hence, the comparison between the two building models needs more investigation.

To conclude, in this paper, only the comparison between the calculated building model and its LiDAR point cloud is considered. In future research, a more thorough investigation will be carried out to compare the constructed models with the reference models. Another question about the improvement of quality as well as the accuracy of the calculated building model will also be handled.

6. Conclusions

The novel proposed approach to modeling atypical landmark buildings at the LoD3 level has significant implications for all applications that rely on 3D building models. The suggested algorithm is based on the hypothesis of the rotating surface form of the target building. One building point cloud will be present by three matrices, X, Y, and Z. Moreover, the visualization will be realized using 3D pixel form. Only buildings that are solids of the revolution were modeled in the present study. A strategy for filling gaps in vertical cross-sections was described for buildings whose unique features prevent laser pulses from reaching the scanned surface. The developed strategy significantly improved the quality of the generated models. The operation of integration of the deviated points into the constructed building model aids in completely fitting the constructed model with the point cloud, but it may generate undesirable deformation in the building model when the deviated points represent noisy points, or their density is not great enough.

The main advantage of the suggested modeling algorithm is that it targets complicated geometric buildings, and the model data volume is light. Further efforts are needed to render building facades in greater detail because the deviated points (outliers) in the calculated models can belong to the façade. These points can result from noise, residual errors in the process of determining the building axis, or even permissible deviations from the wall and roof surfaces stipulated in structural designs. Also, more investigations are needed to estimate the constructed building model with vectorial models.

Finally, the novel suggested modeling strategy can be extended in future work to be employed for most levels of building architectural complexity, especially when a high point density is available. This approach can be extended for tree modeling as well as statues and other solid objects.

Author Contributions: Conceptualization, F.T.K., E.L. and Z.G.; methodology, F.T.K. and E.L.; software, F.T.K.; validation, F.T.K. and J.S.; formal analysis, F.T.K. and J.S.; resources, E.L., data curation, F.T.K. and E.L.; writing—original draft preparation, F.T.K. writing—review and editing, Z.G. and J.S.; visualization, F.T.K., E.L. and J.S. All authors have read and agreed to the published version of the manuscript.

Funding: This research was financed as part of a statutory research project of the Faculty of Geoengi neering of the University of Warmia and Mazury in Olsztyn, Poland, entitled "Geoinformation from the Theoretical, Analytical and Practical perspective" (No. 29.610.008-110_timeline: 2023–2025).

Data Availability Statement: The publication uses LAS measurement data obtained from an open Polish portal run by the Central Office of Geodesy and Cartography, and street view images were used. Water towers were searched based on the portal https://wiezecisnien.eu/en/wieze-cisnien/, accessed on 25 June 2023.

Acknowledgments: We would like to thank the Central Office of Geodesy and Cartography (GUGiK) in Poland for providing Lidar measurement data and for making data from the 3D portal available.

Conflicts of Interest: The authors declare no conflict of interest.

References

1. Richa, J.P.; Deschaud, J.-E.; Goulette, F.; Dalmasso, N. AdaSplats: Adaptive Splatting of Point Clouds for Accurate 3D Modeling and Real-Time High-Fidelity LiDAR Simulation. *Remote Sens.* **2022**, *14*, 6262. [CrossRef]
2. Beil, C.; Ruhdorfer, R.; Coduro, T.; Kolbe, T.H. Detailed Streetspace Modelling for Multiple Applications: Discussions on the Proposed CityGML 3.0 Transportation Model. *ISPRS Int. J. Geo-Inf.* **2020**, *9*, 603. [CrossRef]
3. Biljecki, F.; Lim, J.; Crawford, J.; Moraru, D.; Tauscher, H.; Konde, A.; Adouane, K.; Lawrence, S.; Janssen, P.; Stouffs, R. Extending CityGML for IFC-sourced 3D city models. *Autom. Constr.* **2021**, *121*, 103440. [CrossRef]
4. Jayaraj, P.; Ramiya, A.M. 3D CityGML building modelling from lidar point cloud data. In *The International Archives of Photogrammetry, Remote Sensing and Spatial Information Sciences*; Gottingen Tom XLII-5; Copernicus GmbH: Gottingen, Germany, 2018; pp. 175–180. [CrossRef]
5. Xu, Y.; Stilla, U. Towards Building and Civil Infrastructure Reconstruction From Point Clouds: A Review on Data and Key Techniques. *IEEE J. Sel. Top. Appl. Earth Obs. Remote Sens.* **2021**, *14*, 2857–2885. [CrossRef]
6. Tarsha Kurdi, F.; Awrangjeb, M.; Liew, A.W.-C. Automated Building Footprint and 3D Building Model Generation from Lidar Point Cloud Data. In Proceedings of the 2019 Digital Image Computing: Techniques and Applications (DICTA), Perth, Australia, 2–4 December 2019; pp. 1–8. [CrossRef]
7. Tarsha Kurdi, F.; Gharineiat, Z.; Campbell, G.; Dey, E.K.; Awrangjeb, M. Full Series Algorithm of Automatic Building Extraction and Modelling from LiDAR Data. In Proceedings of the 2021 Digital Image Computing: Techniques and Applications (DICTA), Gold Coast, Australia, 29 November–1 December 2021; pp. 1–8. [CrossRef]
8. Labetski, A.; Vitalis, S.; Biljecki, F.; Ohori, K.A.; Stoter, J. 3D building metrics for urban morphology. *Int. J. Geogr. Inf. Sci.* **2023**, *37*, 36–67. [CrossRef]
9. Pfeifer, N.; Rutzinger, M.; Rottensteiner, F.; Muecke, W.; Hollaus, M. Extraction of Building Footprints from Airborne Laser Scanning: Comparison and Validation Techniques. In Proceedings of the Joint IEEE-GRSS/ISPRS Workshop on Remote Sensing and Data Fusion over Urban Areas, Urban 2007, Paris, France, 11–13 April 2007. [CrossRef]
10. Wang, X.; Luo, Y.-P.; Jiang, T.; Gong, H.; Luo, S.; Zhang, X.-W. A New Classification Method for LIDAR Data Based on Unbalanced Support Vector Machine. In Proceedings of the 2011 International Symposium on Image and Data Fusion, Tengchong, China, 9–11 August 2011; pp. 1–4. [CrossRef]
11. Chen, D.; Zhang, L.; Mathiopoulos, P.; Huang, X. A Methodology for Automated Segmentation and Reconstruction of Urban 3-D Buildings from ALS Point Clouds. *IEEE J. Sel. Top. Appl. Earth Obs. Remote Sens.* **2014**, *7*, 4199–4217. [CrossRef]
12. Sampath, A.; Shan, J. Building Boundary Tracing and Regularization from Airborne Lidar Point Clouds. *Photogramm. Eng. Remote Sens.* **2007**, *73*, 805–812. [CrossRef]
13. Gilani, S.A.N.; Awrangjeb, M.; Lu, G. Segmentation of Airborne Point Cloud Data for Automatic Building Roof Extraction. *GIScience Remote Sens.* **2017**, *55*, 63–89. [CrossRef]
14. Jung, J.; Sohn, G. Progressive modeling of 3D building rooftops from airborne Lidar and imagery. In *Topographic Laser Ranging and Scanning: Principles and Processing*, 2nd ed.; Shan, J., Toth, C.K., Eds.; Taylor & Francis Group; CRC Press: Boca Raton, FL, USA, 2018; pp. 523–562. Available online: https://www.taylorfrancis.com/chapters/edit/10.1201/9781315154381-17/progressive-modeling-3d-building-rooftops-airborne-lidar-imagery-jaewook-jung-gunho-sohn (accessed on 25 June 2023).
15. Dey, E.K.; Tarsha Kurdi, F.; Awrangjeb, M.; Stantic, B. Effective Selection of Variable Point Neighbourhood for Feature Point Extraction from Aerial Building Point Cloud Data. *Remote Sens.* **2021**, *13*, 1520. [CrossRef]
16. Dong, Y.; Hou, M.; Xu, B.; Li, Y.; Ji, Y. Ming and Qing Dynasty Official-Style Architecture Roof Types Classification Based on the 3D Point Cloud. *ISPRS Int. J. Geo-Inf.* **2021**, *10*, 650. [CrossRef]
17. Tarsha Kurdi, F.; Awrangjeb, M.; Munir, N. Automatic filtering and 2D modeling of LiDAR building point cloud. *Trans. GIS* **2020**, *25*, 164–188. [CrossRef]
18. Mahphood, A.; Arefi, H. Density-based method for building detection from LiDAR point cloud. *ISPRS Ann. Photogramm. Remote Sens. Spatial Inf. Sci.* **2023**, *X-4/W1-2022*, 423–428. [CrossRef]
19. Park, S.-Y.; Lee, D.G.; Yoo, E.J.; Lee, D.-C. Segmentation of Lidar Data Using Multilevel Cube Code. *J. Sens.* **2019**, *2019*, 4098413. [CrossRef]
20. Cheng, L.; Zhang, W.; Zhong, L.; Du, P.; Li, M. Framework for Evaluating Visual and Geometric Quality of Three-Dimensional Models. *IEEE J. Sel. Top. Appl. Earth Obs. Remote Sens.* **2014**, *8*, 1281–1294. [CrossRef]
21. Ostrowski, W.; Pilarska, M.; Charyton, J.; Bakuła, K. Analysis of 3D building models accuracy based on the airborne laser scanning point clouds. In *International Archives of the Photogrammetry, Remote Sensing & Spatial Information Sciences*; ISPRS: Vienna, Austria, 2018; p. 42. Available online: https://ui.adsabs.harvard.edu/link_gateway/2018ISPAr.422..797O/ (accessed on 25 June 2023). [CrossRef]
22. Tarsha Kurdi, F.; Awrangjeb, M. Comparison of LiDAR Building Point Cloud with Reference Model for Deep Comprehension of Cloud Structure. *Can. J. Remote Sens.* **2020**, *46*, 603–621. [CrossRef]
23. Tarsha Kurdi, F.; Gharineiat, Z.; Campbell, G.; Awrangjeb, M.; Dey, E.K. Automatic Filtering of Lidar Building Point Cloud in Case of Trees Associated to Building Roof. *Remote Sens.* **2022**, *14*, 430. [CrossRef]
24. Gharineiat, Z.; Tarsha Kurdi, F.; Campbell, G. Review of Automatic Processing of Topography and Surface Feature Identification LiDAR Data Using Machine Learning Techniques. *Remote Sens.* **2022**, *14*, 4685. [CrossRef]
25. Adeleke, A.K.; Smit, J.L. Building roof extraction as data for suitability analysis. *Appl. Geomat.* **2020**, *12*, 455–466. [CrossRef]

26. Yang, W.; Liu, X.; Zhang, Y.; Wan, Y.; Ji, Z. Object-based building instance segmentation from airborne LiDAR point clouds. *Int. J. Remote Sens.* **2022**, *43*, 6783–6808. [CrossRef]
27. Axel, C.; Van Aardt, J. Building damage assessment using airborne lidar. *J. Appl. Remote Sens.* **2017**, *11*, 046024. [CrossRef]
28. Dorninger, P.; Pfeifer, N. A Comprehensive Automated 3D Approach for Building Extraction, Reconstruction, and Regularization from Airborne Laser Scanning Point Clouds. *Sensors* **2008**, *8*, 7323–7343. [CrossRef]
29. Liu, X.; Zhang, Y.; Ling, X.; Wan, Y.; Liu, L.; Li, Q. TopoLAP: Topology Recovery for Building Reconstruction by Deducing the Relationships between Linear and Planar Primitives. *Remote Sens.* **2019**, *11*, 1372. [CrossRef]
30. Li, Z.; Shan, J. RANSAC-based multi primitive building reconstruction from 3D point clouds. *ISPRS J. Photogramm. Remote Sens.* **2022**, *185*, 247–260. [CrossRef]
31. Liu, X.; Zhu, X.; Zhang, Y.; Wang, S.; Jia, C. Generation of concise 3D building model from dense meshes by extracting and completing planar primitives. *Photogramm. Rec.* **2023**, *38*, 22–46. [CrossRef]
32. Matikainen, L.; Hyyppä, J.; Hyyppä, H. Automatic detection of buildings from laser scanner data for map updating. In *International Archives of the Photogrammetry and Remote Sensing, XXXIV, 3/W13*; ISPRS: Dresden, Germany, 2003. Available online: https://www.isprs.org/proceedings/xxxiv/3-W13/papers/Matikainen_ALSDD2003.pdf (accessed on 25 June 2023).
33. Vosselman, G.; Dijkman, S. 3D Building Model Reconstruction from Point Clouds and Ground Plans. In *International Archives of the Photogrammetry and Remote Sensing, XXXIV, 3/W4*; ISPRS: Annapolis, MA, USA, 2001; pp. 37–44.
34. Wen, C.; Yang, L.; Li, X.; Peng, L.; Chi, T. Directionally constrained fully convolutional neural network for airborne LiDAR point cloud classification. *ISPRS J. Photogramm. Remote Sens.* **2020**, *162*, 50–62. [CrossRef]
35. Maltezos, E.; Doulamis, A.; Doulamis, N.; Ioannidis, C. Building Extraction from LiDAR Data Applying Deep Convolutional Neural Networks. *IEEE Geosci. Remote Sens. Lett.* **2018**, *16*, 155–159. [CrossRef]
36. Yuan, J. Learning Building Extraction in Aerial Scenes with Convolutional Networks. *IEEE Trans. Pattern Anal. Mach. Intell.* **2017**, *40*, 2793–2798. [CrossRef] [PubMed]
37. Kuras, A.; Brell, M.; Rizzi, J.; Burud, I. Hyperspectral and Lidar Data Applied to the Urban Land Cover Machine Learning and Neural-Network-Based Classification: A Review. *Remote Sens.* **2021**, *13*, 3393. [CrossRef]
38. Zhou, L.; Geng, J.; Jiang, W. Joint Classification of Hyperspectral and LiDAR Data Based on Position-Channel Cooperative Attention Network. *Remote Sens.* **2022**, *14*, 3247. [CrossRef]
39. Pantoja-Rosero, B.; Achanta, R.; Kozinski, M.; Fua, P.; Perez-Cruz, F.; Beyer, K. Generating LOD3 building models from structure-from-motion and semantic segmentation. *Autom. Constr.* **2022**, *141*, 104430. [CrossRef]
40. Fan, H.; Wang, Y.; Gong, J. Layout graph model for semantic façade reconstruction using laser point clouds. *Geo. Spat. Inf. Sci.* **2021**, *24*, 403–421. [CrossRef]
41. Gui, S.; Qin, R. Automated LoD-2 model reconstruction from very-high-resolution satellite-derived digital surface model and orthophoto. *ISPRS J. Photogramm. Remote Sens.* **2021**, *181*, 1–19. [CrossRef]
42. Peters, R.; Dukai, B.; Vitalis, S.; van Liempt, J.; Stoter, J. Automated 3D Reconstruction of LoD2 and LoD1 Models for All 10 Million Buildings of the Netherlands. *Photogramm. Eng. Remote Sens.* **2022**, *88*, 165–170. [CrossRef]
43. Zhang, Z.; Qian, Z.; Zhong, T.; Chen, M.; Zhang, K.; Yang, Y.; Zhu, R.; Zhang, F.; Zhang, H.; Zhou, F.; et al. Vectorized rooftop area data for 90 cities in China. *Sci. Data* **2022**, *9*, 66. [CrossRef] [PubMed]
44. Pang, H.E.; Biljecki, F. 3D building reconstruction from single street view images using deep learning. *Int. J. Appl. Earth Obs. Geoinform.* **2022**, *112*, 102859. [CrossRef]
45. Lewandowicz, E.; Tarsha, K.F.; Gharineiat, Z. 3D LoD2 and LoD3 Modeling of Buildings with Ornamental Towers and Turrets Based on LiDAR Data. *Remote Sens.* **2022**, *14*, 4687. [CrossRef]

remote sensing

MDPI

Article

Confidence-Guided Planar-Recovering Multiview Stereo for Weakly Textured Plane of High-Resolution Image Scenes

Chuanyu Fu [1], Nan Huang [1], Zijie Huang [1], Yongjian Liao [1], Xiaoming Xiong [2], Xuexi Zhang [1] and Shuting Cai [2,*]

[1] School of Automation, Guangdong University of Technology, Guangzhou 510006, China; zxxnet@gdut.edu.cn (X.Z.)

[2] School of Integrated Circuits, Guangdong University of Technology, Guangzhou 510006, China

* Correspondence: shutingcai@gdut.edu.cn

Abstract: Multiview stereo (MVS) achieves efficient 3D reconstruction on Lambertian surfaces and strongly textured regions. However, the reconstruction of weakly textured regions, especially planar surfaces in weakly textured regions, still faces significant challenges due to the fuzzy matching problem of photometric consistency. In this paper, we propose a multiview stereo for recovering planar surfaces guided by confidence calculations, resulting in the construction of large-scale 3D models for high-resolution image scenes. Specifically, a confidence calculation method is proposed to express the reliability degree of plane hypothesis. It consists of multiview consistency and patch consistency, which characterize global contextual information and local spatial variation, respectively. Based on the confidence of plane hypothesis, the proposed plane supplementation generates new reliable plane hypotheses. The new planes are embedded in the confidence-driven depth estimation. In addition, an adaptive depth fusion approach is proposed to allow regions with insufficient visibility to be effectively fused into the dense point clouds. The experimental results illustrate that the proposed method can lead to a 3D model with competitive completeness and high accuracy compared with state-of-the-art methods.

Keywords: confidence calculation; depth estimation; multiview stereo; plane supplementation; weakly textured regions

Citation: Fu, C.; Huang, N.; Huang, Z.; Liao, Y.; Xiong, X.; Zhang, X.; Cai, S. Confidence-Guided Planar-Recovering Multiview Stereo for Weakly Textured Plane of High-Resolution Image Scenes. *Remote Sens.* **2023**, *15*, 2474. https://doi.org/10.3390/rs15092474

Academic Editors: Jorge Delgado García, Tarig Ali and Fayez Tarsha Kurdi

Received: 20 March 2023
Revised: 25 April 2023
Accepted: 6 May 2023
Published: 8 May 2023

1. Introduction

Multiview stereo (MVS) is an important research topic in photogrammetry and computer vision. Over the last few years, impressive results [1–4] have been achieved in terms of the quality of 3D geometric representation reconstructed from multiview stereo. The reconstructed 3D model is applied in real-scene applications, such as digital cities, unmanned aerial vehicles (UAV), augmented reality (AR), and virtual reality (VR). The MVS, based on the PatchMatch algorithm [5,6], represents the most advanced MVS method. It aims at estimating depth maps using a set of 2D images with multiple views and then merging the dense 3D point clouds of the objects or scenes via depth fusion.

The PatchMatch algorithm can be divided into two main parts, including depth estimation and depth fusion. The depth estimation relies on the cost function based on photometric consistency, which is computed as normalized cross correlation (NCC) of corresponding patches between multiple views. Further, the NCC is expressed as the similarity of luminosity (pixel values or grayscale values) between different images' patches. The PatchMatch algorithm achieves adequately reliable results in strongly textured regions as well as in Lambert surfaces. However, it is mainly faced with the following challenges:

(1) Depending on the photometric consistency, traditional depth estimation [6,7] exhibits the fuzzy matching problem in weakly textured regions. The fuzzy matching problem is that even the erroneous plane hypothesis allows patches to match highly similar regions between multiple views. This makes depth estimation insufficiently reliable in weakly textured regions.

(2) During depth estimation, some views are invisible and cannot accurately reflect a reliable matching relationship due to occlusion and illumination. The matching cost calculated via invisible view would be an outlier in the multiview matching cost, which affects the accuracy of depth estimation.

In response to the above problems, some state-of-the-art methods [8–11] have been proposed. For outliers caused by invisible views in the multiview matching cost, a good idea is to determine the importance of each neighboring view, thereby altering the influence of each view in the multiview matching cost. Refs. [12–15] explored the contribution of neighborhood views to the multiview matching cost to achieve a highly accurate MVS. Ref. [15] designs a view weight to adjust the contribution of neighboring views in the multiview matching cost. It jointly estimates view selection and depth-normal information via a probabilistic graphical model. By using a generalized expectation maximization algorithm, each view would be assigned a view weight. The weighted multiview matching cost function effectively achieves highly accurate depth estimation. However, the view weights never change the essence of the fuzzy matching problem of photometric consistency. It makes [15] suffer from a significant inadequacy in terms of the completeness of the reconstruction, especially in weakly textured regions.

To solve the deficiency of [15] in completeness, we propose a plane supplement module, which is based on plane hypothesis confidence calculation. The generated reliable plane hypothesis is introduced into a confidence-driven depth estimation, which can effectively improve the completeness of the reconstruction. Meanwhile, confidence is embedded into the multiview matching cost as a constraint to overcome the fuzzy matching problem faced in photometric consistency.

In structured scenes, surfaces with weakly textured regions can be approximately characterized as identical planes. This allows the plane-based methodology [16–21] to effectively guide the elimination of the fuzzy matching problem that occurs in weakly textured regions, then improves the completeness of the reconstruction. Following their previous work, the authors of [18,22] introduce the prior plane to help the recovery of weakly textured regions. Firstly, the pixels with extremely small costs are selected for triangulation and interpolation. The generated triangular prior planes can effectively represent the planar structure of the scene. Secondly, the prior planes are introduced into the multiview cost function through a probabilistic graphical model. The new matching cost balances the photometric consistency with planar compatibility, thus improving the quality of reconstruction.

However, the problem with [18] is that the generation of prior planes is overly dependent on the photometric consistency cost, although incorrect prior planes may not be available due to the multiview matching cost. To address the problem, we propose a new confidence calculation method to express the reliability of the plane hypotheses in depth estimation. The calculated confidence consists of multiview consistency and patch consistency. Via the plane hypothesis confidence calculation, a confidence-driven depth estimation combined with the proposed planar supplement is effective in estimating reliable plane hypotheses.

In addition, the quality of the 3D models merged from the 2D depth maps is dependent on the depth fusion. The authors of [6,7] employ a consistent-matching-based depth fusion approach to obtain dense point clouds. A consistent match is defined as satisfying certain consistency constraints. The plane hypothesis would be accepted when allowing at least certain neighboring views satisfying the consistent match (view constraint). The pixels of accepted plane hypothesis are projected into the 3D space and averaged into uniform 3D points. Based on the depth fusion method of consistent matching, [15,18,22] tighten the constraints of consistent matching using the geometric error.

However, the fusion approach used in [7,15,18,22] relies on the fixed parameters of consistency constraints and view constrain. For regions that are only visible in finite neighborhood views, a rigorous view constraint may make it difficult to be fused into the point clouds. The regions are easily fused if the view constraint is loose, however, leading to a decrease in the accuracy of the point clouds. To address this problem, we propose a depth fusion method that adaptively adjusts view constraint and consistency constraints to improve the quality of the 3D point clouds.

In this paper, we propose an MVS pipeline using confidence calculation as guidance to recover reliable planar surfaces for weakly textured planes. To quantifiably express the reliability of the plane hypothesis in depth estimation, we propose a confidence calculation method consisting of multiview consistency and patch consistency. The plane supplement method is applied to additionally provide reliable planes, especially for the planar surfaces in weakly textured regions. Then, the reliable planes selected by the confidence calculation are embedded in the confidence-driven depth estimation. Finally, an adaptive fusion method can efficiently merge invisible regions into dense point clouds, and achieve a good balance between completeness and accuracy of reconstruction.

Our contributions are summarized as follows:

- To quantify the reliability of the plane hypothesis in depth estimation, a plane hypothesis confidence calculation is proposed. The confidence consists of multiview confidence and patch confidence, which provide global geometry information and local depth consistency.
- Based on the confidence calculation, a plane supplement module is applied to generate reliable plane hypotheses and is introduced into the confidence-driven depth estimation to tackle the estimating problem of weakly textured regions to achieve the high completeness of reconstruction.
- An adaptive depth fusion method is proposed to address the imbalance in accuracy and completeness of point clouds caused by fixed parameters. The view constraint and consistency constraints for fusion are adaptively adjusted according to the dependency of each view on different neighboring views. The method achieves a good balance of accuracy and completeness when merging depth maps into dense point clouds.

2. Related Works

According to [23], the pipelines of multiview stereo can be divided into four categories, which are voxel-based methods [24,25], surface evolution-based methods [26,27], feature point growing-based methods [28], and depth map merging-based methods [6,8,11,15,29].

The depth map merging-based approach is divided into two steps, which are depth map estimation and depth map fusion. Depth maps are estimated for all views, and then all depth maps are merged into the point clouds model based on the relationship between multiple views. Ref. [5] innovatively proposes slanted support windows to achieve highly slanted surface reconstruction with subpixel precision for disparity detail. Ref. [6] applies PatchMatch to MVS to estimate depth maps, and fuses them into point clouds by consistency matching.

The invisible neighboring view in multiple views becomes a disturbance to the accuracy of depth estimation. Ref. [12] heuristically selects the best view by minimum cost for accurate depth estimation. Refs. [13,30] model scene visibility and local depth smoothing assumptions by Markov random fields for pixel-level view selection. Ref. [14] jointly models pixel-level view selection and depth map estimation via a probabilistic framework to adaptively determine pixel-level data associations between the current view and all elements of neighboring views. By discussing the support window selection, visibility determination, and outlier detection, Ref. [9] proposes an accurate visibility estimation method to achieve high-accuracy reconstruction. Ref. [15] establishes a pixelwise view selection scheme and jointly estimates the view selection, as well as depth-normal information, by a probabilistic graphical model.

The sequential propagation in the PatchMatch-based MVS method is an important factor affecting the efficiency of depth estimation. Ref. [10] implements a GPU-based parallel propagation of the red-black checkerboard scheme to accelerate the propagation process of MVS. Ref. [22] proposes adaptive checkerboard propagation and multihypothesis joint view selection to obtain efficient and high-quality reconstruction, which is named

ACMH. On this basis, reliable estimation of weakly textured regions at coarse scales is applied to fine scales in combination with multiscale geometric consistency guidance, which is named ACMM. Similarly based on the adaptive checkerboard propagation scheme, Ref. [18] proposes the prior plane generation method and embeds it into the matching cost calculation, utilizing a probabilistic graphical model. Ref. [4] integrates the two aforementioned works to achieve an extremely competitive 3D reconstruction.

The fuzzy matching problem faced in weakly textured regions greatly affects the completeness of the reconstruction. Ref. [1] adaptively adjusts the patch size by curvature model to attenuate the ambiguity of matching. Ref. [31] considers local consistency in the matching cost and completes the MVS with high integrity in the pyramid structure. Ref. [3] combines dynamic propagation and sequential propagation and introduces coarse inference within a universal window to eliminate artifacts to improve the completeness of reconstruction. Ref. [16] proposes a texture-aware MVS and fills the vacant planes by superpixels after filtering outliers. Ref. [17] improves a planar complementation method by growing superpixels to complement the filtered depth map. Ref. [32] combines the relationship between multiple views while using superpixels to make the complementation more robust. Ref. [20] proposes a plane prior generation method by combining mean-shift clustering and superpixel segmentation, then introduces planar priors and smooth constraints into the cost. The image gradient is used to adaptively adjust the weights of different constraints in the cost. Ref. [21] designs a quadtree-guided prior method and embeds it into the matching cost calculation to improve the estimation of weakly textured regions.

The reconstruction of geometric details is also an important research problem. Ref. [33] proposes a selective joint bilateral propagation upsampling method for recovering the depth maps at coarse scales to geometric details at fine scales. Ref. [2] focuses on the geometric details of reconstruction, especially the preservation of geometric details of thin structures. Ref. [34] considers three types of filters to achieve an outlier and artifact removal method for MVS.

Recently, a popular research approach combines the traditional MVS pipeline with deep learning, which is about confidence. Confidence prediction is widely used in stereo problems [35–37]. In multiview stereo, the photometric consistency is stably supported by multiple views but is still not reliable. Ref. [38] proposes a self-supervised learning method to predict the confidence of multiview depth maps and constructs high-quality reconstructions by confidence-driven and boundary-aware interpolation. Ref. [39] proposes a confidence prediction method, which is a network structure where RGB images, normal maps, and scale-robust TSDF are globally fused by U-Net architecture with intermediate loss and refined by an iterative refinement module with a later-fusion layer and LSTM layer. Ref. [40] customizes a confidence prediction network for MVS using DNNs and uses it for depth map outlier filtering and depth map refinement. Ref. [41] uses a pyramid structure to guide the fine-scale MVS process using a grid at coarse scales, and a deep neural network is designed to predict the confidence.

The successful extraction of semantic information contributes to the further development of the quality of 3D reconstruction. Refs. [42,43] explore the possibility of semantic segmentation for application in MVS. Further, Ref. [44] utilizes semantic segmentation-guided prior planes to tackle the weak texture problem in PatchMatch MVS. Ref. [19] combines the MVS with PlaneNet to repair incorrect points by correcting and integrating inaccurate prior information from pretrained CNN models and depth map merging methods, then interpolating in weak support planes.

3. Review of Depth Estimation in ACMH

In this section, we review an advanced PatchMatch-based MVS algorithm, ACMH [22]. It follows the basic four-step PatchMatch-based MVS algorithm [6]. The purpose of this section is to help clearly understand the details of the depth estimation in our framework.

ACMH adopts a propagation scheme of adaptive checkerboard sampling, and ameliorates the calculation of the multiview matching cost function through multihypothesis joint view selection, thus achieving extremely high accuracy while parallelizing. The entire depth estimation can be summarized as follows:

3.1. Initialization

ACMH generates a random initial plane hypothesis for each pixel. Then, the bilateral weighted NCC [15] is calculated as the matching cost between the current view and each neighboring view. The initial multiview matching cost is calculated as the average of the top five best matching costs.

3.2. Propagation

Based on the diffusion-like propagation scheme [10], ACMH modifies the selection of neighborhood plane hypotheses to four V-shaped areas and four long strip areas (Figure 1). According to the multiview matching cost, the plane hypothesis with the minimum cost is selected as the candidate in each of the eight areas.

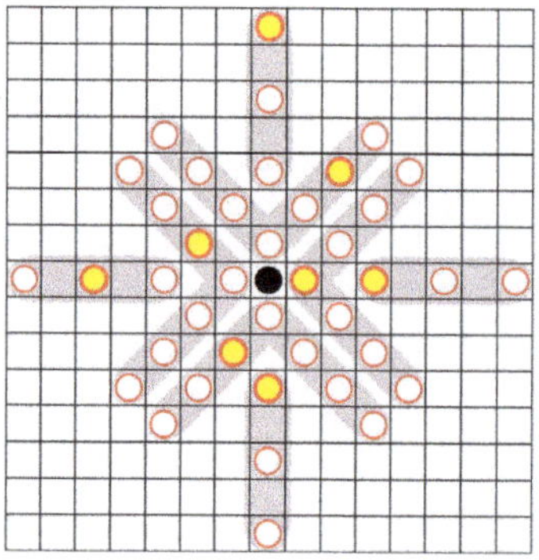

Figure 1. The adaptive checkerboard propagation scheme of ACMH. Each V-shaped area contains 7 sampling pixels, and each long strip area contains 11 sampling pixels.

3.3. Multiview Matching Cost Calculation

For each pixel p, the matching costs between all neighboring views are calculated and embedded into a cost matrix M according to its original plane hypothesis and eight candidate plane hypotheses obtained in propagation,

$$M = \begin{bmatrix} m(\phi_0, 1) & \cdots & m(\phi_0, J) \\ \vdots & \ddots & \vdots \\ m(\phi_8, 1) & \cdots & m(\phi_8, J) \end{bmatrix} \tag{1}$$

where $m(\phi_i, j)$ is the matching cost between the i-th plane hypothesis ϕ_i corresponding to the j-th neighboring view, and J is the total number of neighboring views.

To mitigate the impact of unreliable neighborhood views, ACMH selects an appropriate subset from all neighboring views according to the cost matrix. Then, the reliable neighboring views are given large view weights.

In each column of the cost matrix, a voting decision is adopted to determine the suitability of the view. For the neighboring view V_j of the t-th iteration, the S_{good} is defined as the set whose $m(\phi_i, j) < \tau(t)$, and S_{bad} is the set whose $m(\phi_i, j) > \tau_b$. The parameter τ_b is a constant, and $\tau(t)$ is modeled as

$$\tau(t) = \tau_{init} \cdot e^{-\frac{t^2}{\mu}} \tag{2}$$

where τ_{init} is the initial cost threshold and μ is a constant. The selected subset of neighboring views contains all neighboring views whose $S_{good} > n_1$ and $S_{bad} < n_2$.

Furthermore, to determine the view weight of each neighboring view in the subset, the cost confidence is calculated based on the matching cost,

$$C(m(\phi_i, j)) = e^{-\frac{m(\phi_i,j)^2}{2\beta^2}} \tag{3}$$

Based on the cost confidence, the initial view weight of the neighboring view V_j is calculated as

$$w_{init}\left(V_j\right) = \begin{cases} \frac{1}{|S_{good}|} \sum\limits_{m(\phi_i,j) \in S_{good}} C(m(\phi_i, j)), & V_j \in S_t; \\ 0, & else. \end{cases} \tag{4}$$

After iterative propagation, according to the most important view (the view with the largest weight), the calculation of view weight is modified as

$$w_j' = \begin{cases} (\Lambda(V_j = v_{t-1}) + 1) \cdot w_{init}(V_j), & V_j \in S_t; \\ 0.2 \cdot \Lambda(V_j = v_{t-1}), & else. \end{cases} \tag{5}$$

where $\Lambda(\cdot)$ means that $\Lambda(true) = 1$ and $\Lambda(false) = 0$. According to the calculated view weight, the multiview matching cost is calculated as

$$m(p, \phi_i) = \frac{\sum\limits_{j=1}^{N-1} w'_j \cdot m(\phi_i, j)}{\sum\limits_{j=1}^{N-1} w'_j} \tag{6}$$

The original plane hypothesis of pixel p is updated to the plane hypothesis, with the minimum multiview matching cost calculated in the set of plane hypotheses.

3.4. Refinement

In the refinement, each plane hypothesis is made as close as possible to the global optimal solution after propagation. Random plane hypothesis (d_{rand}, n_{rand}) and perturbed plane hypothesis (d_{pert}, n_{pert}) are generated based on the plane hypothesis (d_p, n_p) of pixel p. The new set of plane hypotheses is combined as (d_p, n_p), (d_{prt}, n_p), (d_{rnd}, n_p), (d_p, n_{prt}), (d_p, n_{rnd}), (d_{rnd}, n_{rnd}), (d_{prt}, n_{prt}). The plane hypothesis of pixel p is updated to the one with the minimum multiview matching cost in the set.

Finally, the steps of propagation, multiview matching cost calculation, and refinement are iterated several times to make the plane hypothesis of each pixel converge to the global optimal solution.

4. Method

4.1. Overview

Given a set of views with known camera parameters, the goal of depth estimation is to obtain each plane hypothesis in views, which contain both depth information and normal information.

Then, all the depth maps are merged into dense point clouds. The whole framework of our method is shown in Figure 2.

Firstly, the sparse point clouds are reconstructed via structure from motion (SFM). Then, the set of views, camera parameters, and sparse point clouds are jointly input into our framework. At the beginning of our framework, we follow the scheme in ACMH [22] to obtain the coarse depth maps with structural details. Via the confidence calculation, reliable plane hypotheses in coarse depth maps are identified and extracted. In plane supplementation, the images with the extracted reliable planar hypotheses are divided into multiple triangular primitives by Delaunay triangulation. For the low-confidence regions in triangular primitives, new reliable planes are generated via triangular interpolation of the extracted reliable plane hypotheses. Afterward, the most accurate plane hypotheses between the supplement planes and coarse depth maps are retained and embedded into the confidence-driven depth estimation. Finally, the depth maps are converted into dense point clouds according to the adaptive fusion approach.

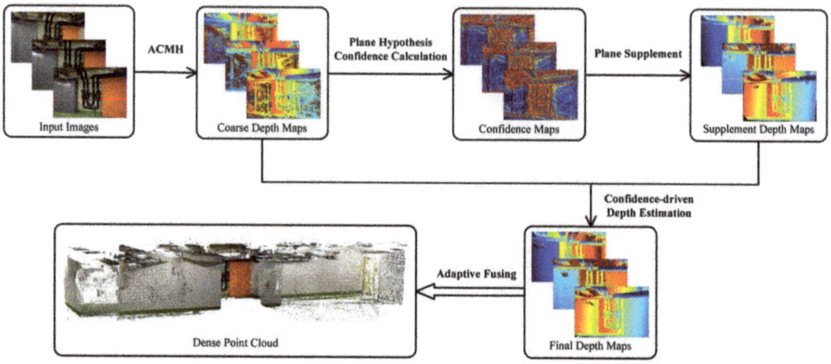

Figure 2. Overview of our method. The framework is divided into five parts, namely depth estimation of ACMH, confidence calculation module, plane supplementation module, confidence-driven depth estimation, and adaptive fusion. Further, both the depth estimation of ACMH and the confidence-driven depth estimation follow the basic four steps of the PatchMatch-based MVS algorithm, namely initialization, propagation, multiview matching cost calculation, and refinement.

4.2. Plane Hypothesis Confidence Calculation

During the original depth estimation, photometric consistency appears as the problem of fuzzy matching in weakly textured regions. The problem is demonstrated by the fact that the depth of the incorrect plane hypothesis can make it possible to match highly similar regions between multiple views, making the multiview matching cost lack credibility. Ref. [15] attempts to add geometric consistency constraints to the multiview matching cost to reduce the erroneous plane hypotheses in weakly textured regions. Ref. [31] tries to add local consistency constraints to eliminate incorrect plane hypotheses.

However, photometric consistency would perfectly characterize the structure of the objects or scenes in structured scenes. Adding constraints to the matching cost certainly allows the fuzzy matching problem that occurs in weakly textured regions to be solved to some extent. However, the new constraints may blur the geometric details in the object or scene.

In contrast, we would like to capture which plane hypotheses are accurate enough to be represented the real objects or scenes after each depth estimate. Therefore, we propose a new confidence calculation method. The confidence expresses the degree of reliability of each plane hypothesis. For the plane hypothesis with large confidence, we consider that the plane hypothesis would accurately indicate the real surface of the scene or object. The plane hypothesis with low confidence is considered to be an incorrect estimation, and these incorrect plane hypotheses need to be filtered or upgraded. In the confidence calculation, the confidence is divided into two parts, including the multiview confidence and the patch confidence.

In multiview stereo, an assumption is that a reliable plane hypothesis should be geometrically stable between multiple views. Thus, based on the relationship of multiple views, a measure of multiview confidence is established firstly, which means the consistency degree of plane hypotheses among multiple views.

Note that the multiview confidence is calculated based on all neighboring views. The component of multiview confidence, which is calculated between the current view and one neighboring view, is defined as the view confidence.

Given a pixel p, its plane hypothesis ϕ_p is (d_p, n_p). The multiview geometry is obtained as shown in Figure 3. The camera projection matrix from the current view V_i to the neighboring view V_j is calculated according to d_p. The pixel p is projected into the neighboring view V_j to obtain the projected pixel q, and the plane hypothesis ϕ_q of q would be obtained in the neighboring view. The pixel q is reprojected back into the current view to obtain the reprojection point p' according to the camera projection matrix from the neighboring view to the current view, which is calculated with d_q. By reprojecting the point p', the plane hypothesis $\phi_{p'}$ of p' corresponding to the current view V_i can be obtained.

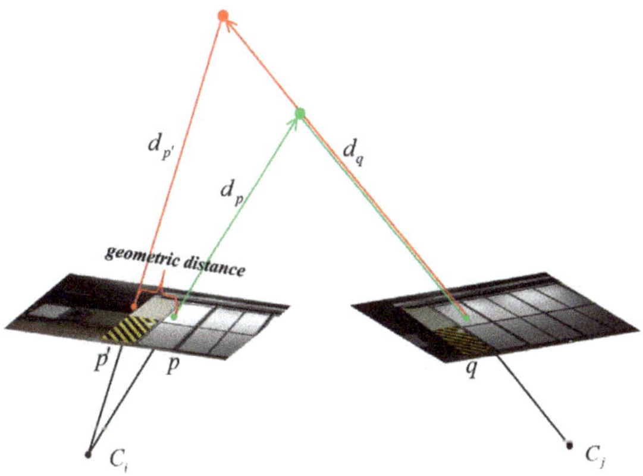

Figure 3. The diagram of multiview geometry, which is using a neighboring view of multiple views as an example. C_i and C_j are the camera center of the current view and neighboring view, respectively.

The view relationship between the current view and j-th neighboring view can be described via the reprojection distance ξ_{geo}, depth relative error ξ_d, normal pinch error ξ_n, and the matching cost $m(\phi_p, j)$. Therefore, ξ_{geo}, ξ_d, ξ_n is calculated according to the pixel p, p' and the plane hypothesis (d_p, n_p), $(d_{p'}, n_{p'})$. Then, the view confidence consists of the geometric confidence C_{geo}, the depth confidence C_d, the normal confidence C_n, and the cost confidence C_c, which are calculated via the Gaussian function,

$$C_{geo} = e^{-\frac{\xi_{geo}^2}{2\sigma_{geo}^2}} = e^{-\frac{\left(\left\|x_p - x_{p'}\right\|\right)^2}{2\sigma_{geo}^2}} \tag{7}$$

$$C_d = e^{-\frac{\xi_d^2}{2\sigma_d^2}} = e^{-\frac{1}{2\sigma_d^2}\left[\frac{\left(d_p - d_{p'}\right)}{d_p}\right]^2} \tag{8}$$

$$C_n = e^{-\frac{\xi_n^2}{2\sigma_n^2}} = e^{-\frac{\left[\arccos\left(n_p \cdot n_{p'}\right)\right]^2}{2\sigma_n^2}} \tag{9}$$

$$C_c = e^{-\frac{m(\phi_p, j)^2}{2\sigma_c^2}} \tag{10}$$

where $\sigma_{geo}, \sigma_d, \sigma_n$ and σ_c are constants in the C_{geo}, C_d, C_n and C_c, respectively.

In a set of neighboring views $\mathcal{V} = \{V_j \mid j = 1, 2, \ldots, J\}$, there exist J sets of confidence relations for neighboring view. The multiview confidence is calculated as the average of the view confidence with all neighboring views:

$$\bar{C}_g = \frac{\sum_{j=1}^{J}\left(C_{geo}^j \cdot C_d^j \cdot C_n^j \cdot C_c^j\right)}{J} \tag{11}$$

A good plane hypothesis should be supported by multiple neighboring views. When the number of neighboring views that maintain consistency is increased, the trustworthiness of the planar hypothesis is improved. Meanwhile, the geometric stability in multiple views is increased. However, the camera's pose variation and the presence of occlusion determine that not all regions in a view can be consistent with multiple views. Specifically, some regions in a view are only visible in a limited number of neighboring views. Thus, in these regions, calculating the view confidence with all neighboring views may cause the correct plane hypothesis to be judged as unreliable. To this end, the multiview confidence calculation is modified to be the average of the best K neighboring views among all neighboring views.

$$C_g = \frac{\sum_{j=1}^{K} \left(C_{geo}^j \cdot C_d^j \cdot C_n^j \cdot C_c^j \right)}{K} \tag{12}$$

The global spatial information is fully considered in the multiview confidence, which is based on the consistency of multiple measurements. The multiview confidence calculation makes most plane hypotheses easy to calculate as reliable estimates with high confidence. However, for some plane hypotheses that are correctly estimated in current view, erroneous multiview confidences are calculated because of wrong plane hypotheses in neighboring views. In addition, because of similar plane hypotheses in multiple views, some noise in the current view may be calculated as high-confidence and retained, especially in weakly textured regions.

In order to reduce the calculation of error confidence, a patch confidence measure based on depth local consistency is added, which only relies on the information in the current view. In the PatchMatch algorithm [5,45], a key statement is that relatively large regions of pixels can be modeled by an approximately 3D plane. It allows the same plane hypotheses to be shared within the pixel regions. The statement can be beneficial to help exploit the local information in a view. To this end, the patch confidence is structured as a calculation based on the consistency of local planes. For each pixel in the current view, a cruciform patch is constructed, centered on the pixel. Firstly, a 3D local plane is constructed in the camera coordinate via the central pixel's 3D point X_c and its corresponding plane hypothesis. Secondly, neighboring pixels in a cruciform patch are projected into the same camera coordinate to obtain 3D points X_n. The average Euclidean distance from the 3D points of neighboring pixels to the local 3D plane is calculated (refer to Figure 4),

$$\overline{\xi} = \frac{1}{N} \cdot \sum_{n=1}^{N} \xi_n = \frac{1}{N} \cdot \sum_{n=1}^{N} \frac{n_c \cdot (X_c - X_n)}{\sqrt{n_{cx}^2 + n_{cy}^2 + n_{cz}^2}} \tag{13}$$

where N is the number of pixels in a cruciform patch and n_c is the normal of patch center pixel. n_{cx}, n_{cy}, n_{cz} are the three components of normal n_c.

Based on the calculated average Euclidean distance, patch confidence is constructed by the Gaussian function as well:

$$C_l = e^{-\frac{\overline{\xi}^2}{2\sigma_p^2}} \tag{14}$$

where σ_p is a constant parameter of patch confidence.

Finally, via the calculated multiview confidence and patch confidence, the confidence of the pixel p can be expressed as

$$C(p, \phi_p) = C_g \cdot C_l \tag{15}$$

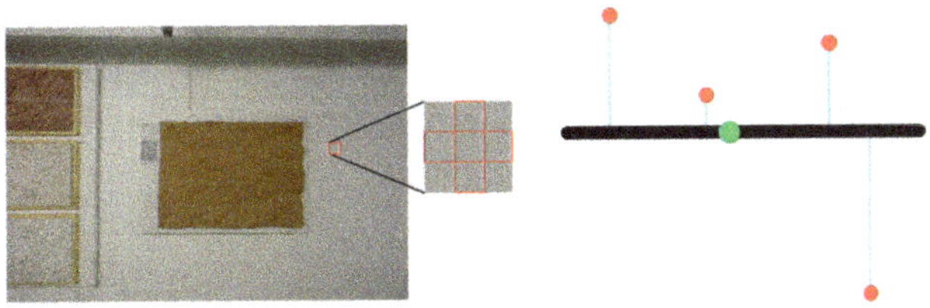

(a) (b)

Figure 4. The diagram of patch confidence. (**a**) Building a cruciform patch with the local window of 3 × 3, which contains four neighborhood pixels. (**b**) Calculation of Euclidean distance in the camera coordinate. The green point is the 3D point of the center pixel, the black line is the plane hypothesis of the center pixel, the red points are 3D points corresponding to neighborhood pixels in the patch, and the blue line is the Euclidean distance.

4.3. Plane Supplement and Confidence-Driven Depth Estimation

The purpose of the propagation scheme is that the reasonable plane hypotheses can be propagated to other pixels in the same plane, making the estimation accurate and reliable. However, in the weakly textured regions, it is difficult to select the correct plane hypothesis using the matching cost function based on luminosity consistency, because the weakly textured regions usually do not contain discriminative information. This makes it difficult for incorrect plane hypotheses to be replaced by correct neighborhood candidate planes via the propagation scheme, and these incorrect plane hypotheses may be propagated to other pixels due to the propagation scheme.

This means that relying on existing plane hypotheses cannot help the reconstruction of weakly textured regions. Ref. [18] chooses the base point via photometric consistency cost to generate the prior plane and introduces them into the calculation of multiview matching cost. However, the photometric consistency is not reliable in weakly textured regions, giving prior planes wrong plane hypotheses. In addition, the photometric consistency cost of incorrect planar hypothesis may be sufficiently small in weakly textured regions. Despite using prior planes as a constraint in the calculation of multiview matching cost, it does not allow these errors to recompute an aggregation cost large enough to be replaced by correct plane hypotheses contained in prior planes. It keeps the wrong plane hypotheses in weakly textured regions of depth maps.

In Section 4.2, the confidence is proposed to discriminate the accuracy and reliability of plane hypotheses to avoid misjudgment of photometric consistency in weakly textured regions. After the plane hypothesis confidence calculation, pixels with high confidence (we set the confidence threshold δ_c to 0.8) are extracted from the coarse depth map. The key observation behind this is that these pixels with high confidence mostly contain the structure of 3D scenes. Meanwhile, the planar hypotheses of extracted pixels are accurate and reliable, because they are supported by multiple views and are consistent in local planes. Using the extracted pixels as base points, the images are divided into multiple triangular primitives with different sizes using Delaunay triangulation [46]. Then, based on the depths of the three base points in the triangular primitives, a local 3D plane where the triangular primitives are located is constructed. For the low-confidence pixels contained in each triangular primitive, they are projected into the local 3D plane to obtain new depths, resulting in additional supplemental depth maps.

The supplemental planes perform well in weakly textured regions, especially those with large planes. However, some edge regions are blurred, which is contrary to photometric consistency. The coarse depth map that depends on photometric consistency is calculated with higher confidence than the supplemental depth map in these edge regions. Conversely, the confidence of the supplemental depth map is better than the coarse depth map in weakly textured regions.

Thus, the coarse depth map and supplemental depth map are jointly fed to a comparison module. Specifically, after obtaining the supplemental depth maps, the plane hypothesis confidence calculation module is reapplied to calculate the confidence for each plane hypothesis in the supplemental depth map. The confidences calculated in the supplemental depth map are compared with the coarse depth map, and the planar hypotheses with higher confidence are retained.

Subsequently, the retained plane hypotheses are used as the initial values for the confidence-driven depth estimation. An important reason for confidence-driven depth estimation is that there are still some erroneous plane hypotheses mixed in with the retained plane hypotheses. These noises tend to exhibit low confidence in both the supplemental depth maps and the coarse depth maps. These noises can be effectively reduced with the help of the propagation mechanism and the modified cost function. The results obtained by combining the supplemental depth maps and the coarse depth maps lose partial structural details. Since the photometric consistency cost has a significant result in textured regions, it is possible to exploit this advantage to help the recuperation of these textured regions. In addition, plane supplementation has a significant recovery for planar surfaces in weakly textured regions. However, there is a subtle variation in the plane hypotheses in curved surfaces of weakly textured regions, which causes a slight decrease in the accuracy of our plane supplement. The propagation step and the refinement step in the depth estimation can effectively help these curved surfaces to produce the correct variations of plane hypotheses instead of keeping them in the same plane.

In the confidence-driven depth estimation, the processes of propagation and refinement are kept in line with the ACMH [22], which are reviewed in Section 3. In particular, for the multiview matching cost calculation, confidence is used as a constraint to limit the propagation of incorrect plane hypotheses with low confidence to other pixels. Meanwhile, planar hypotheses with high confidence can be easily propagated to other pixels of the same plane with the help of the propagation scheme. According to the Equation (6), the confidence-driven multiview matching cost function is modeled as

$$m(p,\phi_p) = \frac{\sum_{j=1}^{N-1} w'_j \cdot m(\phi_p, j) + \lambda \cdot (1 - C(p,\phi_p))}{\sum_{j=1}^{N-1} w'_j} \tag{16}$$

where $C(p,\phi_p)$ is the confidence of pixel p, which is calculated with the plane hypothesis ϕ_p, and λ is a weight constant.

For weakly textured regions, the matching cost of photometric consistency computed by different plane hypotheses is usually similar because of the lack of distinguishability information. It causes the propagation mechanism in traditional MVS to easily transmit erroneous plane hypotheses to other pixels in these regions, and is difficult to replace. According to the modified confidence-driven multiview matching cost, the determining factor for propagation mechanism to judge the reliability of the candidate plane hypotheses is confidence. Because the confidence level calculated in the noise is small, the multiview matching cost calculated in the noise is larger than the correct plane hypothesis. Thus, the confidence-driven multiview matching cost would be helpful to address the problem of propagation mechanism for candidate plane hypothesis selection. Meanwhile, plane hypotheses with high confidence can be easily transmitted to pixels in the same plane because they are computed at a low cost. For the structural detail regions, the important factor that dominates the propagation mechanism's selection of candidate plane hypotheses changes to the photometric consistency matching cost. The reason is that the calculated confidences are all great in these regions. Thus, the detail regions that were previously blurred and erroneous would be improved.

To avoid the complexity of repeated confidence calculations due to changes in plane hypotheses, the confidence-driven depth estimation is restricted to obtaining the final depth maps with one propagation. Via this confidence-driven depth estimation, the final depth maps preserve the structural details well and improve the estimation quality of weakly textured regions.

4.4. Adaptive Fusion

After depth estimation, all the depth maps of views are obtained. In the depth map fusion step, all the depth maps are merged into the dense point clouds. In [6,7], all the depth maps are fused by consistent matching with a fixed threshold. Specifically, for each pixel, it is projected into each neighboring view via its depth of plane hypothesis. Then, it is reprojected back to current view by the depth of hypothesis, which is in the neighboring view. The corresponding matching relationship can be obtained based on the reprojected point and the pixel in the current view. A consistent matching is defined as satisfying the consistent constraints, including depth difference $\delta_d \leq 0.01$ and normal angle $\delta_n \leq 10$. For all neighboring views, if there exist $n \geq \delta$ neighboring views (defining δ as the view constraint) satisfying the consistent matching, the hypothesis is accepted. Finally, all pixels that satisfy the consistency matching are projected into the 3D space and averaged into uniform 3D points, thus becoming part of the dense 3D point clouds. Refs. [15,18,22] further tighten the consistent constraints of consistency matching on this depth fusion approach; the reprojection geometry error $\delta_{geo} \leq 2$ should be satisfied.

However, we observe that such a depth map fusion approach relies on fixed consistent constraints and fixed view constrain. There are always situations where some regions of the current view are only visible in a limited number of neighboring views. Then, too large a view constraint will cause these regions cannot to be fused into the dense point clouds, resulting in a lack of completeness. Too small a view constraint ensures the fusion of these areas, but leads to a decrease in the overall reconstruction quality, especially in terms of accuracy.

To solve this problem, an adaptive depth fusion approach is developed. Specifically, the view weight is added to each neighboring view when calculating the multiview matching cost. Such view weights can reflect the visibility relationships of pixels in multiple views. At the end of the last depth estimation, the view weights corresponding to all neighboring views of all pixels are retained. In the depth map fusion step, firstly, all neighborhood view weights corresponding to each pixel are sorted from large to small. Secondly, based on the distribution changes of neighborhood view weights, the view constraints $\delta(V_i)$ can be adjusted adaptively.

$$\delta(V_i) = \begin{cases} j, & w'_j > \delta_w \cap w'_j < \delta_w \cap j \leq 4; \\ 4 & j > 4. \end{cases} \tag{17}$$

where w'_j denotes the j-th sorted view weight of neighboring views. δ_w is the threshold of the view weight. After sorting, the comparison starts from the largest neighborhood view weight to the threshold of view weight. For the view weight $w'_j \geq \delta_w$, we consider the pixel to be visible in corresponding neighboring view. The view constraint is adaptively adjusted to the number of neighboring views accumulated. Until the j-th view weight $w'_j \leq \delta_w$, we consider that the pixel's visibility starts to be insufficient. In addition, the main goal of our adaptive fusion is to ensure accuracy while improving the integrity of invisible regions. The increase in view constraint indicates that the visibility of the regions is satisfied in multiple neighborhood views, but it becomes difficult to satisfy the consistency of plane hypotheses between multiple views. To prevent the influence of excessive view constraint on the completeness of these regions, the view constraint is phased at the maximum value of 4.

Simultaneously, to ensure as much as possible that the adaptive view constraints are adjusted by visibility judgments rather than resulting in incorrect plane hypotheses, the consistency constraints of consistent matching are adaptively adjusted according to the size of the view constraint. For pixels with small view constraints, the consistency constraints are tightened to ensure that their plane hypotheses are accurate enough. For pixels with large view constraints, the consistency constraints are relaxed appropriately, allowing the pixels supported via multiple neighboring views to be easily merged into point clouds to improve the completeness of reconstruction.

$$\delta_d^{new} = [\ln(2 \cdot \delta(V_i) - 1) + 1] \cdot \delta_d \tag{18}$$

$$\delta_n^{new} = [\ln(2 \cdot \delta(V_i) - 1) + 1] \cdot \delta_n \tag{19}$$

$$\delta_{geo}^{new} = [\ln(2 \cdot \delta(V_i) - 1) + 1] \cdot \delta_{geo} \tag{20}$$

where δ_d, δ_n, δ_{geo} is the strictest consistency constraint when the view constraint $\delta(V_i)$ is 1. With the view constraint increased, the consistency constraints become loose, making pixels easy to be merged into dense point clouds when they are visible among multiple views.

In addition, for outdoor scenes, the sky regions become redundant in the dense point clouds, because the sky regions lack true depth. Through a guided-filter-based mask refinement method, Ref. [47] uses a neural network and weighted guided upsampling to create accurate sky alpha masks at high resolution, resulting in the segmentation of sky regions. Thus, before the beginning of the depth fusion step, the method in [47] is applied to filter out the plane hypotheses of sky regions contained in the depth maps. The sky-filtering step has almost no effect on the calculation of the quantifiers. However, we can obtain clean depth maps as well as dense point clouds.

With the depth map fusion approach described above, we can obtain dense 3D point clouds with high completeness and accuracy.

5. Experiments

The proposed CGPR-MVS is implemented in C++ with CUDA. To evaluate the proposed pipeline, we perform quantitative and qualitative evaluations on the published dataset ETH3D benchmark [48]. In addition, the qualitative evaluation is performed on the sensefly dataset. The experiments were conducted on a machine equipped with Intel Xeon E5-1630 v4 CPU, 64G RAM, and NVIDIA Quadro K2200 GPU.

The ETH3D benchmark contains both high-resolution datasets and low-resolution datasets for the MVS task. Further, the high-resolution dataset is divided into the training branch and the test branch, with all images having a resolution of 6048 × 4032. The training branch dataset contains 13 sequences of indoor and outdoor scenes, with additional ground truth point clouds and ground truth depth maps. The test branch dataset contains 12 sequences of indoor and outdoor scenes, and the evaluation is only available by uploading to the online website, and the ground truth data are not publicly available. During the evaluation, the quality of the reconstruction results is quantified in three metrics as accuracy, completeness, and F_1 score. The accuracy is the percent fraction of the reconstruction, which is closer to the ground truth than the evaluation tolerance. The completeness is the percent fraction of the ground truth, which is closer to the reconstruction than the evaluation tolerance. The F_1 score is the harmonic mean of accuracy and completeness. For details of the whole evaluation, please refer to [48].

The sensefly datasets are collected from real remote sensing images captured by various drones with different cameras from AgEagle, a company that provides fixed-winged drones and aerial imagery-based data collection and analytics solutions. The datasets contain several different scenes, each with different flight heights and applied in different practical applications. A challenging scene is selected to test in our experiment. The dataset of Thammasat University campus in Bangkok, Thailand was collected by an eBee X drone carrying an Aeria X photogrammetry camera. The drone flew at a height of 285 m, photographed scenes covering 2.1 square kilometers, and captured high-resolution images of 6000 × 4000. Usually, these collected datasets are used for 3D mapping, regular updating of city maps, inspecting infrastructure, monitoring construction projects, and studying architectural aspects.

5.1. Parameter Settings

Firstly, the undistorted images are downsampled to the resolution of 3200 × 2130 for reconstruction. For all datasets, the same set of parameters is used in the experiments. The specific parameter settings are shown in Table 1.

Table 1. Parameter settings of experiments and their meaning.

Parameter	Meaning	Value
σ_{geo}	constant of geometric confidence	5.0
σ_d	constant of depth confidence	0.05
σ_n	constant of normal confidence	0.8
σ_c	constant of cost confidence	0.5
K	best K neighboring views	2
σ_p	constant of patch confidence	1.0
δ_c	confidence threshold	0.8
λ	constant of confidence constraint in multiview matching cost	2.0
δ_w	threshold of view weight	0.6
δ_d	the strictest depth difference	0.01
δ_n	the strictest normal angle	0.15
δ_{geo}	the strictest geometry error	1.5

5.2. Quantification

Some state-of-the-art MVS methods and our method were compared by quantitative evaluation on the high-resolution of ETH3D benchmark, including Gipuma [10], COLMAP [15], ACMH [22], OpenMVS [7], ACMP [18], CLD-MVS [38], QAPM [21]. The quantitative evaluation performance of the training branch dataset and the test branch dataset are shown in Tables 2 and 3 respectively. Note that the evaluation tolerance is 2 cm, as defaulted by the ETH3D benchmark. The quantitative evaluation of other approaches is dependent on the published results on the online website of the ETH3D benchmark.

As shown in Table 2, the proposed method achieves the best performance of F_1 score and completeness compared with other methods in the training branch dataset, except for outdoor scenes, where completeness is slightly inferior to QAPM. The main contribution of Gipuma is the parallelized red–black checkerboard propagation, which brings a huge improvement in the efficiency of depth estimation. However, its performance is far inferior to other schemes in the quality of reconstruction, because it simply selects the top-k minimum matching cost for averaging to represent the multiview matching cost. Both COLMAP and ACMH are based on the matching cost of photometric consistency. It makes them suffer from fuzzy matching problems in weakly textured regions and perform poorly in completeness, which in turn affects the F_1 score. On the contrary, they possess an extremely high accuracy attributed to their contribution to view selection strategy and the check of geometric consistency. OpenMVS also uses the matching cost based on photometric consistency. By relaxing the view constraint on depth fusion, it performs poorly in terms of accuracy, but the increase in completeness results in an improvement in the F_1 score. ACMP introduces planar priors to improve completeness. However, the generation of planar priors relies excessively on the multiview matching cost based on photometric consistency, which allows erroneous planar priors to be generated and mislead the computation of the improved cost function. CLD-MVS utilizes a boundary-aware interpolation method, which improves completeness while decreasing its accuracy, and it results in an inferior performance to our method. QAPM extracts pixel information with the same plane by constructing the quadtree, then the plane priors are generated by a plane fitting algorithm. However, the plane fitting algorithm is not implemented completely for all quadtree blocks, which leads to a lot of vacancies in the generated prior planes. In addition, nearly but not sufficiently accurate prior planes would affect the accuracy of the reconstruction. The great success in completeness makes our method ahead of other SOTA methods in the F_1 score, while the accuracy is not overly behind the most accurate method

COLMAP. The improvement in completeness is attributed to the fact that after utilizing the confidence calculation method, the supplemental depth maps are generated based on it and combined with the coarse depth maps. It allows for an effective recovery of weakly textured regions, especially those planes in weakly textured regions, without blurring the structural detail regions.

Table 2. Quantitative evaluation comparative results (F_1 score, accuracy, completeness) at default tolerance of 2 cm on high-resolution training dataset of ETH3D benchmark.

Method	All			Indoor			Outdoor		
	F_1	Acc.	Comp.	F_1	Acc.	Comp.	F_1	Acc.	Comp.
Gipuma [10]	36.38	86.47	24.91	35.80	89.25	24.61	37.07	83.23	25.26
COLMAP [15]	67.66	**91.85**	55.13	66.76	**95.01**	52.90	68.70	**88.16**	57.73
ACMH [22]	70.71	88.94	61.59	70.00	92.62	59.22	71.54	84.65	64.36
OpenMVS [7]	76.15	78.44	74.92	76.82	81.39	73.91	75.37	74.99	76.09
ACMP [18]	79.79	90.12	72.15	80.53	92.30	72.25	78.94	87.58	72.03
CLD-MVS [38]	79.35	82.75	77.36	81.23	87.22	77.29	77.16	77.54	77.45
QAPM [21]	78.47	80.43	77.50	80.22	84.34	77.43	76.43	75.86	**77.59**
OURS	**82.64**	86.66	**79.39**	**85.03**	88.52	**82.13**	**79.86**	84.48	76.19

The best results are marked in bold black.

Table 3. Quantitative evaluation comparative results (F_1 score, accuracy, completeness) at default tolerance of 2 cm on high-resolution test dataset of ETH3D benchmark.

Method	All			Indoor			Outdoor		
	F_1	Acc.	Comp.	F_1	Acc.	Comp.	F_1	Acc.	Comp.
Gipuma [10]	45.18	84.44	34.91	41.86	86.33	31.44	55.16	78.78	45.30
COLMAP [15]	73.01	**91.97**	62.98	70.41	**91.95**	59.65	80.81	**92.04**	72.98
ACMH [22]	75.89	89.34	68.62	73.93	91.14	64.81	81.77	83.96	80.03
OpenMVS [7]	79.77	81.98	78.54	78.33	82.00	75.92	84.09	81.93	86.41
ACMP [18]	81.51	90.54	75.58	80.57	90.60	74.23	84.36	90.35	79.62
CLD-MVS [38]	82.31	83.18	82.73	81.65	82.64	82.35	84.29	84.79	83.86
QAPM [21]	80.88	82.59	79.95	79.50	82.59	77.39	85.03	82.58	**87.64**
OURS	**85.76**	86.17	**85.71**	**85.29**	85.54	**85.46**	**87.17**	88.05	86.46

The best results are marked in bold black.

The performance of the proposed method is further demonstrated by the comparison results of the test branch dataset shown in Table 3. Except for the outdoor scenes, where the completeness is slightly inferior to QAPM, our method ranks first in both completeness and F_1 score. In addition, the comparison results of different scenes in the test branch dataset at different distance tolerances are shown in Figure 5. It can be seen that our method almost achieves the most competitive f1 scores for different sequences at each distance tolerance, except for the 'exhibition hall' sequence. It means that our method is robust for different scene sequences, although at different evaluation tolerances. In addition, the accuracy requirements for the reconstructed point clouds vary greatly in practical applications, which gives all of the reconstruction results for different thresholds a reference significance of comparison. Through the comparison between different sequences, it can be seen that while the scenes contain a lot of weakly textured planes, our method achieves excellent results for resolving the problems of luminosity consistency in these regions, resulting in the most competitive F_1 score. The 'lounge' scene sequence is noteworthy among them all. The 'lounge' sequence is the one scene where all methods perform poorly because of the presence of large reflective floor areas in this indoor scene. It causes a failure in photometric consistency and makes depth estimation difficult. An important reason for the best competitiveness of our method in this sequence is the proposed confidence calculation method, then the planar supplement based on our confidence. ACMP and QAPM are both planar-based methods, but their planar generation is based on photometric consistency, which makes them perform worse than us on this sequence. The interpolation method of CLD-MVS blurs the geometric details and results in reduced accuracy, which also affects the F_1 score. In contrast, we perform plane supplementation and then combine the supplemental depth maps with the coarse depth maps, which effectively improves the deficiency of plane supplementation in geometric detail regions and provides reliable planes in weakly textured regions.

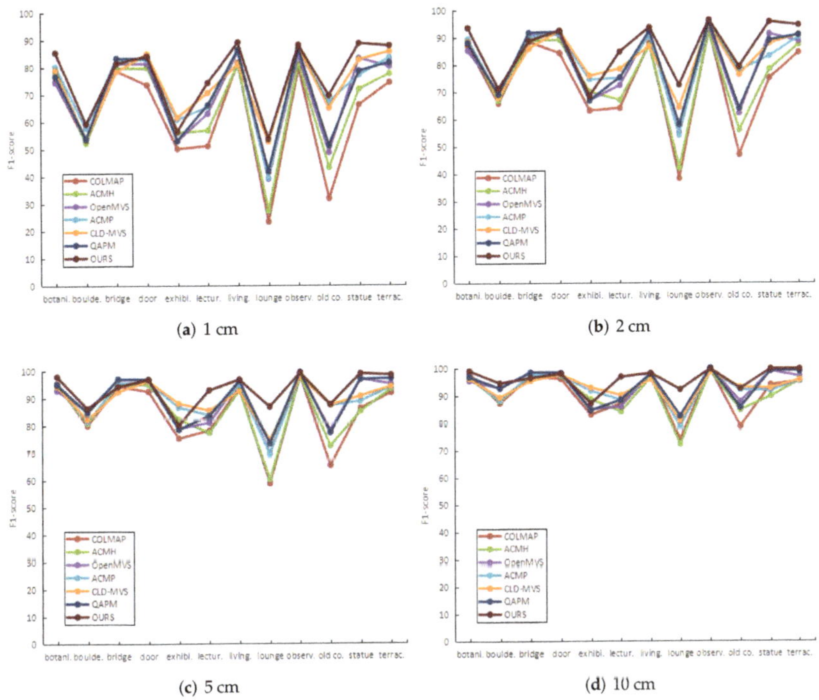

Figure 5. Quantitative evaluation comparison results (F_1 score) of different tolerances for all sequences (botanical garden[30], boulders[26], bridge[110], door[7], exhibition hall[68], lecture room[23], living room[65], lounge[10], observatory[27], old computer[54], statue[11], terrace2[13]) of the ETH3D benchmark's high-resolution **test** branch dataset.

5.3. Qualification

The qualitative evaluation is compared with some state-of-the-art PatchMatch-based MVS methods in terms of both depth maps as well as dense point clouds. For the ETH3D benchmark, all the dense point clouds are obtained from the results submitted on the online website to fairly compare the reconstruction quality of all methods. The depth map results with other methods are implemented in our machine via their open-source code. For the sensefly dataset, both the dense point clouds and the depth map results are implemented through open-source code.

For partial sequences of the ETH3D benchmark's high-resolution dataset, the comparison of qualitative depth maps is shown in Figure 6. The challenges in the ETH3D benchmark arise from a huge variation in camera angles between different images and the magnification of weakly textured regions in the high-resolution images, while the former leads to an increase in occlusion. The second aggravates the difficulty of reconstruction of weakly textured regions, because the images in the benchmark contain a large number of weakly textured surfaces (e.g., walls, floors, roads, ceilings). It can be seen that the depth maps of OpenMVS and ACMH contain a large amount of noise, and these incorrectly estimated depths are detrimental to both accuracy and completeness. COLMAP also contains a lot of noise; most of the noise is filtered out after the geometric consistency check, resulting in accurate but extremely incomplete depth maps. ACMP improves the quality of the depth maps via the plane priors. However, it is inferior to the depth maps of our method, which is due to the generation of prior planes relying on the cost function of photometric consistency. In contrast, the proposed confidence calculation would address well the unreliability of photometric consistency in weakly textured regions. Thus, the depth maps with high quality are estimated in combination with the plane supplement module.

A comparison of the qualitative depth maps for the university scene of the sensefly dataset is shown in Figure 7. The challenges of the sensefly dataset are the poor overlap of the images and the absence of common viewing areas. In addition, the weakly textured regions in these remote sensing images are mostly concentrated on roads and building roofs. As shown in Figure 7, the depth map of the COLMAP exhibits large vacancies. Besides the weakly textured regions that fail to estimate the correct depths, the poor overlap of the images leads to the misuse of geometric consistency in the COLMAP. The reason is that some regions are invisible in partial—or even all—neighborhood views. This removes the depth of these regions in the geometric consistency check, resulting in large vacancies. OpenMVS and ACMH still perform poorly in the weakly textured planes, while ACMP improves. In contrast, the depth maps of our method are most intact in these planar regions, which indicates the successful recovery of our method in the weakly textured planes.

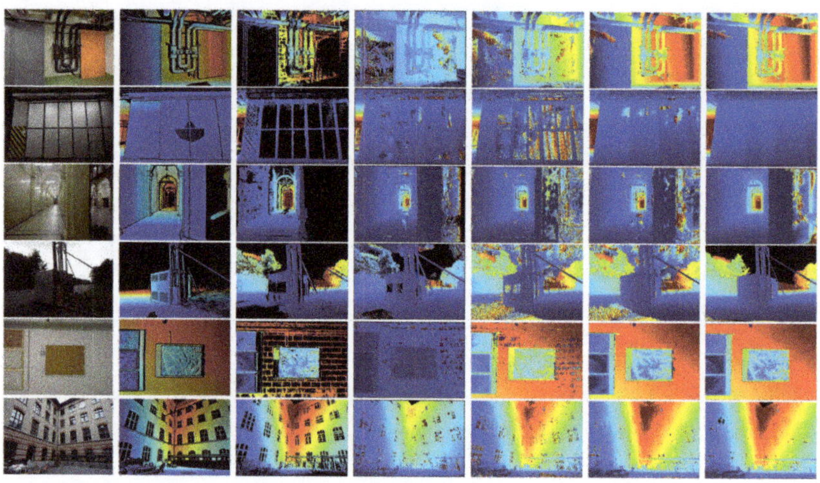

(**a**) Images (**b**) GT (**c**) COLMAP (**d**) OpenMVS (**e**) ACMH (**f**) ACMP (**g**) OURS

Figure 6. Comparative results of qualitative depth map with other methods on partial sequences (pipes^{14}, delivery^{44}, relief^{31}, electro^{45}, terrains^{42}, and courtyard^{38}) of ETH3D benchmark's high-resolution training dataset. The black regions indicate no depth.

Through the proposed adaptive depth fusion approach, the obtained dense point clouds are compared with other MVS methods. For the high-resolution training and test datasets of the ETH3D benchmark, qualitative comparisons of the dense point clouds for some sequences are shown in Figures 8 and 9, respectively. It can be seen that COLMAP and ACMH exhibit large vacancies in weakly textured regions, which makes their point clouds sparse. OpenMVS, which also utilizes the photometric consistency matching cost, sacrifices significant accuracy for an increase in completeness by decreasing the view constraint in depth fusion. Therefore, the point clouds of OpenMVS seem to be dense. However, they contain a lot of redundancy, which severely reduces the accuracy of the point clouds. The planar priors of ACMP bring a great improvement in completeness while maintaining high accuracy. However, it can be seen that the point cloud still appears sparse and vacant in weakly textured regions, especially in weakly textured planar surfaces. This is due to the incorrect prior planes generated in these regions, which are guided by the matching cost of photometric consistency. In contrast, our method recovers completely in these regions, especially in indoor scenes, which usually contain more weakly textured planar surfaces (e.g., walls, floors). Due to the effect of adaptive fusion, although the regions with insufficient visibility are effectively recovered in dense point clouds, some erroneous redundancies inevitably appear. These redundancies are one reason why the accuracy of our quantitative evaluation does not outperform other methods. However, the slight decrease in accuracy is worth it compared to our great improvement in completeness.

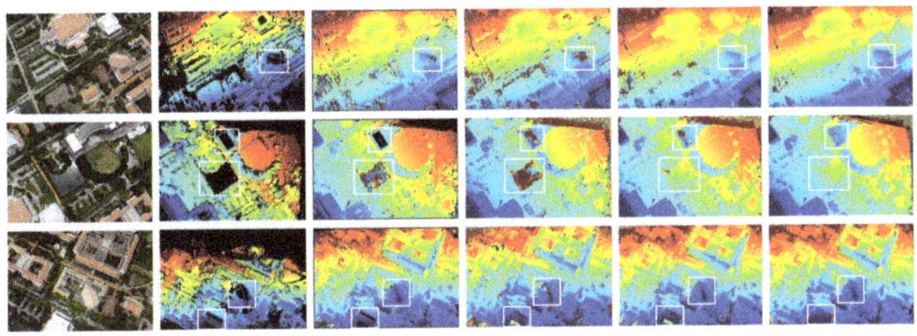

(**a**) Images (**b**) COLMAP (**c**) OpenMVS (**d**) ACMH (**e**) ACMP (**f**) OURS

Figure 7. Comparative results of qualitative depth map with other methods on the university^{443} scene of the sensefly dataset.

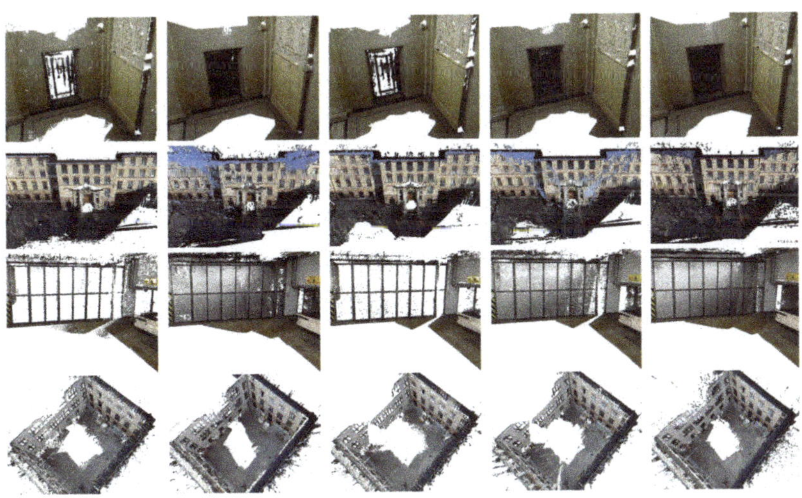

(**a**) COLMAP (**b**) OpenMVS (**c**) ACMH (**d**) ACMP (**e**) OURS

Figure 8. Comparative results of qualitative point clouds with other methods on partial sequences (relief2^{31}, facade^{76}, delivery^{44}, and courtyard^{38}) of ETH3D benchmark's high-resolution training dataset.

(a) COLMAP (b) OpenMVS (c) ACMH (d) ACMP (e) OURS

Figure 9. Comparative results of qualitative point clouds with other methods on partial sequences (door[7], terrace2[13], observatory[27], and statue[11]) of ETH3D benchmark's high-resolution test dataset.

In addition, the comparison of point clouds on the sensefly dataset is shown in Figure 10. It can be seen that the point clouds reconstructed by COLMAP and ACMH are both sparse. ACMP has a better reconstruction in the weakly textured regions, but the overall denseness is not as good as that of OpenMVS and ours. The point clouds of OpenMVS, which are only based on the photometric consistency, show the densest point clouds in the comparison. An important reason for the above observation is attributed to the difference in depth fusion methods. COLMAP, ACMH, and ACMP all require a high view constraint for depth fusion, which makes their reconstructed point clouds sparse, although ACMP performs well in depth maps. In contrast, OpenMVS uses the most relaxed view constraint for depth fusion, which results in the densest reconstructed point clouds. Our adaptive fusion method dynamically adjusts the view constraint, which results in a far denser point cloud than COLMAP, ACMH, and ACMP, but slightly sparser than OpenMVS. However, the advantage of our method is that the reconstruction is more integral in the weakly textured regions, especially the planar surfaces of these regions, such as the building roofs and the water surface shown in the red box of Figure 10.

To further illustrate the effectiveness of our pipeline in weakly textured regions, the comparison results of completeness visualizations are shown individually in Figure 11. The completeness visualizations are provided on the online website of the ETH3D benchmark and are only available in the training branch dataset. It can be clearly seen that the results of our method have more green parts (meaning the parts that are reconstructed successfully) of point clouds compared to other methods. Most of the green parts of point clouds belong to weakly textured regions. For the successful recovery in weakly textured regions, our method can outperform other methods in terms of completeness.

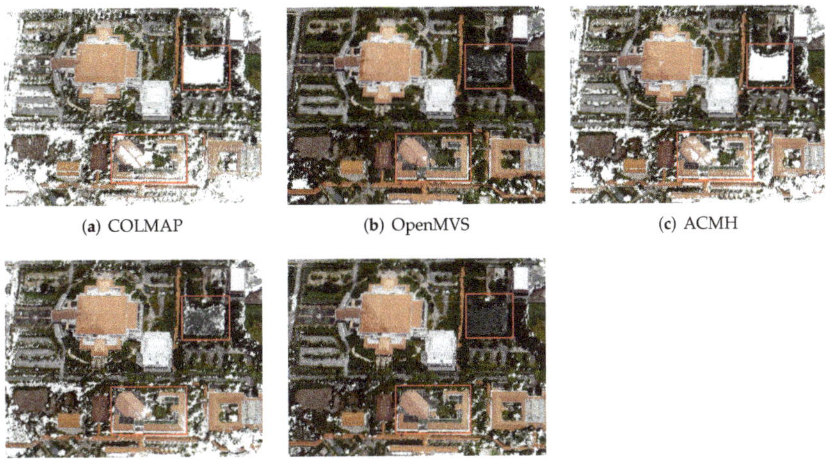

(**a**) COLMAP (**b**) OpenMVS (**c**) ACMH (**d**) ACMP (**e**) OURS

Figure 10. Comparative results of qualitative point clouds with other methods on the university scene of sensefly dataset.

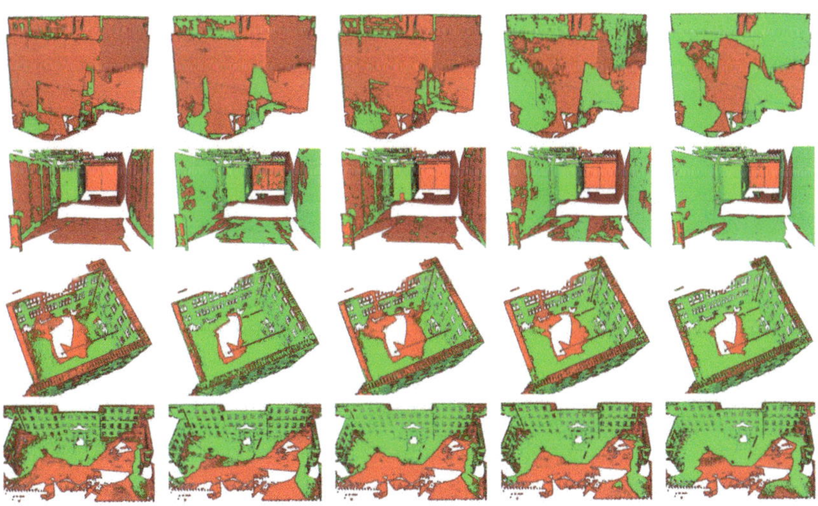

(**a**) COLMAP (**b**) OpenMVS (**c**) ACMH (**d**) ACMP (**e**) OURS

Figure 11. Comparative results of completeness visualizations at default tolerance of 2 cm on partial sequences (office, pipes, courtyard, and facade) of ETH3D benchmark's high-resolution training dataset. The green areas of point clouds are the parts that are less than the distance tolerance between the reconstruction result and the ground truth. The red regions of point clouds are the ground truth that cannot be accepted within the distance tolerance.

5.4. Ablation Study

To evaluate the effectiveness of each part of our proposed method, we conducted ablation experiments on the high-resolution training dataset of the ETH3D benchmark. The evaluation results are presented in Table 4. In the table, we list the results of removing different modules from our proposed CGPR-MVS, including without all modules proposed (baseline), without confidence calculation (CGPR-MVS/C), without plane supplement (CGPR-MVS/S), without adaptive fusing (CGPR-MVS/A), and the whole pipeline in

CGPR-MVS. For CGPR-MVS/C, we use the matching cost as a substitute for completing the plane supplement module. For CGPR-MVS/S, we filter the unreliable estimation by confidence after the confidence is computed in the confidence calculation module. For CGPR-MVS/A, we set fixed-view constraint and fixed-consistency constraints for depth fusion, like other pipelines [7,15,18,22].

Firstly, a quantitative comparison between CGPR-MVS and CGPR-MVS/A shows that the adaptive fusion approach greatly balances the accuracy and completeness of the reconstruction results. After removing adaptive fusion, the pipeline achieves extremely high accuracy. Nonetheless, both accuracy and completeness are increased compared to the baseline method. In CGPR-MVS/C, the proposed confidence calculation is replaced by the multiview matching cost of photometric consistency to help the subsequent implementation of the plane hypothesis supplement. By comparing CGPR-MVS and CGPR-MVS/C, the result without confidence calculation is that the completeness and accuracy of the quantitative evaluation are significantly reduced. The results further illustrate the failure of photometric consistency in the weakly textured regions, and prove that the proposed confidence calculation is extremely effective for the improvement in reconstruction quality. Compared to the quantitative evaluation results of CGPR-MVS and CGPR-MVS/S, there is essentially no excessive change in accuracy, but there is a significant decrease in completeness after removing the planar hypothesis supplement. Based on the implementation of confidence calculation, the planar hypothesis supplement provides reliable planar hypotheses for the weakly textured region, which helps converge to the global optimal solution. In contrast, after losing the planar hypothesis supplement, the plane hypotheses of weakly textured regions are limited to the wrong local optimal solution, resulting in the failure of reconstruction.

Table 4. Ablation study results (F_1 score, accuracy, completeness) at default tolerance of 2 cm on high-resolution training dataset of ETH3D benchmark.

Method	F_1 Score	Accuracy	Completeness
Baseline	72.77	90.65	62.46
CGPR-MVS/C	74.40	76.59	73.70
CGPR-MVS/S	78.41	85.68	73.40
CGPR-MVS/A	79.71	**90.72**	71.87
CGPR-MVS	**82.64**	86.66	**79.39**

The best results are marked in bold black.

5.5. Time Evaluation

For each 3200 × 2130 resolution view in the high-resolution training dataset of the ETH3D benchmark, the runtimes of each proposed section and the total runtime are listed in Table 5. It can be seen that both the plane hypothesis confidence calculation module and plane supplement module do not impose an excessive runtime burden for depth estimation. Moreover, since the machine configuration is not expensive, the runtime results show that the proposed pipeline can be equipped on low-performance machines without consuming excessive computational resources.

In addition, the comparison results between the proposed method and some GPU-based methods are shown in Table 6. The running times of all the methods are experimentally obtained on our machines equipped with a single GPU. It can be seen that even with the GPU, COLMAP [15] still has the longest running time, because of the sequential propagation in depth estimation. Our pipeline takes more time than ACMP [18], but still has a shorter running time than ACMM [22] and COLMAP [15]. The comparison results further show that the proposed method is efficient enough and does not require more computational resources.

Table 5. Different modules' running times for one image of 3200 × 2130 resolution on high-resolution training dataset of ETH3D benchmark.

Module	Time(s)	Ratio (%)
depth estimation of ACMH	18.79	49.70
plane hypothesis confidence calculation	2.36	6.24
plane supplement	3.52	9.31
confidence-driven depth estimation	13.14	34.75
Total	37.81	-

Table 6. Comparison of running times for one image of 3200 × 2130 resolution on high-resolution training dataset of ETH3D benchmark.

Method	COLMAP	ACMM	ACMP	OURS
Time(s)	129.9	43.0	**23.7**	37.8

The best results are marked in bold black.

6. Conclusions

In this work, we propose a novel MVS method, which is called confidence-guided planar recovering multiview stereo (CGPR-MVS). After depth estimation, the confidence calculation module is applied to depth maps to produce pixel-wise confidence, which contains multiview consistency and patch consistency. Based on the plane hypothesis confidence calculation, a Delaunay triangle based plane supplement module additionally provides reliable plane information. The supplemental depth map and coarse depth map are fed into a confidence-driven depth estimation to achieve high-integrity recovery without losing the structural detail regions. Via adaptive fusion, invisible regions can be merged into dense point clouds. Qualitative and quantitative evaluations of high-resolution MVS datasets demonstrate the efficiency and effectiveness of our method, especially in the reconstruction quality of weakly textured planes. In future work, we will focus on improving the accuracy of texture detail regions while maintaining reconstruction completeness.

Author Contributions: Conceptualization, C.F. and N.H.; methodology, C.F.; software, C.F. and Z.H.; validation, C.F., N.H., and S.C.; formal analysis, C.F., N.H., Z.H., and Y.L.; investigation, C.F., Z.H., and Y.L.; resources, C.F.; data curation, C.F.; writing—original draft preparation, C.F.; writing—review and editing, C.F., N.H., and S.C.; visualization, C.F.; supervision, X.X., X.Z., and S.C.; project administration, X.X., X.Z., and S.C. All authors have read and agreed to the published version of the manuscript.

Funding: This work was supported by the Guangdong-Hong Kong-Macao Joint Innovation Field Project (No.2021A0505080006), Research and Development Project in Key Field of Guangdong Province, China (No.2022B0701180001), the Science Technology Planning Project of Guangdong Province, China (Nos. 2019B010140002, 2020B111110002), and the National Natural Science Foundation of China (61801127).

Data Availability Statement: Not applicable.

Conflicts of Interest: The authors declare no conflict of interest.

References

1. Xu, Z.; Liu, Y.; Shi, X.; Wang, Y.; Zheng, Y. MARMVS: Matching Ambiguity Reduced Multiple View Stereo for Efficient Large Scale Scene Reconstruction. In Proceedings of the 2020 IEEE/CVF Conference on Computer Vision and Pattern Recognition (CVPR), Seattle, WA, USA, 13–19 June 2020; pp. 5980–5989. [CrossRef]
2. Zhou, L.; Zhang, Z.; Jiang, H.; Sun, H.; Bao, H.; Zhang, G. DP-MVS: Detail Preserving Multi-View Surface Reconstruction of Large-Scale Scenes. *Remote Sens.* **2021**, *13*, 4569. [CrossRef]
3. Zhang, Q.; Luo, S.; Wang, L.; Feng, J. CNLPA-MVS: Coarse-Hypotheses Guided Non-Local PatchMatch Multi-View Stereo. *J. Comput. Sci. Technol.* **2021**, *36*, 572–587. [CrossRef]
4. Xu, Q.; Kong, W.; Tao, W.; Pollefeys, M. Multi-Scale Geometric Consistency Guided and Planar Prior Assisted Multi-View Stereo. *IEEE Trans. Pattern Anal. Mach. Intell.* **2023**, *45*, 4945–4963. [CrossRef] [PubMed]

5. Bleyer, M.; Rhemann, C.; Rother, C. PatchMatch Stereo-Stereo Matching with Slanted Support Windows. In Proceedings of the British Machine Vision Conference (BMVC), Dundee, UK, 29 August–2 September 2011; pp. 14.1–14.11. [CrossRef]
6. Shen, S. Accurate Multiple View 3D Reconstruction Using Patch-Based Stereo for Large-Scale Scenes. *IEEE Trans. Image Process.* **2013**, *22*, 1901–1914. [CrossRef] [PubMed]
7. Cernea, D. OpenMVS: Multi-View Stereo Reconstruction Library. Available online: https://cdcseacave.github.io/openMVS (accessed on 4 August 2022).
8. Fuhrmann, S.; Langguth, F.; Goesele, M. MVE—A Multi-View Reconstruction Environment. In Proceedings of the Eurographics Workshop on Graphics & Cultural Heritage, Darmstadt, Germany, 6–8 October 2014.
9. Zhu, Z.; Stamatopoulos, C.; Fraser, C.S. Accurate and occlusion-robust multi-view stereo. *ISPRS J. Photogramm. Remote Sens.* **2015**, *109*, 47–61. [CrossRef]
10. Galliani, S.; Lasinger, K.; Schindler, K. Massively Parallel Multiview Stereopsis by Surface Normal Diffusion. In Proceedings of the 2015 IEEE International Conference on Computer Vision (ICCV), Santiago, Chile, 7–13 December 2015; pp. 873–881. [CrossRef]
11. Kuhn, A.; Hirschmüller, H.; Scharstein, D.; Mayer, H. A TV Prior for High-Quality Scalable Multi-View Stereo Reconstruction. *Int. J. Comput. Vis.* **2016**, *124*, 2–17. [CrossRef]
12. Kang, S.B.; Szeliski, R.; Chai, J. Handling occlusions in dense multi-view stereo. In Proceedings of the 2001 IEEE Computer Society Conference on Computer Vision and Pattern Recognition, CVPR 2001, Kauai, HI, USA, 8–14 December 2001; Volume 1, p. I. [CrossRef]
13. Strecha, C.; Fransens, R.; Van Gool, L. Combined Depth and Outlier Estimation in Multi-View Stereo. In Proceedings of the 2006 IEEE Computer Society Conference on Computer Vision and Pattern Recognition (CVPR'06), New York, NY, USA, 17–22 June 2006; Volume 2, pp. 2394–2401. [CrossRef]
14. Zheng, E.; Dunn, E.; Jojic, V.; Frahm, J.M. PatchMatch Based Joint View Selection and Depthmap Estimation. In Proceedings of the 2014 IEEE Conference on Computer Vision and Pattern Recognition, Columbus, OH, USA, 23–28 June 2014; pp. 1510–1517. [CrossRef]
15. Schnberger, J.L.; Zheng, E.; Pollefeys, M.; Frahm, J.M. Pixelwise View Selection for Unstructured Multi-View Stereo. In Proceedings of the European Conference on Computer Vision (ECCV), Amsterdam, The Netherlands, 11–14 October 2016.
16. Romanoni, A.; Matteucci, M. TAPA-MVS: Textureless-Aware PAtchMatch Multi-View Stereo. In Proceedings of the 2019 IEEE/CVF International Conference on Computer Vision (ICCV), Seoul, Republic of Korea, 27 October–2 November 2019.
17. Kuhn, A.; Lin, S.; Erdler, O. Plane Completion and Filtering for Multi-View Stereo Reconstruction. In Proceedings of the GCPR, Dortmund, Germany, 10–13 September 2019.
18. Xu, Q.; Tao, W. Planar Prior Assisted PatchMatch Multi-View Stereo. In Proceedings of the AAAI Conference on Artificial Intelligence, New York, NY, USA, 7–12 February 2020; Volume 34, pp. 12516–12523.
19. Yan, S.; Peng, Y.; Wang, G.; Lai, S.; Zhang, M. Weakly Supported Plane Surface Reconstruction via Plane Segmentation Guided Point Cloud Enhancement. *IEEE Access* **2020**, *8*, 60491–60504. [CrossRef]
20. Huang, N.; Huang, Z.; Fu, C.; Zhou, H.; Xia, Y.; Li, W.; Xiong, X.; Cai, S. A Multi-View Stereo Algorithm Based on Image Segmentation Guided Generation of Planar Prior for Textureless Regions of Artificial Scenes. *IEEE J. Sel. Top. Appl. Earth Obs. Remote Sens.* **2023**, *16*, 3676–3696. [CrossRef]
21. Stathopoulou, E.K.; Battisti, R.; Cernea, D.; Georgopoulos, A.; Remondino, F. Multiple View Stereo with quadtree-guided priors. *ISPRS J. Photogramm. Remote Sens.* **2023**, *196*, 197–209. [CrossRef]
22. Xu, Q.; Tao, W. Multi-Scale Geometric Consistency Guided Multi-View Stereo. In Proceedings of the 2019 IEEE/CVF Conference on Computer Vision and Pattern Recognition (CVPR), Long Beach, CA, USA, 15–20 June 2019.
23. Seitz, S.; Curless, B.; Diebel, J.; Scharstein, D.; Szeliski, R. A Comparison and Evaluation of Multi-View Stereo Reconstruction Algorithms. In Proceedings of the 2006 IEEE Computer Society Conference on Computer Vision and Pattern Recognition (CVPR'06), New York, NY, USA, 17–22 June 2006; Volume 1, pp. 519–528. [CrossRef]
24. Goesele, M.; Curless, B.; Seitz, S. Multi-View Stereo Revisited. In Proceedings of the 2006 IEEE Computer Society Conference on Computer Vision and Pattern Recognition (CVPR'06), Seattle, WA, USA, 14–19 June 2006; Volume 2, pp. 2402–2409. [CrossRef]
25. Kostrikov, I.; Horbert, E.; Leibe, B. Probabilistic Labeling Cost for High-Accuracy Multi-view Reconstruction. In Proceedings of the 2014 IEEE Conference on Computer Vision and Pattern Recognition, Washington, DC, USA, 23–28 June 2014; pp. 1534–1541. [CrossRef]
26. Hiep, V.H.; Keriven, R.; Labatut, P.; Pons, J.P. Towards high-resolution large-scale multi-view stereo. In Proceedings of the 2009 IEEE Conference on Computer Vision and Pattern Recognition, Miami, FL, USA, 20–25 June 2009; pp. 1430–1437. [CrossRef]
27. Cremers, D.; Kolev, K. Multiview Stereo and Silhouette Consistency via Convex Functionals over Convex Domains. *IEEE Trans. Pattern Anal. Mach. Intell.* **2011**, *33*, 1161–1174. [CrossRef]
28. Lhuillier, M.; Quan, L. A quasi-dense approach to surface reconstruction from uncalibrated images. *IEEE Trans. Pattern Anal. Mach. Intell.* **2005**, *27*, 418–433. [CrossRef] [PubMed]
29. Furukawa, Y.; Ponce, J. Accurate, Dense, and Robust Multiview Stereopsis. *IEEE Trans. Pattern Anal. Mach. Intell.* **2010**, *32*, 1362–1376. [CrossRef]

30. Strecha, C.; Fransens, R.; Van Gool, L. Wide-baseline stereo from multiple views: A probabilistic account. In Proceedings of the 2004 IEEE Computer Society Conference on Computer Vision and Pattern Recognition, CVPR 2004, Washington, DC, USA, 27 June–2 July 2004; Volume 1, p. I. [CrossRef]
31. Liao, J.; Fu, Y.; Yan, Q.; Xiao, C. Pyramid Multi-View Stereo with Local Consistency. *Comput. Graph. Forum* **2019**, *38*, 335–346. [CrossRef]
32. Jung, W.K.; Han, J.k. Depth Map Refinement Using Super-Pixel Segmentation in Multi-View Systems. In Proceedings of the 2021 IEEE International Conference on Consumer Electronics (ICCE), Las Vegas, NV, USA, 10–12 January 2021; pp. 1–5. [CrossRef]
33. Wei, M.; Yan, Q.; Luo, F.; Song, C.; Xiao, C. Joint bilateral propagation upsampling for unstructured multi-view stereo. *Vis. Comput.* **2019**, *35*, 797–809. [CrossRef]
34. Yodokawa, K.; Ito, K.; Aoki, T.; Sakai, S.; Watanabe, T.; Masuda, T. Outlier and Artifact Removal Filters for Multi-View Stereo. In Proceedings of the 2018 25th IEEE International Conference on Image Processing (ICIP), Athens, Greece, 7–10 October 2018; pp. 3638–3642. [CrossRef]
35. Egnal, G.; Mintz, M.; Wildes, R.P. A stereo confidence metric using single view imagery with comparison to five alternative approaches. *Image Vis. Comput.* **2004**, *22*, 943–957. [CrossRef]
36. Pfeiffer, D.; Gehrig, S.; Schneider, N. Exploiting the Power of Stereo Confidences. In Proceedings of the 2013 IEEE Conference on Computer Vision and Pattern Recognition, Portland, OR, USA, 23–28 June 2013; pp. 297–304. [CrossRef]
37. Seki, A.; Pollefeys, M. Patch Based Confidence Prediction for Dense Disparity Map. In Proceedings of the British Machine Vision Conference (BMVC), York, UK, 19–22 September 2016; pp. 23.1–23.13. [CrossRef]
38. Li, Z.; Zuo, W.; Wang, Z.; Zhang, L. Confidence-Based Large-Scale Dense Multi-View Stereo. *IEEE Trans. Image Process.* **2020**, *29*, 7176–7191. [CrossRef]
39. Li, Z.; Zhang, X.; Wang, K.; Jiang, H.; Wang, Z. High accuracy and geometry-consistent confidence prediction network for multi-view stereo. *Comput. Graph.* **2021**, *97*, 148–159. [CrossRef]
40. Kuhn, A.; Sormann, C.; Rossi, M.; Erdler, O.; Fraundorfer, F. DeepC-MVS: Deep Confidence Prediction for Multi-View Stereo Reconstruction. In Proceedings of the 2020 International Conference on 3D Vision (3DV), Fukuoka, Japan, 25–28 November 2020; pp. 404–413.
41. Wang, Y.; Guan, T.; Chen, Z.; Luo, Y.; Luo, K.; Ju, L. Mesh-Guided Multi-View Stereo With Pyramid Architecture. In Proceedings of the 2020 IEEE/CVF Conference on Computer Vision and Pattern Recognition (CVPR), Seattle, WA, USA, 13–19 June 2020; pp. 2036–2045. [CrossRef]
42. Stathopoulou, E.E.K.; Remondino, F. Semantic photogrammetry—Boosting image-based 3D reconstruction with semantic labeling. *Int. Arch. Photogramm. Remote Sens. Spat. Inf. Sci.* **2019**, *42*, 685–690. [CrossRef]
43. Stathopoulou, E.K.; Remondino, F. Multi view stereo with semantic priors. *arXiv* **2020**, arXiv:2007.02295. [CrossRef]
44. Stathopoulou, E.K.; Battisti, R.; Cernea, D.; Remondino, F.; Georgopoulos, A. Semantically Derived Geometric Constraints for MVS Reconstruction of Textureless Areas. *Remote Sens.* **2021**, *13*, 1053. [CrossRef]
45. Barnes, C.; Shechtman, E.; Finkelstein, A.; Goldman, D.B. PatchMatch: A randomized correspondence algorithm for structural image editing. *ACM Trans. Graph.* **2009**, *28*, 24. [CrossRef]
46. Delaunay, B. Sur la sphère vide. A la mémoire de Georges Voronoï. *Bull. De L'académie Des Sci. De L'urss. Cl. Des Sci. Mathématiques Et Na* **1934**, *6*, 793–800.
47. Liba, O.; Movshovitz-Attias, Y.; Cai, L.; Pritch, Y.; Tsai, Y.T.; Chen, H.; Eban, E.; Barron, J.T. Sky Optimization: Semantically aware image processing of skies in low-light photography. In Proceedings of the 2020 IEEE/CVF Conference on Computer Vision and Pattern Recognition Workshops (CVPRW), Seattle, WA, USA, 14–19 June 2020; pp. 2230–2238. [CrossRef]
48. Schöps, T.; Schönberger, J.L.; Galliani, S.; Sattler, T.; Schindler, K.; Pollefeys, M.; Geiger, A. A Multi-View Stereo Benchmark with High-Resolution Images and Multi-Camera Videos. In Proceedings of the Conference on Computer Vision and Pattern Recognition (CVPR), Honolulu, HI, USA, 21–26 July 2017.

remote sensing

MDPI

Article

A Quantitative Assessment of LIDAR Data Accuracy

Ahmed Elaksher [1,*], Tarig Ali [2] and Abdullatif Alharthy [3]

1 College of Engineering, New Mexico State University, Las Cruces, NM 88003-0001, USA
2 Department of Civil Engineering, College of Engineering, American University of Sharjah, Sharjah 26666, United Arab Emirates
3 Ministry of National Guard, Riyadh 11173, Saudi Arabia
* Correspondence: elaksher@nmsu.edu

Abstract: Airborne laser scanning sensors are impressive in their ability to collect a large number of topographic points in three dimensions in a very short time thus providing a high-resolution depiction of complex objects in the scanned areas. The quality of any final product naturally depends on the original data and the methods of generating it. Thus, the quality of the data should be evaluated before assessing any of its products. In this research, a detailed evaluation of a LIDAR system is presented, and the quality of the LIDAR data is quantified. This area has been under-emphasized in much of the published work on the applications of airborne laser scanning data. The evaluation is done by field surveying. The results address both the planimetric and the height accuracy of the LIDAR data. The average discrepancy of the LIDAR elevations from the surveyed study area is 0.12 m. In general, the RMSE of the horizontal offsets is approximately 0.50 m. Both relative and absolute height discrepancies of the LIDAR data have two components of variation. The first component is a random short-period variation while the second component has a less significant frequency and depends on the biases in the geo-positioning system.

Keywords: assessment; LIDAR; GPS; surveying; RMSE; accuracy

Citation: Elaksher, A.; Ali, T.; Alharthy, A. A Quantitative Assessment of LIDAR Data Accuracy. *Remote Sens.* **2023**, *15*, 442. https://doi.org/10.3390/rs15020442

Academic Editor: Gianluca Groppelli

Received: 20 October 2022
Revised: 20 December 2022
Accepted: 23 December 2022
Published: 11 January 2023

1. Introduction

Light Detection and Ranging (LIDAR) technology is a cost-effective, reliable, and a fast source to collect dense and accurate elevation data for many surveying and mapping applications [1–3]. Aerial LIDAR mapping systems are equipped with an active laser sensor, a Global Positioning System (GPS), and an Inertial Navigation System (INS) [4]. The laser sensor emits pulses at a near infrared wavelength of about 1000 nm [5]. Pulses travel through the atmosphere [6] and are reflected from ground features back to the detector. Their travel time multiplied by the speed of light is used to calculate the distance between the laser scanner and the ground [7].

Manned and unmanned LIDAR systems are widely employed in different applications including urban planning, damage assessment, natural resources, geomorphology, and archaeology [8–14]. The assessment of LIDAR data is becoming increasingly important in the geospatial and mapping communities. Several studies have been conducted to evaluate the quality of LIDAR data and products from such data [15–17].

In [18], two LIDAR datasets were acquired from about 2000 m above ground over two heavy tree coverage areas during leaf-off and snow-free seasons with an Optech GEMINI Airborne Laser Terrain Mapper. Two DEMs were then interpolated from the LIDAR datasets at one-meter ground resolution and resampled to three-, five-, and ten-meter ground resolution using mean cell aggregation. Each DEM underwent a simple low-pass smoothing filter using a mean filtering technique. Ground checkpoints were collected through total station ground surveys. Total stations have a typical accuracy of 1–3 mm ± 1–3 parts per million. Discrepancies of about 0.70 m to 2.10 m were discovered.

In [19], a Leica ALS50 Phase II was used to map an archaeological site that is ecologically and topographically diverse with alluvial terraces, foothills, mountains, and vegetation. Data were collected at 15 points/m^2 ground density with flight lines overlapping more than 50%. Two control points surveyed with a Trimble R8 GNSS rover and a Trimble R7 base were used to geospatially register the LiDAR data through a 3-D shift of the LIDAR point cloud. Data were classified into bare-ground and manmade and natural features points. A Digital Terrain Model (DTM) was generated from the bare-ground points and a Digital Surface Model (DSM) was generated from all points. Results showed an average accuracy of 0.5 m for the produced DTM and DEM.

LIDAR data popularity is increasing driven mainly by the demand for 3D data for different applications ranging from 3D mapping to earth surface and city models. Common earth surface representations created from LIDAR data are the digital elevation and surface models [20–22]. In [23], the accuracy of LiDAR-driven DEMs interpolated from two LIDAR point clouds collected during the leaf-off season with a Leica ALS60, flying at about 3000 m above ground, for two watershed areas was evaluated. The DEMs were interpolated at one-meter ground resolution using the "LAS dataset to Raster" tool in ArcMap. Four interpolation methods in ArcMap were assessed including the inverse distance weighting and nearest neighbor interpolation techniques. Reference data were collected with a GPS ground survey. The 3-D coordinates of 45 LIDAR ground points were compared with those collected with GPS. The Root Mean Square Error (RMSE) for elevation differences between the LIDAR and GPS coordinates was 0.75 m.

In [24], the accuracy of LIDAR data collected with an Optech™ ALTM 3100 EA LIDAR sensor, flying at an average altitude of 400 m, over four sites in forestry areas was assessed. Original data was collected with an average point sampling density of 10 points/m^2 then thinned to two datasets of five and one points/m^2. DTMs and DSMs with a grid size of one meter were produced using the TerraScan software. Field surveys for 100 trees were conducted with differentially corrected GPS data collected with a GeoXM 2005 receiver. Mean differences and Root mean square errors (RMSEs) between LIDAR data and field surveys were about 0.10 m and 1.25 m, consecutively.

In [25], the accuracy and usefulness of point clouds obtained by Airborne Laser Scanners (ALS) and Unmanned Aerial Systems (UASs) in detecting and measuring individual tree heights were compared. ALS data were collected with a Leica ALS80-HP laser scanner over an area of 100 hectares. The average altitude during the scanning was 2700 m above mean sea level. The UAS images were collected with an RGB SONY ILCE-5000 camera mounted on a custom-built fixed wind UAS and processed in pix4D. Analysis of the discrepancies in elevation data showed that elevations measured from the ALS data were lower than those estimated from the UAS data by approximately 1.15 m with a standard deviation of about 1.95 m.

The reliability and of LIDAR data for coastal geomorphological changes was investigated [26]. Data was provided by the U.S. Geological Survey (USGS) and was collected with the Experimental Advanced Airborne Research Lidar (EAARL) sensor from 300 m above the ground. The LIDAR point cloud contained had a ground point density of 0.65 points/m^2. Ground truth data consisted of 734 points and was surveyed with two Trimble R10 and one Leica GS12 geodetic GPS receivers configured for Real Time Kinematic (RTK) differential correction. In vegetated dune topography areas, dissimilarities of about 0.5 m RMSE were unveiled horizontally. However, better accuracies were attained in non-vegetated flat areas. RMSE was reduced to 0.25 m after systematic offsets between LIDAR data and ground truth were compensated for.

In [27], a point cloud obtained with a Riegl LMS-Q780 scanner, flying at an average flight height of 1240 m above mean ground, was evaluated. The point cloud had an average density of approximately 15 points/m^2. Ground truth was collected with a DJI Phantom 4 Pro UAS equipped with 5472 pixels × 3648 pixels digital camera. Captured images were processed with Agisoft Metashape Professional® using structure-from-motion photogrammetric techniques. Geodetic control was established with a Leica TS02 total station. Height comparisons between the LIDAR and UAS point clouds reviled that 98% of the dissimilarities in elevations were in the range from −150 to +150 mm with an absolute maximum of 0.75 m for the whole area.

A Riegl LMS-Q560, flown at an altitude of 750 m, was employed in [28] to assess the accuracy of nine interpolation methods using cross-validation. The interpolation methods examined were Delauney Triangulation, Natural Neighbor, Nearest neighbor, Ordinary Kriging, Inverse Distance Weighted, and four spline-based interpolators. Data was collected as a point cloud with a point density of 5.13 points/m^2 and a subset was held back for validation. Several DTMs were generated at 0.5-, 1.0-, 1.5-, and 2.0-meter ground resolutions. For most of the algorithms, RMSE values between 0.11 and 0.28 m were reported. Spline-based methods and the Natural Neighbor algorithm had vertical errors of less than 0.20 m for over 90 percent of validation points.

The same dataset was used in another study to map a mountainous terrain area covered by dense forest vegetation with a mixture of spruce and pine trees [29]. Ground filtering was carried out to distinguish ground points from non-ground points using the LasTools software and then manually checked. Data were resampled at different ground point densities ranging from 0.89 points/m2 to 0.09 points/m^2. It was found that DTM accuracy is highly dependent on the ground point density as the RMSE ranged from 0.15 m for the 0.89 points/m^2 ground point density to 0.62 m for the 0.09 points/m^2 ground point density.

In [30], LIDAR data from the Golden Gate LiDAR Project was evaluated. The flying altitude was about 2500 m above mean sea level. The data minimum point density was 2 points/m^2 and was filtered using TerraScan providing a set of points classified as bare-earth ground. In addition to the LiDAR data, a set of 753 points was surveyed by the USGS using the RTK differential GPS technique. The distribution of the vertical dissimilarities between the LIDAR elevation values and those observed by RTK was 19% less than 50 cm, 50% between 50 cm and 1 m, 30% between 1 cm and 2 cm, and only about 1% greater than 2 m.

From the presented literature, we can conclude that most of the previous research used GPS ground surveys to assess the quality of the LIDAR data. Despite the fact that GPS in general has higher accuracy than LIDAR data, the reliability of GPS is affected by distances to the base point, time of observation, and satellite geometry. In addition, the lack of a clear line of sight is occasionally found in urban geographic environments and this could harm the quality of the GPS signals. Therefore, in our study we used traditional ground surveying techniques, a total station, to assess the quality of the LIDAR data. Moreover, most of the previous research focused on either relative or absolute assessment of LIDAR data. In our study, we evaluated both the relative and absolute quality of the LIDAR data. The LIDAR data set was collected with an Optech LIDAR system and ground control was surveyed using conventional ground surveying techniques.

2. Data Acquisition and Dataset Characteristics

2.1. LIDAR Dataset

The data used in this research was collected in November 2014 using an Optech ALTM 3100 EA LiDAR flown at 600 m above the mean ground. Under optimal conditions, the system is capable of achieving elevation accuracies as high as ±3 cm, 2-σ at 500 m elevation, 33 kHz Pulse repetition frequency (PRF) with ±10° scan angles, [31]. The instrument operated with a scan angle of 18° and a laser pulse rate of 100 kHz. An Applanix POS/AV 510 system was co-mounted with LiDAR to record the aircraft's 3-D position and attitude (pitch, roll, and yaw). The reported positional accuracy of the system is 0.05–0.3 meter with velocity uncertainty of 0.005 m/s, roll and pitch accuracies of 0.005°, and heading accuracy of 0.008° [32]. The accuracy of the GPS unit in RTK mode is 2 cm + 2 ppm. The reported range resolution of the lidar is 1 centimeter with a resolution of 0.01° and beam divergence of 0.1 mrad. The overlap between swaths was 30%. The point cloud on the ground had

9.54 points/m^2. The data consists of fourteen strips flown in the north–south direction. Figure 1 below shows an across-band view, 500 m wide, of the fourteen strips.

Figure 1. Area coverage of a portion of the 14 strips.

Imprecision in the GPS positioning, the INS orientation, component misalignments, errors in the angle encoder and other sources of error, cause the same ground spot in two adjacent overlap strips to have two different coordinates. This leads to systematic errors in position and attitude and may produce a tilt in the resulting scanned surface. In order to reduce discrepancies between adjacent strips, we applied the mathematical model presented in [33] for strip adjustment. In this model, errors are modeled by three offsets, three rotations, and three time-dependent rotations. Tie points in adjacent strips were identified and least squares matching was applied to laser scanner data points to achieve higher relative accuracy between adjacent strips.

2.2. Reference Dataset

In order to evaluate the data set used in this research, a ground survey was conducted on an area with a large athletic area, which contains tennis courts and soccer and basketball fields representing the selected test area. The study area has two main useful characteristics. First, the area is flat and horizontal with almost no significant slope over the tennis courts or sports fields. This enables the examination of pure height accuracy because the flatness and the lack of slope of the surface rule out any planimetric uncertainty effects. Second, the presence of drainage ditches around the area facilitates the computation of planimetric accuracy.

The fieldwork was conducted by surveying 440 points using a TOPCON (GTS-303) total station. Three measurements were made for each point and the reported horizontal and vertical RMSEs were less than one centimeter. Figure 2 shows a part of the surveyed area. Those points were then used to establish the correspondence with the laser data based on the spatial positions and consequently compute the absolute accuracy.

Figure 2. The athletic area where the topographical survey was conducted.

3. Relative Accuracy of LIDAR Data

Airborne laser scanning data are acquired in a strip-wise pattern with a strip width varying depending on the chosen scan angle and the flying height. Usually, those strips are flown in parallel and overlapping until the entire region of interest has been covered. Overlap between strips, as shown in Figure 3, provides a means to verify the relative consistency between them. It is evident that imprecision in system positioning, orientation, and ranging may cause the same point to have two different heights if scanned at two different times, which always happens in neighboring overlapping data strips. These points can be considered as tie points in strip adjustment to adjust strips and eliminate or at least minimize relative error between them. However, the discrepancy between tie points from adjacent strips gives an indication of the relative offsets without any strong conclusion of the absolute error. In this section, the height discrepancy is examined between adjacent strips.

Relative Height Offset

In the test data in this research, the percentage overlap between adjacent strips was designed to be about 30% of the total swath width. A nominal swath width of 200 m would then have an overlap region on either side of 60 m. However, due to the actual conditions during the data collection, such as wind, overlap areas between strips ranged from less than 60 m to as wide as 120 m as shown in Figure 4. Since the data consists of 14 strips, 13 overlap regions were examined to quantify the relative height discrepancies. The fact that the data is dense and the overlap regions included in the testing are large increases the likelihood of having coincident and nearly coincident data points. Therefore, the relative height accuracy was obtained directly by computing the differences between such nearly coincident data points.

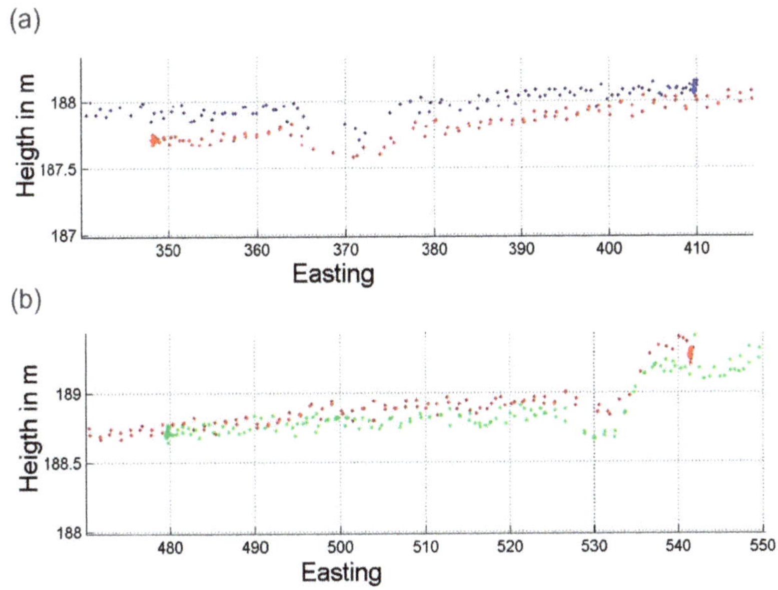

Figure 3. Profile of the overlap area (2 m width) between (**a**) strip 1 and 2, (**b**) strip 2 and 3.

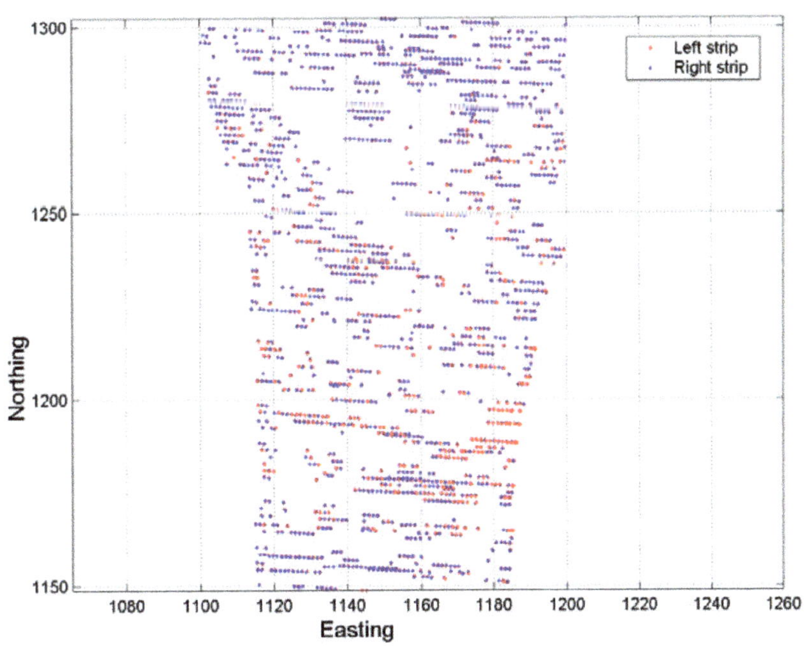

Figure 4. Closer View of the Overlap Region.

Three experiments were conducted during the testing. In order to minimize the effect of slope and spatial position on the collected heights, the planimetric distance between tested points was limited to be less than 0.05 m in the first test. In spite of these limitations, there were some misinterpreted differences (more than a half meter) due to the height jump between the tested points. Height jumps on building edges is one example. On building edges, two points within a few centimeters could have a great height jump since one might be on the ground and the other one is on the roof. Such outliers were detected based on the statistical interpretation of the computed discrepancy. Consequently, they were deleted from the data set to eliminate their influences. A manual examination verified the stated assumption that these outliers coincided with abrupt height transitions. Then, to include more points in the computation and strengthen the statistics, the planimetric-distance constraint was increased to include points within 0.10 m and 0.25 m. Table 1 summarizes the results for all of the 13 overlap regions with 0.05 m, 0.10 m, and 0.25 m planimetric distance allowance. However, changing the planimetric distance and including more points did not significantly change the discrepancy average in all overlap regions as shown in Table 1. This is because the planimetric distance is less than the point density. In the rest of the study, we used the 0.05 m planimetric distance allowance.

Table 1. Relative height discrepancy between adjacent data strips.

		1,2	**2,3**	**3,4**	**4,5**	**5,6**	**6,7**	**7,8**	**8.9**	**9,10**	**10,11**	**11,12**	**12,13**	**13,14**
0.05 m	#pts	521	474	587	552	658	576	0.10	539	725	626	649	517	492
	Mean ΔH(m)	0.04	0.00	0.03	−0.04	−0.10	−0.12	−0.07	−0.11	−0.09	−0.10	−0.10	−0.06	−0.07
	STD m	0.12	0.14	0.14	0.14	0.10	0.10	0.10	0.09	0.11	0.11	0.12	0.12	0.13
0.10 m	#pts	2017	2019	2193	2108	2667	2304	3172	2087	2880	2583	2616	2085	2121
	Mean ΔH m	0.04	0.00	0.03	−0.04	−0.10	−0.11	−0.07	−0.11	−0.09	−0.09	−0.10	−0.06	−0.07
	STD m	0.12	0.15	0.14	0.14	0.11	0.10	0.10	0.10	0.11	0.11	0.11	0.11	0.12
0.25 m	#pts	10,891	11,017	11,944	11,872	14,761	11,889	17,702	11,014	14,768	13,986	13,951	11,242	11,142
	Mean ΔH m	0.04	0.00	0.03	−0.03	−0.10	−0.10	−0.07	−0.10	−0.08	−0.09	−0.10	−0.07	−0.06
	STD m	0.12	0.15	0.14	0.15	0.11	0.11	0.11	0.11	0.11	0.12	0.12	0.12	0.10

Figure 5 summarizes the behavior of the mean relative height offset between adjacent strips. As shown in Table 1, the discrepancy did not show a well-defined behavior. However, the discrepancy shows a trend with time (North–South direction) since the strips where ordered on the time they were scanned (strip 1 was the first one to scan and strip 14 was the last). In the first four overlap regions, the discrepancy was within ±0.04 m. Then, the average discrepancy between strips (5–6, 6–7, 7–8, 8–10, 10–11, and 11–12) seems to increase up to 0.10 m with a negative sign (since the discrepancy was computed by subtracting the height of the right strip from the height of the left strip). Then, in the last two overlap regions (12–13 and 13–14) the discrepancy dropped to −0.06 m.

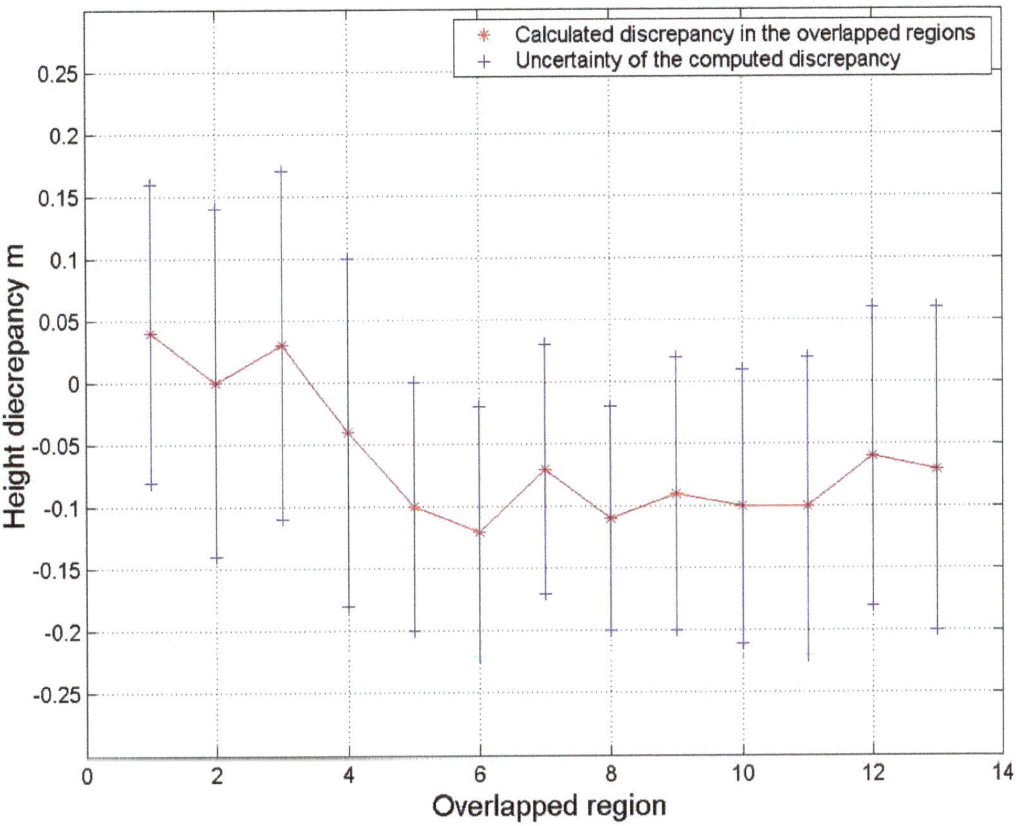

Figure 5. Mean Relative discrepancy behavior between adjacent strips, where the x-axis represents the overlap regions (where 1 is the overlap region between strip one and two).

Figure 6a shows the discrepancy behavior with respect to time (North–South direction) and Figure 6b shows the histogram of the computed discrepancy for the overlap region between strips 1 and 2. In this overlap region, the average relative offset was 0.04m, which means the left data strip (strip 1) was higher in average than the right-hand strip (strip 2). On the other hand, Figure 7a,b show the relative height offset between strips 7 and 8 where the left strip (strip 7) seems to be below the right-hand one (strip 8) over most of the overlap region.

In general, the mean of the discrepancies does not equal zero and is not consistent with all strips. In addition, offsets are not purely a result of height differences at exactly corresponding points in the two strips since they are not error-free in planimetric positioning. Consequently, due to this planimetric uncertainty, the compared points might not have the same exact planimetric position and this mis-correspondence may contribute to the height differences. This correlation between height and planimetric offsets is to be expected in sloping surfaces.

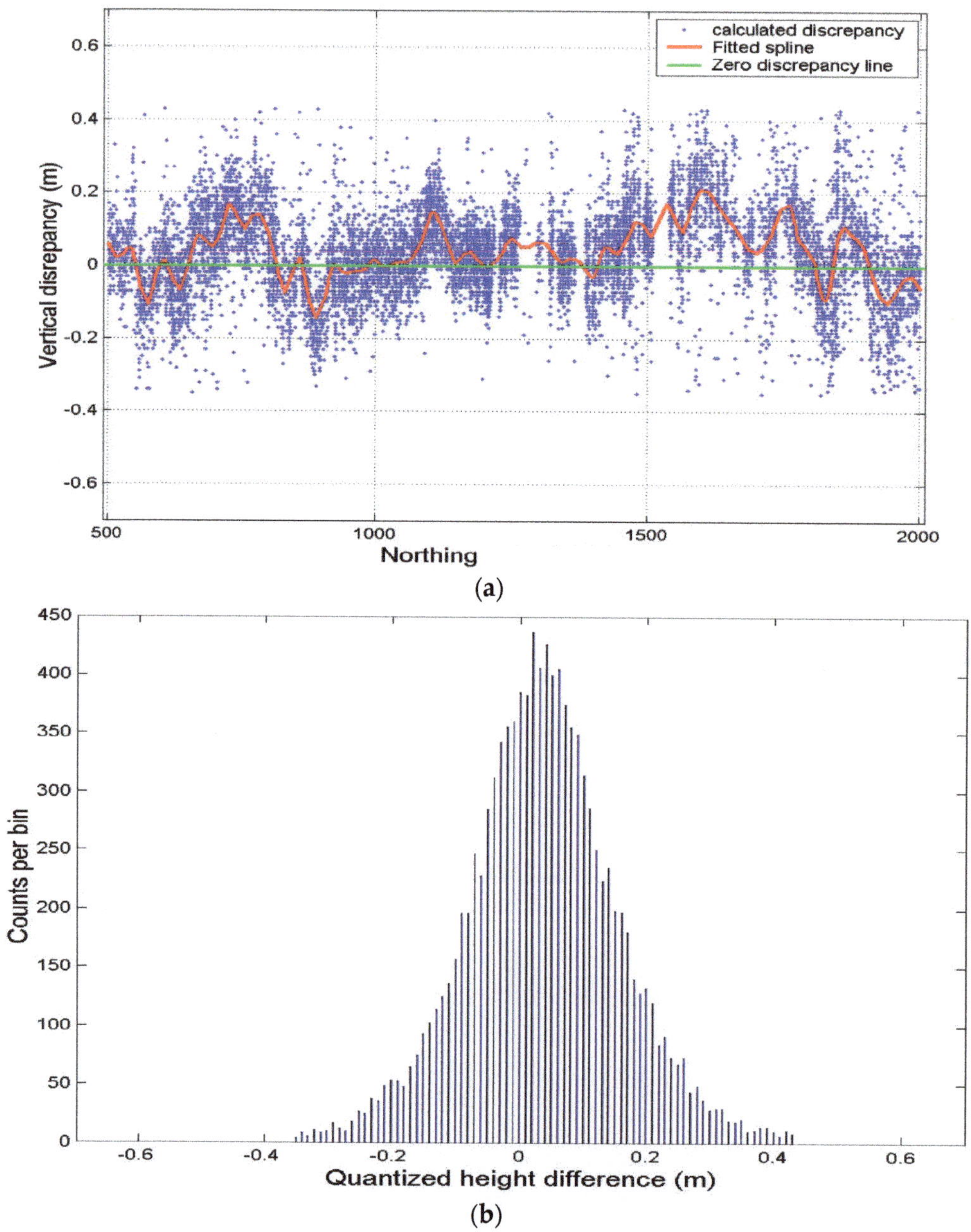

Figure 6. (**a**): Relative height discrepancy along the North–South direction in the overlap region between the 1st and the 2nd strip. (**b**): The histogram of relative height discrepancy between the 1st strip and the 2nd strip.

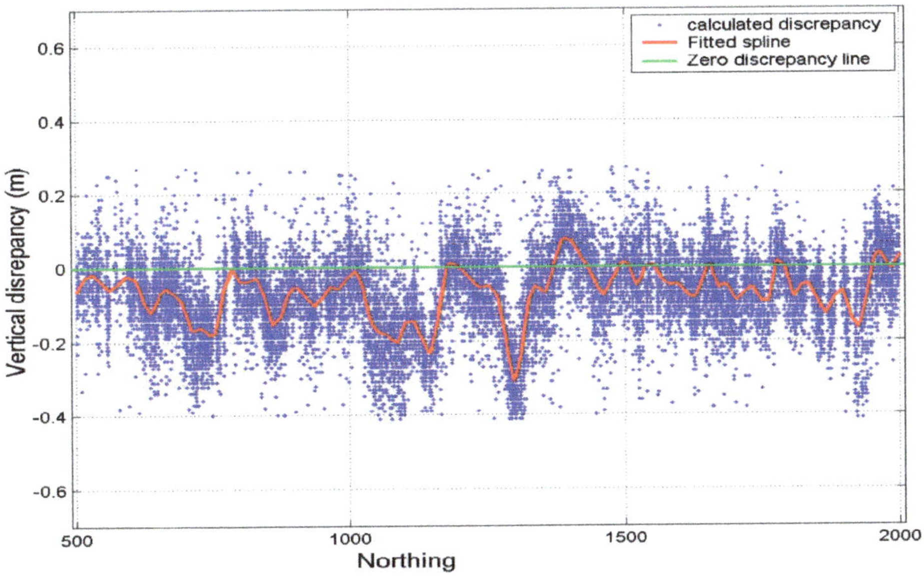

(**a**)

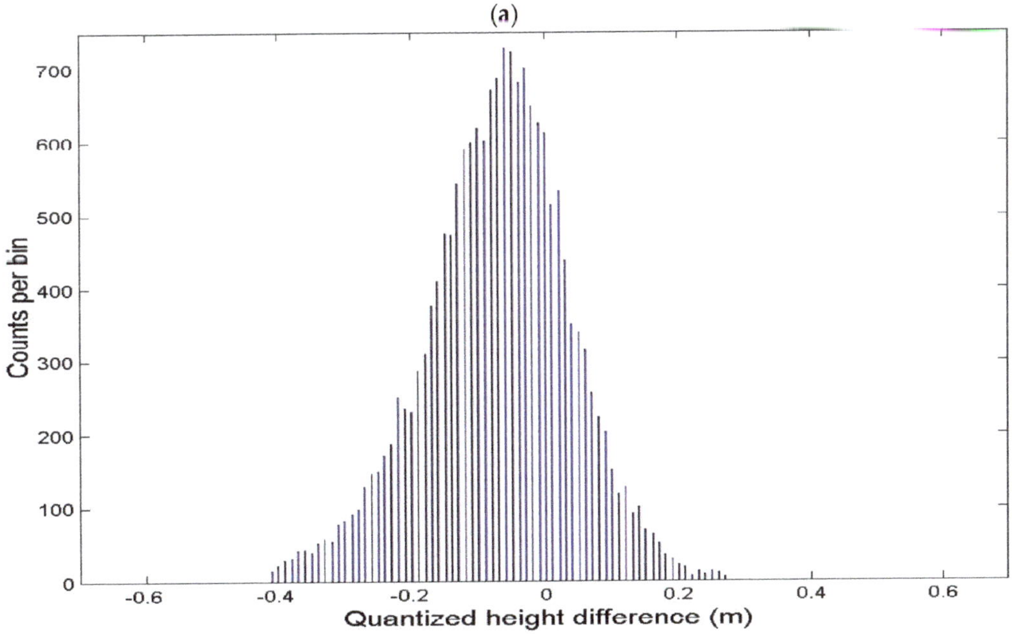

(**b**)

Figure 7. (**a**): Relative height discrepancy along the North–South direction in the overlap region between the 7th and the 8th strip. (**b**): The histogram of the relative height discrepancy between the 7th strip and the 8th strip.

4. Data Absolute Accuracy

Many sources of error affect the quality of airborne laser scanning data. Companies that work in collecting such datasets usually publish a fixed number for the uncertainty of their data. However, these numbers are usually not verified by the users because this is not a simple task. Correspondence between laser data points and ground points is not a straightforward matter. The absolute displacements can be found by measuring the location on the ground of a point or feature of known coordinates in the data set and comparing the two measurements.

4.1. Absolute Height Accuracy

The ground points were collected in a specific pattern except when on distinct features (ditches and curves on the ground). This pattern (lines) facilitates establishing a correspondence between the two data types. A buffer zone (with a width of one meter) was constructed around each ground data line and the data points inside that buffer zone were identified. Then, for each laser data point that lies within a meter or less from two ground data points, a corresponding height point with the same planimetric position was interpolated from the surrounding ground-surveyed points. The absolute accuracy was then estimated as the absolute difference between the laser height and the ground height. Since the points are nearby and the surface is flat, a linear interpolation was used. In order to minimize the effect of the planimetric uncertainty of the laser data, the interpolation was limited to totally flat areas. The slope of those areas was restricted to not exceed 10%. The slope was calculated as the rate of change of elevation for each LIDAR point using the Surface Slope tool in ArcMap.

Direct differencing was applied to compute the discrepancy between the two data points from the laser and ground, i.e., the laser height was subtracted from the ground interpolated height. More than a thousand such differences were computed. Figure 8 shows the histogram of those differences. The histogram shows a bias in the differences of −0.088 m with a standard deviation of ±0.082 m. Hence, the average height of the laser data points included in the test is higher than the estimated ground by 0.088 m. Consequently, the total RMSE of the LIDAR heights from the surveyed ground (zero in the histogram) was 0.12 m with a mean value of zero. The calculated statistical characteristics represent only a small sample of the data. The tested area covers less than 2 s of collection time (about 19,000 data points since the designed scanning rate is 10,000 points/s). The 19,000 data points represent a small portion of the whole data set which is more than 3,000,000 data points. However, only 1008 points were included in the real computation since it is not practical to survey more points on the ground.

4.2. Absolute Planimetric Accuracy

Planimetric offsets are more complicated to determine since they require identifiable spatial features to establish the correspondence between the two data sets. Such features and locations for estimating the offsets might not be available, or when they exist, are usually limited. Moreover, identifying these locations is costly in time and requires great care in order to be reliable. Drainage ditches, terrain features, and building gable roofs are some examples of such features. Procedures and results of estimating the planimetric accuracy in both directions (Easting and Northing) are presented in this section.

Eight positions were identified and used in obtaining the planimetric accuracy. Six of those locations were used to compute the offset in the X (Easting) direction and the other two were utilized to compute the offset in the Y (Northing) direction. In each of these locations the data points from both data sets, LIDAR and ground, were identified as shown in part (a) of Figures 9–11. Parts (a) of these figures show the plan view of the points. From each set, an estimate of the two shifts was made to realize the coincidence. The idea here is to translate these two curves and obtain the shift that will maximize the match. Prior to that, the height bias should be removed in order to exclude its effect in the matching. In part (b) of the figures, the green dashed curve represents the LIDAR data after removing the height bias between the two data curves.

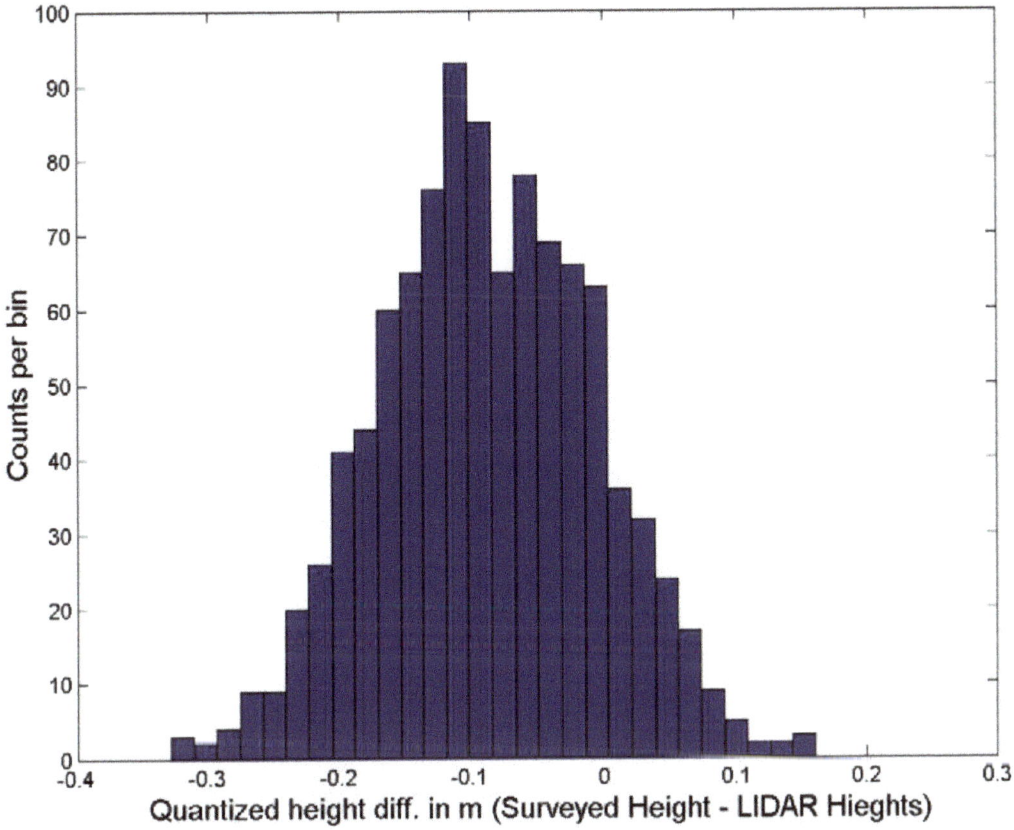

Figure 8. Height differences histogram between the laser height and the ground height.

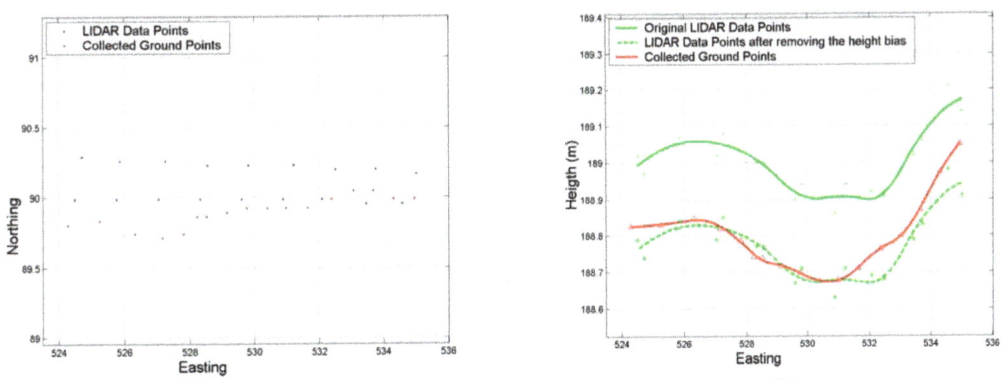

(a) (b)

Figure 9. *Cont.*

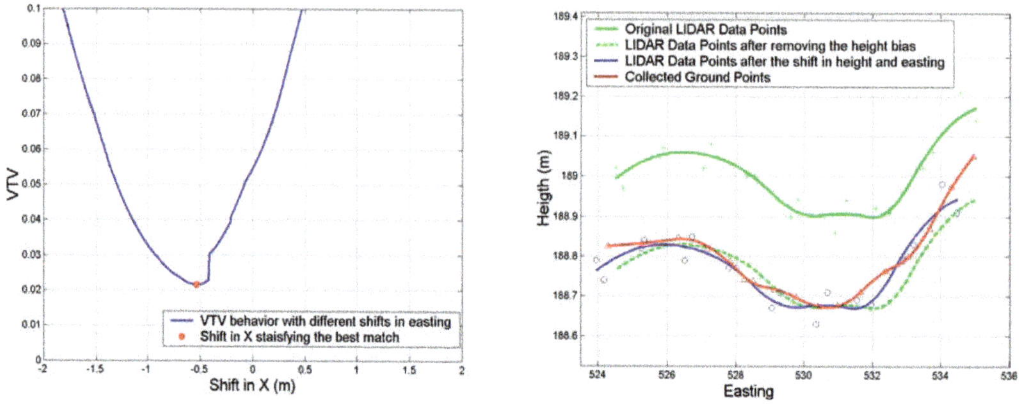

(c) (d)

Figure 9. (**a**–**d**) Planimetric offset in Easting (X) direction at location A.

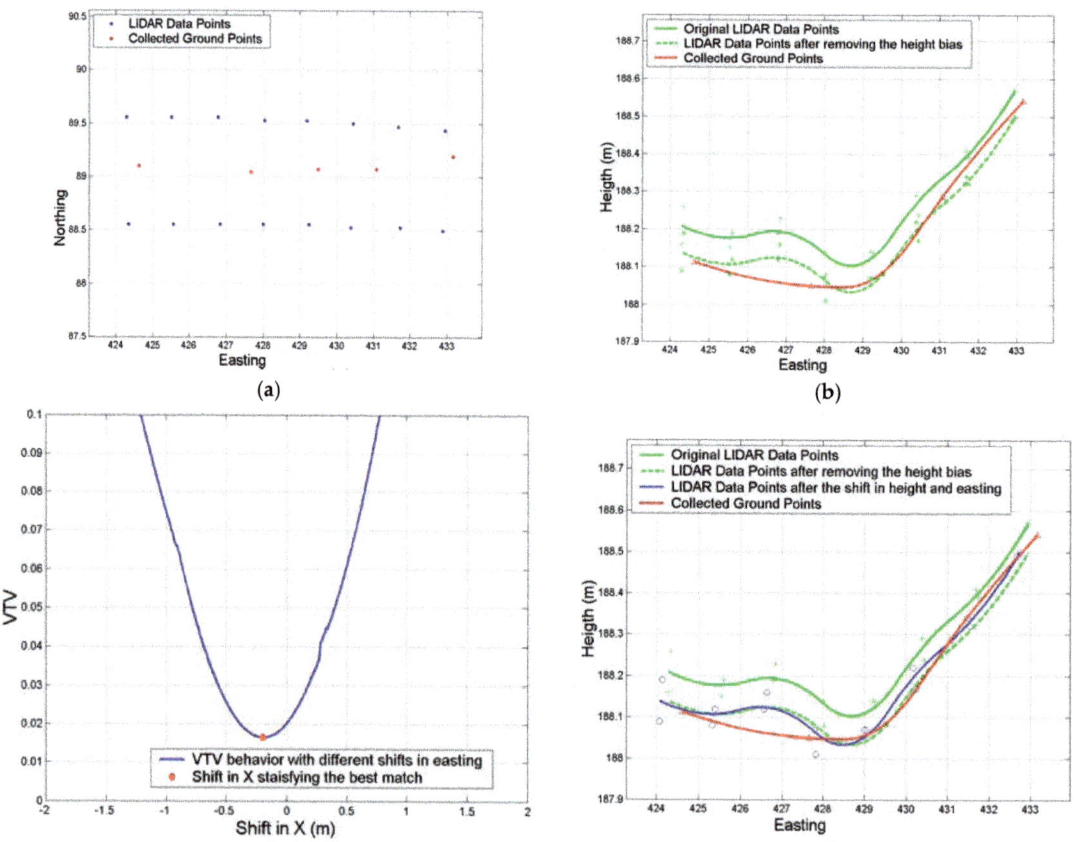

Figure 10. (**a**–**d**) Planimetric offset in Easting (X) direction at location B.

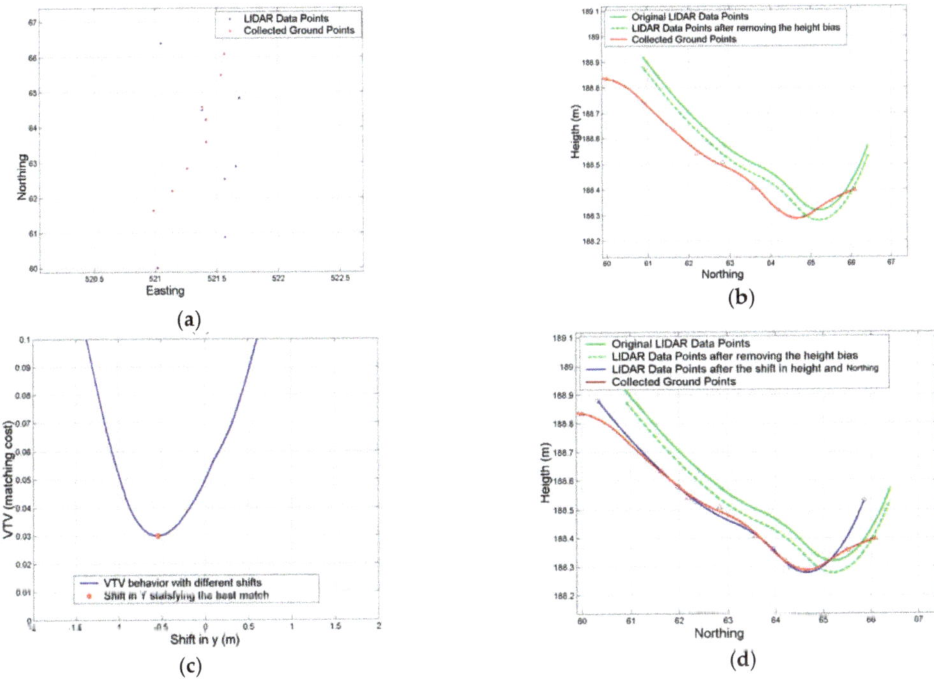

Figure 11. (**a–d**) Planimetric offset in Northing (Y) direction at location G.

To get the best match between the two curves, the LIDAR data curve will be shifted gradually around the ground data set in the direction of the computed offset (X as in Figures 9 and 10, and Y as in Figure 11). The shift search window ranged from −2 m to 2 m with an increment of 0.01m. At each increment, the discrepancy in Z of each ground point and its interpolated-correspondence point from the LIDAR curve data is computed. The sum of the squares of these offsets is considered as the matching cost at each location. Parts (c) of the figures show the matching cost function behavior with respect to different shift values. At the minimum matching cost, which is associated with the best match between the two curves, the corresponding shift is obtained. Parts (d) show the original data curves, after removing the height bias, and after the planimetric shift.

As stated above, eight locations were tested to obtain the planimetric shift. Table 2 summarizes the results at those selected locations. Regarding the offset in the X (Easting) direction, which is across the flight direction and coincides with the scanning direction, six locations were selected, three at the edge of the swath width of strip two and the rest at the middle of the strip. As expected at the strip edge, the offset in the scanning direction (−0.60 m as an average) was larger than at the middle of the strip (−0.30 m as an average). However, those shifts were in the same direction. The same thing could be said for the height bias, height offsets seem to be larger in magnitude at the edge of the strip. On the other hand, two locations were selected to test the accuracy along the flight direction Y (Northing), one at the middle of the strip and the other at the edge. The two locations had the same direction.

Table 2. Planimetric accuracy results.

Location	Height Bias (m)	Direction	Plan. Offset	Location Description
A	−0.12	X	−0.67	At the right edge of the swath width of strip 2
B	−0.03	X	−0.29	At the middle of the swath width of strip 2
C	−0.23	X	−0.54	At the right edge of the swath width of strip 2
D	−0.07	X	−0.20	At the middle of the swath width of strip 2
E	−0.23	X	−0.62	At the right edge of the swath width of strip 2
F	−0.08	X	−0.43	At the middle of the swath width of strip 2
G	−0.04	Y	−0.55	At the middle of the swath width of strip 2
H	−0.07	Y	−0.40	At the middle of the swath width of strip 2

4.3. *Error Behavior with Respect to Time*

To show how errors changed while the scanner was advancing, the LIDAR data points over the test area were sorted based on the time they were scanned. Then, the calculated height differences (1008 biases) were sorted by time. The test area covered only less than two seconds of the scanning time, which contains about 19,000 data points. Figure 12 shows the behavior of the absolute height error with respect to the time they were scanned.

In Figure 12, the height differences between the LIDAR points and their corresponding ground-surveyed points show two types of variations. The two types can be divided in two components: a bias in the geo positioning system and a random variation around that bias. The first type of variation is called short-period variation. This variation seems to be random and has a high frequency. This variation between two consecutive points could reach 0.30 m as a maximum within 0.001 s. This random variation has a standard deviation of 0.082 m. The short period variation, Figure 13, of the uncertainty of LIDAR heights gives an indication of the system precision since the consecutive points are so close to each other in the time domain and the test was conducted on a flat surface where the height is approximately the same.

On the other hand, as shown in Figure 12, a trend (green spline) of the differences is observed which is the second form of the variation. A trend is defined as the component of a random phenomenon which has a period larger than the recorded data sample. The trend in this test data represents the biases in the geopositioning system. Although the test data represents only a small sample, this trend is very noticeable. The relative height accuracy between data strips confirms this inference regarding the two forms of random variation since the computation of the relative accuracy covers most of the data.

The total computed uncertainty of the LIDAR heights over the test area is about ±0.121 m. This uncertainty consists of two components: biases in the geopositioning of the platform and random variation around these biases due to the ranging system uncertainty. Regarding the planimetric accuracy, the computed offsets in the scanning direction over the test area show two main outcomes. First, the planimetric uncertainty is larger (almost double) at the end of the swath than at the middle. So, in general, the planimetric accuracy seems to vary based on the location along the scan (cross-track) direction. At the edge of the swath, the average offset was about 0.60 m, while at the middle it was about 0.30 m. The second outcome is that the whole strip seems to be shifted in the east direction since all the computed offsets have the same direction.

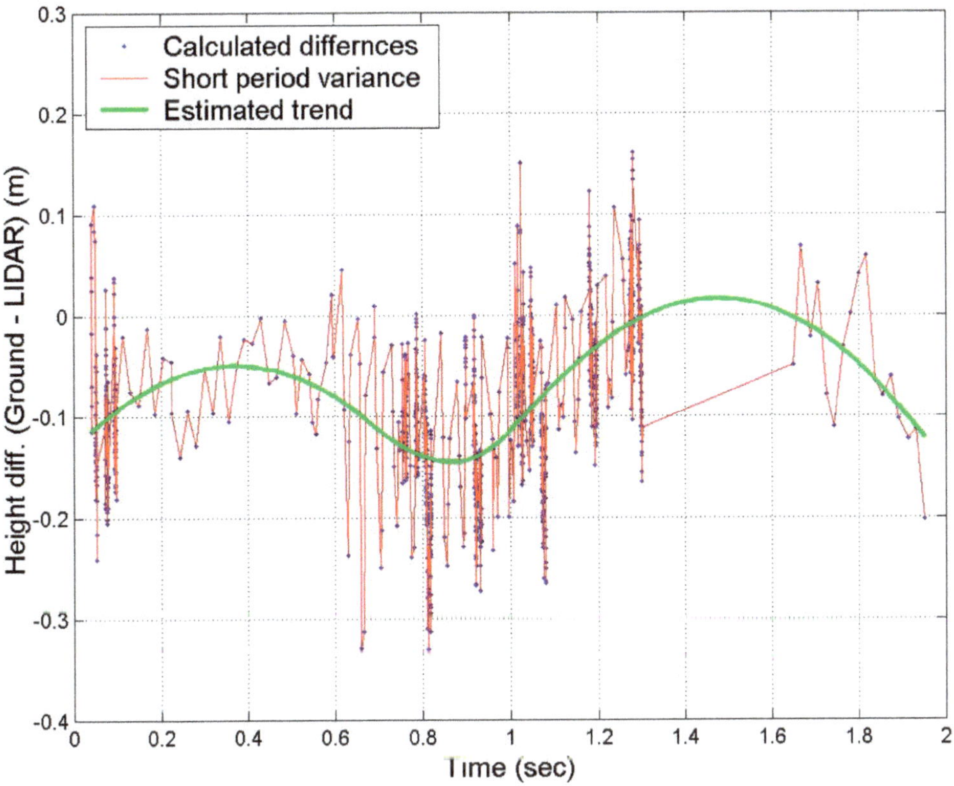

Figure 12. Height accuracy (Ground—LIDAR) with respect to time.

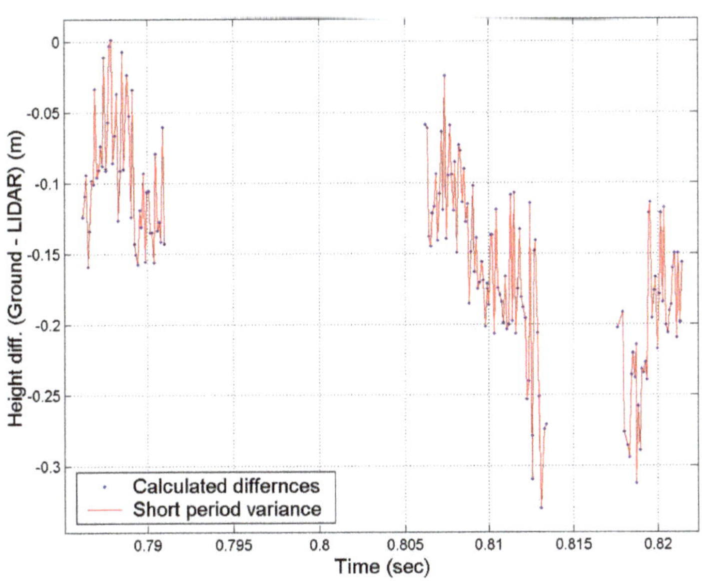

Figure 13. Short period variation of the height biases.

Both accuracies, relative and absolute, show two types of variation. The first variation form is the short period random variation which has a high frequency with time. This random variation seems to represent the actual precision of the LIDAR subsystem. The other variation form is the long period variation or a trend. This trend has a much lower frequency and seems to be a result of errors in the positioning/altitude subsystem. If this trend is modeled and the data is adjusted accordingly, the accuracy of data will be improved. Testing more samples that cover different parts with a longer period of time would be needed to strengthen these conclusions.

5. Results Analysis and Discussion

The relative height offsets between LIDAR data strips were examined based on the height offsets between coincident or partially coincident data points. More than 16,600 height offsets from all the 13 test overlap regions were computed. The results show a variation in the height offset between adjacent strips. Among the 13 overlap regions between adjacent strips, the average relative height offset ranges from 0 to 0.12 m. The maximum offset between coincident laser measurements from two different strips is about 0.40 m. Two variation types are observed in the discrepancy between adjacent strips, short- and long-period variations. The short-period variation is attributed with a high frequency with time and has a standard deviation of about ±0.10 m. The long-period variation has a much lower frequency than the first type. The source of the short-period variation appears to be the uncertainty in the LIDAR subsystem. Ranging and scanning observation uncertainty are some examples of error sources in the LIDAR subsystem. The short-period variation appears to be random and represents the LIDAR subsystem precision. On the other hand, the uncertainty in the navigation subsystem appears to be the source of the long-period variation, which can be modeled and corrected and consequently improve the data quality.

More than 1000 height discrepancies between the LIDAR data and collected ground points were calculated in order to examine the height absolute accuracy of the LIDAR data. The histogram of the computed differences shows a bias of about 0.088 m from the surveyed ground with a standard deviation of ±0.082 m around the middle of the histogram. The average height of the laser data points included in the test is higher than the estimated ground by 0.088 m. However, the total variation of the LIDAR heights from the surveyed ground is 0.120 m. This variation contains both the biases and its random differences. This number (0.120 m) represents a sample of the real system height precision of the LIDAR data. The computed discrepancies were then sorted based on the time they were scanned in order to examine the error behavior with respect to time. The height differences between the LIDAR points and their corresponding ground-surveyed points showed the two types of variation that were shown in the relative height accuracy. However, this time the tested sample is much smaller than in the relative height accuracy.

The variation between two consecutive points in time could reach 0.30 m as a maximum. The time span between two consecutive points in the test data is about 0.001 s since the pulse rate is 10 kHz. This short-period variation of the uncertainty of LIDAR heights gives an indication of the system precision since the consecutive points are so close to each other in the time domain and the test was conducted on a generally flat surface where the height is approximately the same.

In the scanning direction (across track), the planimetric offset of the LIDAR data from the collected ground points varies based on the location in the swath. As expected, at the strip edge, the offset in the scanning direction (−0.60 m as an average) was larger than at the middle of the strip (−0.30 m as an average). However, those shifts were in the same direction, which show the presence of a trend or bias. In the other direction (along track), the average offset from the ground was about 0.47 m. In general, the root mean square error of the planimetric offsets in both directions was estimated to be 0.50 m.

6. Conclusions

Coincident LIDAR data points in overlap regions between adjacent strips show a relative discrepancy in height. This discrepancy varies from one position to another and varies with time. Regarding the absolute height accuracy, the data points show a systematic departure from the collected ground reference and a random variation of −0.088 m. The total computed RMSE of the test data from the ground reference was about 12 cm. Planimetric offset of the LIDAR data from the ground reference in the across flight direction varies with the location in the swath. At the edge of the swath, the average planimetric offset was almost double its value in the middle.

Both relative and absolute height discrepancies of the LIDAR data have two components of variation. The first component is the short-period variation, which is random and has high frequency with time. This short-period random variation runs over the second component, which is the long-period variation. The long-period variation has a much lower frequency than the first type and represents the biases in the geopositioning system. The short-period variation is attributed to the LIDAR subsystem measurement uncertainty. System ranging resolution and scanning system-pointing precision are some examples of the error sources in the LIDAR subsystem. On the other hand, the uncertainty in the navigation subsystem appears to be the source of the long period variation, which can be modeled and accounted for in order to improve the quality of the data.

The LIDAR data accuracy evaluation is still an open area for more research and analysis, particularly with the growing application of new UAV-LIDAR systems in many applications. Such systems can provide centimeter-level accuracy and the procedure outlined in this research has the capability to evaluate the accuracy of these systems with a high degree of certainty. This will extend the use of LIDAR data in many applications such as building information models (BIMs), 3D Geographic Information Systems (GIS), mobile communications, disaster management, and city planning. Therefore, future research will focus on assessing the quality of LIDAR data produced from UAV systems and best practices of collecting LIDAR data for different applications.

Author Contributions: Methodology, A.A.; Validation, A.A.; Formal analysis, A.E.; Investigation, T.A. and A.A.; Resources, T.A.; Data curation, A.E.; Writing—original draft, A.E.; Writing—review & editing, T.A.; Project administration, A.E.; Funding acquisition, A.A. All authors have read and agreed to the published version of the manuscript.

Funding: This research received no external funding.

Data Availability Statement: Not applicable.

Conflicts of Interest: The authors declare no conflict of interest.

References

1. Wani, Z.M.; Nagai, M. An approach for the precise DEM generation in urban environments using multi-GNSS. *Measurement* **2021**, *177*, 109311. [CrossRef]
2. Gargoum, S.A.; El Basyouny, K. A literature synthesis of LiDAR applications in transportation: Feature extraction and geometric assessments of highways. *GISci. Remote Sens.* **2019**, *56*, 864–893. [CrossRef]
3. Kulawiak, M. A Cost-Effective Method for Reconstructing City-Building 3D Models from Sparse Lidar Point Clouds. *Remote Sens.* **2022**, *14*, 1278. [CrossRef]
4. Lv, D.; Ying, X.; Cui, Y.; Song, J.; Qian, K.; Li, M. Research on the technology of LIDAR data processing. In Proceedings of the First International Conference on Electronics Instrumentation and Information Systems (EIIS), Harbin, China, 3–5 June 2017.
5. Elaksher, A.F. Co-registering satellite images and LIDAR DEMs through straight lines. *Int. J. Image Data Fusion* **2016**, *7*, 103–118. [CrossRef]
6. Li, X.; Liu, C.; Wang, Z.; Xie, X.; Li, D.; Xu, L. Airborne LiDAR: State-of-the-art of system design, technology and application. *Meas. Sci. Technol.* **2020**, *32*, 032002. [CrossRef]
7. Wehr, A. LiDAR systems and calibration. In *Topographic Laser Ranging and Scanning*, 2nd ed.; Shan, J., Toth, C., Eds.; CRC Press: Boca Raton, FL, USA, 2018; pp. 159–200.
8. Elaksher, A.; Bhandari, S.; Carreon-Limones, C.; Lauf, R. Potential of UAV lidar systems for geospatial mapping. In Proceedings of the SPIE Optical Engineering and Applications, San Diego, CA, USA, 6–10 August 2017.

9. Cao, S.; Weng, Q.; Du, M.; Li, B.; Zhong, R.; Mo, Y. Multi-scale three-dimensional detection of urban buildings using aerial LiDAR data. *GISci. Remote Sens.* **2020**, *57*, 1125–1143. [CrossRef]
10. Megahed, Y.; Shaker, A.; Yan, W. Fusion of airborne LiDAR point clouds and aerial images for heterogeneous land-use urban mapping. *Remote Sens.* **2021**, *13*, 814. [CrossRef]
11. Wang, X.; Li, P. Extraction of urban building damage using spectral, height and corner information from VHR satellite images and airborne LiDAR data. *ISPRS J. Photogramm. Remote Sens.* **2020**, *159*, 322–336. [CrossRef]
12. Muhadi, N.A.; Abdullah, A.F.; Bejo, S.K.; Mahadi, M.R.; Mijic, A. The use of LiDAR-derived DEM in flood applications: A review. *Remote Sens.* **2021**, *12*, 2308. [CrossRef]
13. Görüm, T. Landslide recognition and mapping in a mixed forest environment from airborne LiDAR data. *Eng. Geol.* **2019**, *258*, 105155. [CrossRef]
14. Štular, B.; Lozić, E.; Eichert, S. Airborne LiDAR-derived digital elevation model for archaeology. *Remote Sens.* **2021**, *13*, 1855. [CrossRef]
15. Mesa-Mingorance, J.L.; Ariza-López, F.J. Accuracy assessment of digital elevation models (DEMs): A critical review of practices of the past three decades. *Remote Sens.* **2021**, *12*, 2630. [CrossRef]
16. Szypuła, B. Quality assessment of DEM derived from topographic maps for geomorphometric purposes. *Open Geosci.* **2019**, *11*, 843–865. [CrossRef]
17. Montealegre, A.L.; Lamelas, M.T.; Riva, J.D. Interpolation routines assessment in ALS-derived digital elevation models for forestry applications. *Remote Sens.* **2015**, *7*, 8631–8654. [CrossRef]
18. Gillin, C.P.; Bailey, S.W.; McGuire, K.J.; Prisley, S.P. Evaluation of LiDAR-derived DEMs through terrain analysis and field comparison. *Photogramm. Eng. Remote Sens.* **2015**, *81*, 387–396. [CrossRef]
19. Von Schwerin, J.; Richards-Rissetto, H.; Remondino, F.; Spera, M.; Auer, M.; Billen, N.; Reindel, M. Airborne LiDAR acquisition, post-processing and accuracy-checking for a 3D WebGIS of Copan, Honduras. *J. Archaeol. Sci. Rep.* **2016**, *5*, 85–104. [CrossRef]
20. Ali, T. On the selection of an interpolation method for creating a terrain model (TM) from LIDAR data. In Proceedings of the American Congress on Surveying and Mapping (ACSM) Conference, Nashville, TN, USA, 16–21 April 2004.
21. Ali, T.; Mehrabian, A. A novel computational paradigm for creating a triangular irregular network (TIN) from LiDAR data. *Nonlinear Anal. Theory Methods Appl.* **2009**, *71*, e624–e629. [CrossRef]
22. Ali, T. Building of robust multi-scale representations of LiDAR-based digital terrain model based on scale-space theory. *Opt. Lasers Eng.* **2010**, *48*, 316–319. [CrossRef]
23. Contreras, M.; Staats, W.; Yiang, J.; Parrott, D. Quantifying the accuracy of LiDAR-derived DEM in deciduous eastern forests of the Cumberland Plateau. *J. Geogr. Inf. Syst.* **2017**, *9*, 339–353. [CrossRef]
24. Sibona, E.; Vitali, A.; Meloni, F.; Caffo, L.; Dotta, A.; Lingua, E.; Garbarino, M. Direct measurement of tree height provides different results on the assessment of LiDAR accuracy. *Forests* **2016**, *8*, 7. [CrossRef]
25. Guerra-Hernández, J.; Cosenza, D.N.; Rodriguez, L.C.E.; Silva, M.; Tomé, M.; Díaz-Varela, R.A.; González-Ferreiro, E. Comparison of ALS-and UAV (SfM)-derived high-density point clouds for individual tree detection in Eucalyptus plantations. *Int. J. Remote Sens.* **2018**, *39*, 5211–5235. [CrossRef]
26. Schmelz, W.J.; Psuty, N.P. Quantification of Airborne Lidar Accuracy in Coastal Dunes (Fire Island, New York). *Photogramm. Eng. Remote Sens.* **2019**, *85*, 133–144. [CrossRef]
27. Kovanič, Ľ.; Blistan, P.; Urban, R.; Štroner, M.; Blišťanová, M.; Bartoš, K.; Pukanská, K. Analysis of the Suitability of High-Resolution DEM Obtained Using ALS and UAS (SfM) for the Identification of Changes and Monitoring the Development of Selected Geohazards in the Alpine Environment—A Case Study in High Tatras, Slovakia. *Remote Sens.* **2020**, *12*, 3901. [CrossRef]
28. Căţeanu, M.; Ciubotaru, A. Accuracy of Ground Surface Interpolation from Airborne Laser Scanning (ALS) Data in Dense Forest Cover. *ISPRS Int. J. Geo-Inf.* **2020**, *9*, 224. [CrossRef]
29. Căţeanu, M.; Ciubotaru, A. The effect of lidar sampling density on DTM accuracy for areas with heavy forest cover. *Forests* **2021**, *12*, 265. [CrossRef]
30. Liu, X. A new framework for accuracy assessment of LIDAR-derived digital elevation models. *ISPRS Ann. Photogramm. Remote Sens. Spat. Inf. Sci.* **2022**, *4*, 67–73. [CrossRef]
31. Ussyshkin, R.V.; Smith, B. Performance analysis of ALTM 3100EA: Instrument specifications and accuracy of lidar data. In Proceedings of the ISPRS Commission I Symposium, Paris, France, 4–6 July 2006.
32. Mostafa, M.; Hutton, J.; Reid, B.; Hill, R. GPS/IMU products—The Applanix approach. In *Photogrammetric Week*; Wichmann: Berlin/Heidelberg, Germany, 2001; Volume 1, pp. 63–83.
33. Vosselman, G.; Maas, H.G. Adjustment and filtering of raw laser altimetry data. In Proceedings of the OEEPE Workshop on Airborne Laser Scanning and Interferometric SAR for Detailed Digital Terrain Models, Stockholm, Sweden, 1–3 March 2001.

remote sensing

MDPI

Article

New Structural Complexity Metrics for Forests from Single Terrestrial Lidar Scans

Jonathan L. Batchelor [1,*], Todd M. Wilson [2], Michael J. Olsen [3] and William J. Ripple [4]

1 School of Environmental and Forest Sciences, University of Washington, Seattle, WA 98195, USA
2 USDA Forest Service, Pacific Northwest Research Station, Corvallis, OR 97331, USA
3 School of Civil and Construction Engineering, Oregon State University, Corvallis, OR 97331, USA
4 Department of Forest Ecosystems and Society, Oregon State University, Corvallis, OR 97331, USA
* Correspondence: jonbatch@uw.edu

Abstract: We developed new measures of structural complexity using single point terrestrial laser scanning (TLS) point clouds. These metrics are depth, openness, and isovist. Depth is a three-dimensional, radial measure of the visible distance in all directions from plot center. Openness is the percent of scan pulses in the near-omnidirectional view without a return. Isovists are a measurement of the area visible from the scan location, a quantified measurement of the viewshed within the forest canopy. 243 scans were acquired in 27 forested stands in the Pacific Northwest region of the United States, in different ecoregions representing a broad gradient in structural complexity. All stands were designated natural areas with little to no human perturbations. We created "structural signatures" from depth and openness metrics that can be used to qualitatively visualize differences in forest structures and quantitively distinguish the structural composition of a forest at differing height strata. In most cases, the structural signatures of stands were effective at providing statistically significant metrics differentiating forests from various ecoregions and growth patterns. Isovists were less effective at differentiating between forested stands across multiple ecoregions, but they still quantify the ecological important metric of occlusion. These new metrics appear to capture the structural complexity of forests with a high level of precision and low observer bias and have great potential for quantifying structural change to forest ecosystems, quantifying effects of forest management activities, and describing habitat for organisms. Our measures of structure can be used to ground truth data obtained from aerial lidar to develop models estimating forest structure.

Keywords: terrestrial lidar; TLS; forest structure; depth; openness; viewshed; Research Natural Areas; forest vegetation

Citation: Batchelor, J.L.; Wilson, T.M.; Olsen, M.J.; Ripple, W.J. New Structural Complexity Metrics for Forests from Single Terrestrial Lidar Scans. *Remote Sens.* **2023**, *15*, 145. https://doi.org/10.3390/rs15010145

Academic Editors: Jorge Delgado García, Fayez Tarsha Kurdi and Tarig Ali

Received: 19 November 2022
Revised: 21 December 2022
Accepted: 23 December 2022
Published: 27 December 2022

1. Introduction

1.1. Importance of Structure

The structural complexity of a forest (e.g., arrangement and amount of above-ground, biophysical components) has profound influences on its ecological function and health [1–3]. Structural complexity influences everything from canopy closure and connectivity that determines the amount of light infiltration to the forest floor, which influences the type and amount of understory and midstory plant species. Biophysical components also heavily influence the foraging and movement patterns of wildlife [4–7].

Quantifying structure is becoming increasingly important for resolving contemporary forest management issues. In the Pacific Northwest (PNW) of the United States, efforts to address some of these issues include developing late seral forest characteristics in young, managed forests to promote habitat for threatened northern spotted owls (Strix occidentalis caurina), landscape-level efforts to reduce fire fuel loads in dry forests after more than a century of fire suppression, and developing programs to monitor the long-term effects of climate change on natural and managed forests [8–11]. Each of these efforts requires, in part, the ability to quantify changes in the physical arrangement and amount of forest structure over time.

Historical measures of structural complexity have included simple measures of structural elements (e.g., tree diameter, height, spacing, basal area, variance in diameter at breast height (DBH)) and indices combining multiple measures (e.g., Shannon-Weiner Index, old growth index, & diameter diversity index [DDI]) [12–15]. These measures have been used successfully for a range of ecological applications, from single species habitat analysis at a localized level, to broad landscape-level planning [16,17]. Given the importance of forest structure, new methods are constantly being developed to quantify more nuanced and subtle differences and changes as new technologies emerge [18–20]. While most historical measures have relied on rudimentary hand tools, such as spherical densitometers or wedge prisms that introduce observer bias [21,22], many of the new methods to quantify forest structure leverage the more objective measurements derived from remote sensing platforms such as lidar.

1.2. Lidar for Forest Structure

Lidar technology is increasingly being used for quantifying forest structure [23–25]. Lidar data can be acquired from a variety of platforms, depending on the accuracy, resolution, and spatial coverage needed. Airborne lidar can efficiently capture forest structure across broad landscapes and has been used effectively to measure tree heights, leaf area index, above ground biomass, and forest heterogeneity [25–28]. Aerial lidar is so ingrained in current forest inventory assessments that it is a standard tool employed by many management organizations [29–31]. However, airborne lidar can miss much of the structural complexity in understory and midstory layers, particularly in dense forests; hence, efforts to quantify understory vegetation from ALS are limited in their characterizations [32–34].

Terrestrial laser scanning (TLS), also referred to as terrestrial lidar, is used to collect structural data from the ground and provides much more detailed information about the lower and middle layers of a forest than airborne lidar, especially in structurally complex forests where tree canopies can obscure aerial detection of understory and mid-story vegetation layers. Inversely, the upper canopy can be occluded from TLS in these same structurally complex and dense forests. The resolution of TLS is also much finer, with several hundred to many thousands of point returns captured per square meter (compared to a few dozen for airborne lidar). Although limited in area covered, TLS only requires a single operator and can be collected at any time weather conditions are favorable. TLS is increasingly being used to measure structural elements such as amount of woody biomass and basal area [35–37]. TLS has also been used to assess fire fuel loads and help predict fire behavior [38–40]. Further, TLS has been used to quantify forest structure for ecological analyses, including measuring forest stratification, wildlife habitat and prey cover, and structural density [41–44].

Typically, studies using TLS for forestry applications use multiple scans stitched together to create a comprehensive three-dimensional model. TLS point clouds are highly susceptible to occlusion as close objects block everything located behind them relative to the scanner position. The effects of this occlusion on derived forest metrics varies depending on what metrics are being obtained [45,46]. In relatively open forest stands, collecting multiple scans and creating a larger composite is a simple task of matching shared features between the scans (e.g., specific TLS targets or natural elements such as rocks or tree branches). In forest stands that have dense understory, stitching together multiple scans can be extremely difficult and time consuming due to the extreme amount of occlusion at each sampling location. Because of these limitations with using stitched together TLS scans, there is increasing interest in the use of single TLS scans to derive forest structural metrics. While single point TLS scans have been shown to be able to derive typical forest metrics such as DBH or stem maps [47–49], typically, multiple scans are required to replicate typical forest structure measurements [50]. The key to using single point TLS scans for forest structure characterization is likely to focus on deriving different types of metrics that still relate to forest structure beyond trying to replicate conventional sampling techniques (e.g., basal area and mean diameter). Richardson et al. [51] proposed the Three-dimensional Vegetation Density Index (3DI), which is a metric based on median distance traveled of each pulse between defined scan angles from the scanner. This method can quantify differences between stands with markedly different forest structure such as an area of dense understory vegetation compared to a stand where heavy cattle grazing removed much of the understory vegetation. However, there are some major limitations to the 3DI, namely that it does not account for slope effects on lidar point distances and thus can only be used effectively in relatively flat areas.

Studies that have used single TLS scans for novel ecological characterizations include using voxelization and point height statistics to quantify burn severity [52,53], using relative point location, geometry, and intensity to differentiate between tree stems and leaves [54], and correlating point height statistics with species richness [55]. While there has been work deriving forest metrics from single terrestrial lidar scans, only a small portion explicitly deal with how far individual pulses travel [56] or requires multiple scans to assess view area [57].

1.3. Objectives

We propose three different sets of metrics derived from single point TLS scans; depth, openness, and isovists. All three of these metrics are created in a manner that contours to localized plot topography and can be derived at sites that are both level and on a significant slope. Depth is a measurement of the distance traveled by each individual laser pulse stratified by the angle of the pulse leaving the scanner and corrected for slope. Openness accounts for pulses that did not return to the scanner (i.e., did not interact with any vegetation within the range of the scanner). Isovists, quantify the area visible approximately at 1.4 m above ground level with a 360° view from a center location. This metric is expressed as a percentage of total area visible at differing distances from the scanner. We generated our metrics from TLS scans collected in natural (no previous timber harvesting or other major human disturbance) forests that varied in structural complexity and composition to explore whether the expected variation in forest structure could be distinguished using our proposed metrics.

We then outline potential applications of these new metrics and discuss the potential advantages and disadvantages of using single TLS scans to quantify forest structure parameters.

2. Materials and Methods

2.1. Site Selection

We selected twenty-seven forested stands across six ecoregions (Coast Range, Klamath Mountains, Willamette Valley, West Cascades, and East Cascades [58]) within a 300 km radius of Corvallis, Oregon for study. We chose stands within Forest Service and Bureau of Land Management, Research Natural Areas (RNAs) and Areas of Critical Environmental Concern (ACECs) because they represent relatively pristine forests with minimal anthropogenic influence and allowed us to explore a wide range of natural variation in forest structure [59] (Figure 1).

We used DDI as the primary metric for selecting stands that collectively represented a gradient in structural complexity across the study area. DDI is a measure based on counts of trees in four different size classes, with 0 indicating the absence of trees (non-forest) and 10 indicating maximum representation of 4 size classes [13]. DDI values were derived using gradient nearest neighbor (GNN) raster maps developed from LandSat satellite data [60]. Site selection was done remotely using ESRI GIS software [61] and relevant data layers. DDI raster values were smoothed using focal statistics, generating mean values within a 3 × 3 cell dimension neighborhood. We divided forested DDI values (i.e., values from 1–10) into nine classes (Table 1) and delineated patches of forest (i.e., stands) large enough to include 9 plots (3 × 3 grid, with 100 m spacing between grid points). From this pool of potential stands, we randomly selected three stands for each DDI class that could accommodate our plot grid to determine plot coordinates. We reselected a few stands due to access issues.

Figure 1. Distribution of 27 natural areas in Oregon and Washington used to evaluate metrics of structural complexity derived from Terrestrial LIDAR scans collected between August and October, 2014.

Table 1. The ecoregion, name, and Landsat image gradient nearest neighbor diameter diversity index (GNN DDI) for our 27 study stands.

Ecoregion	Name	GNN DDI
Coast Range	North Spit	1–2
	New River	3–4
	Mary's Peak	8–9
	Port Orford Cedar	8–9
East Cascades	Wechee Butte	1–2
	Mokst Butte	2–3
	Bluejay	3–4
	Mill Creek	4–5
	Smith Butte	5–6
	Monte Cristo	7–8
Klamath Mountains	North Bank	2–3
	Ashland	4–5
	Crooks Creek	5–6
	Hunter Creek Bog	7–8
	Grayback Glades	8–9
	French Flat	9–10

Table 1. *Cont.*

Ecoregion	Name	GNN DDI
West Cascades	Goat Marsh	2–3
	Sherwood Butte	3–4
	Limpy Rock	4–5
	Cultus River	6–7
	Katsuk Butte	6–7
	Three Creek	7–8
	Carolyn's Crown	9–10
	Steamboat Mountain	9–10
Willamette Valley	Coburg Hills	1–2
	Little Sink	5–6
	Camas Swale	6–7

2.2. Scan Acquisition

One scan was taken at each plot location for a total of 243 scans across the 27 stands (Figure 2). Scans were taken between 20 August and 9 October 2014, prior to leaf fall. A Faro Focus 3D s120 terrestrial lidar scanner (Faro Technologies Inc., Laker Mary, FL, USA) was used. Plot centers were located with a Garmin GPSmap handheld GPS unit (Garmin Ltd., Olathe, KS, USA) with 5 m to 20 m accuracy depending on canopy closure and weather conditions. TLS unit was set up at the plot center at approximately 1.4 m above ground. Some variation in scanner height was due to topography. If the plot center was too heavily occluded (e.g., within a dense shrub or directly next to a large tree) then the closest location to plot center that offered a full 360° of the stand within 10 m of plot center was used. This was to maximize area captured in the scan and to minimize occlusion by trees and shrubs. The scanner reliably received returns for objects ≤60 m away within a panoramic scan capturing a horizontal window from 0 to 360° and a vertical window from −60 to 90°. Vertical and horizontal scan lines were spaced every 0.035°, resulting in a 10,266 horizontal × 4267 vertical resolution per scan. Each scan required approximately 10 min to complete once onsite.

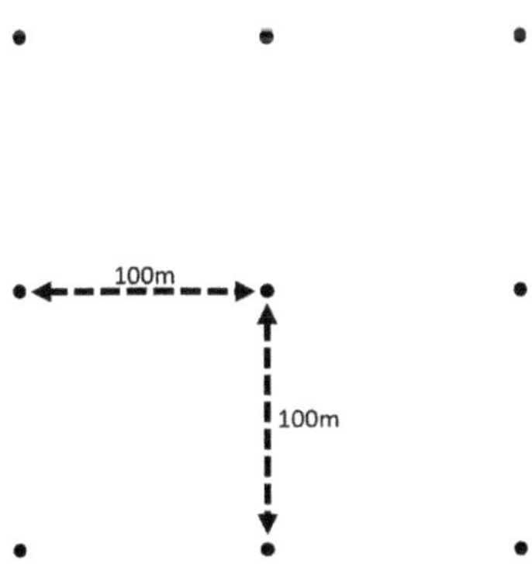

Figure 2. Nine single point scans (represented by the points) were collected in each forest stand placed at 100 m spacing. This was to get adequate representation of each forest type while limiting spatial autocorrelation between scans.

2.3. Scan Processing

Initial scan processing consisted of filtering artifacts and noise present in the scan data using preset filters in Faro Scene version 5.3.3 [62]. Dark scan points were isolated using an intensity (return signal strength) threshold of 200. Stray or isolated scan points were removed using a grid size of 3 px, distance threshold of 0.02 m, and allocation threshold of 33.3%. Noise and edge effects are inherent in all lidar data due to edge effects and back scattering [63].

Processed scans were exported into Leica PTX format, which preserved scanning acquisition structure with fixed angular increments between scan pulses [64]. We used PTX reader [65] to create two-dimensional intensity (Online Supplement: Supplement A) and depth (range) raster images using first-return point values, where each column of pixels represented an individual scan line and each pixel represented an angular location where the laser pulse was fired. Each scan resulted in a 10,266 × 4267 pixel raster, matching the scanline resolution settings of the scanner. Because the rasters are 2D representations of a 3D point cloud, there are distortions in the image. This is essentially a cylindrical projection of the point cloud. As long as pixels within the image are only compared to other pixels in the same relative horizontal position, the distortion will be equivalent and allow for valid comparisons. In these planar projections, if a scan is taken on a steep slope, the ground will take on a wave-like appearance with the uphill slope appearing taller in the image and the downhill slope appearing lowering the image (Figure 3).

Figure 3. Conceptual illustration of the process used to identify vertical layers and remove the ground from the raster images. The raster data had 10,266 columns and each column was divided into 100 sections between the top of the image and where the column hit the ground. In this figure there are 8 sections drawn in magenta to illustrate how each column was subdivided into 100 sections contoured to the hill slope, accounting for trees extending below the hillslope horizon, and excluding returns from the ground. The upper, middle, and lower sections of the scan are denoted to help with conceptualizing structural signatures.

This process of using raster depth maps to account for slope has significant advantages over traditional lidar normalization processes for single TLS scans. Traditional lidar normalization uses detailed ground models to remove elevation from each points Z value leaving only the height above ground for each point. Using a single scan, only the ground relatively near the scanner is captured. Slope, density of vegetation and other factors can rapidly occlude the ground beyond a few meters. Normalization is dependent on having a robust ground model which cannot be generated beyond the area directly adjacent to the scanner. Digital terrain models (DTMs) from airborne lidar can be used to normalize TLS scans but this relies on there being ALS data available for all locations and accurately geolocated scans. For our methods we only used the data collected from the single scan and were not reliant on additional data.

Further, normalizing changes the viewshed from the scan location. Branches that were visible can become occluded if normalized, and gaps form where occluded areas before normalizing are moved.

2.4. *Depth and Openness Metric Calculation*

The ground, small herbaceous vegetation, and small woody debris was visually identified and removed in each depth raster. This was accomplished by manually tracing the intersection where the trees and larger vegetation met the ground in the image. Conceptually this is depicted in Figure 3. Bézier curves were used to fit the contours of the "wavey" hillslope when appropriate. The line created through this process was a hillslope horizon line that accounted for vegetation. The ground and small vegetation pixels were simply deleted, and a null value added to the raster. The number of pixels above ground in each vertical scan line in the depth raster was divided into 100 equally spaced vertical increments using a custom MATLAB [66] script (Figure 3).

This discretization process minimized potential slope effects within and across plots. It also ensured that the same relative location on the trees present at a plot were being captured within the same vertical layer. The lowest of the 100 sections uniformly sampled the base of the trees while the highest of the 100 sections uniformly sampled the upper canopy. We averaged both depth and percent of "no returns" (openness) for each increment across the entire 360° horizontal view. Structural signatures were comprised of the average value across all 9 scan plots at each of the 100 angular increments, at each stand. One standard deviation from the mean included as a shaded band that is an indicator of the forest structural heterogeneity. Additionally, overall means were calculated for the upper, middle, and lower forest layers for both the depth and openness plots as structural indices.

2.5. *Isovists*

Isovists (polygons that enclose the visible area from a single vantage point) were generated for each plot. First, a polyline (4 pixels wide) was digitized on each intensity raster at 1.4 m above the ground (Figure 4A). Height above ground was determined by measuring a subset of trees (3–8, depending on slope) in the scan using FARO Scene. Bézier curves were fitted connecting the 1.4 m marks on the trees and contoured to the hill slope. The number of trees varied depending on the number of visible trees in the scan. For a tree to be used, the base needed to be visible. A cross section of points that fell along this line was extracted using PTX reader and subsequently flattened to the XY plane by ignoring Z (elevation) values.

Points within each cross section were imported into ESRI GIS software [61]. A circular buffer was drawn around each point to account for the inherent gaps between sample points. Given that the scanner operated on fixed angular increments rather than fixed sampling distances, we increased the buffer size for each point as range from scanner increased, using the following equation:

$$D\_Bi = R\_i \cdot \tan\varnothing + \in \quad (1)$$

where:

DB = the diameter of the buffer (m) at point i,

R = the 3D Euclidean range from the scanner at the sample point i,

$\varnothing$ = the angular sampling incremental (0.035° for this study), and

$\in$ = a small tolerance value to provide overlap to avoid small data gaps (0.003 m for this study).

The points in each scan were clipped to a horizontal radius of 55 m using circular buffers from each cross section (Figure 4B). Isovist polygons were created by extending 10,266 rays (at 0.035° increments to match scan line intervals) from the scanner location on the cross section and creating a point where each ray intersected a point buffer. The percentage of visible area was calculated at 5 m radius increments from the plot center resulting in 11 values per scan. Percentage was used rather than the actual area value to normalize the isovist metric across all radii.

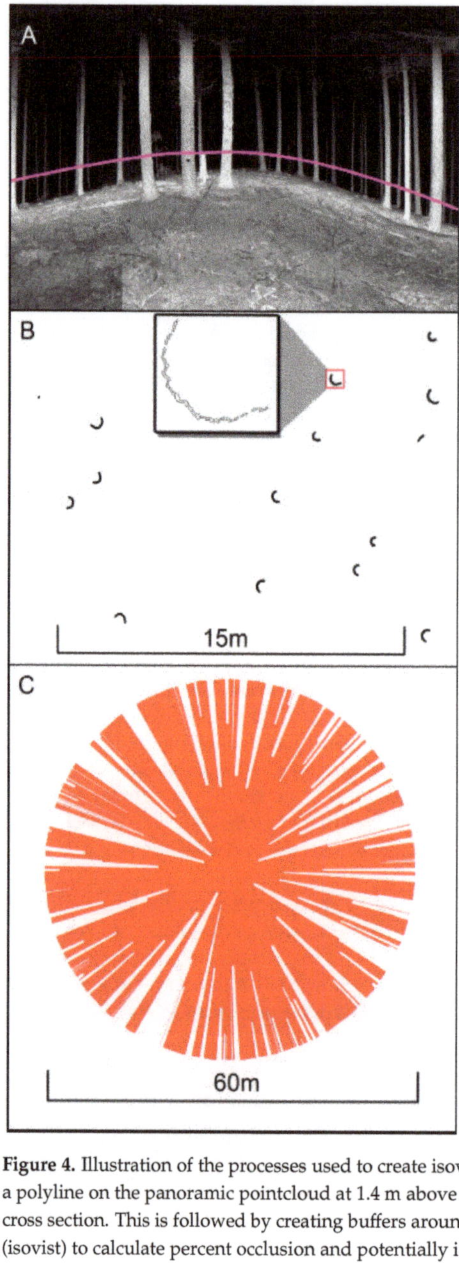

Figure 4. Illustration of the processes used to create isovist polygons at each plot: (**A**) Digitization of a polyline on the panoramic pointcloud at 1.4 m above the ground; (**B**) Planar view of the extracted cross section. This is followed by creating buffers around each point; (**C**) Generate visibility polygon (isovist) to calculate percent occlusion and potentially identify tree locations and DBH.

2.6. Comparisons

2.6.1. Depth and Openness Statistical Tests

Univariate linear regression was used to test for correlation (significant at 95% confidence) between depth and openness (mean and variance between plots in each stand) to ensure these variables were independent.

Principle componence analysis (PCA) [67] was used to visualize how well individual plots in each ecoregion clustered, to determine which depth and openness angular increments most contributed to PC 1 and PC 2, and to visualize the relationships between depth, openness, and GNN DDI vectors. Ordination was performed using all 243 plots with each plot having all depth and openness values assigned to it, resulting in a 243 × 200 main matrix and then simplified to only use the depth and openness angular increments that most contributed to PC 1 and PC 2, and the GNN DDI values. Data was scaled to unit variance before PCA was performed.

Analysis of variance (ANOVA) tests were used on depth and openness metrics at every angular increment with plots grouped by stands to determine if our new metrics differentiated between stands. For this step, 100 ANOVA tests were performed to test if there was a significant difference at each of the 100 angular increment sections between stands.

ANOVA tests using Tukey's honestly significant difference (HSD) [68] was performed on plots using all 100 angular increment sections and grouping stands based on ecoregion to determine if our new metrics differentiated between forest structure based on ecoregion.

2.6.2. Isovist Statistical Tests

ANOVA tests were conducted for each isovist radius grouping isovists by stand. Stands were then aggregated into ecoregion groups to perform an ANOVA test with Tukey's HSD post hoc intergroup comparisons.

3. Results

3.1. Structural Signatures

Each of the 27 stands exhibited distinctly different structural signatures for both depth and openness (Online Supplement: Supplement B). Some stands exhibited large amounts of structural variation at all vertical levels (e.g., Limpy Rock; Figure 5) while others exhibited relatively little variation (e.g., Hunter Creek Bog; Figure 5).

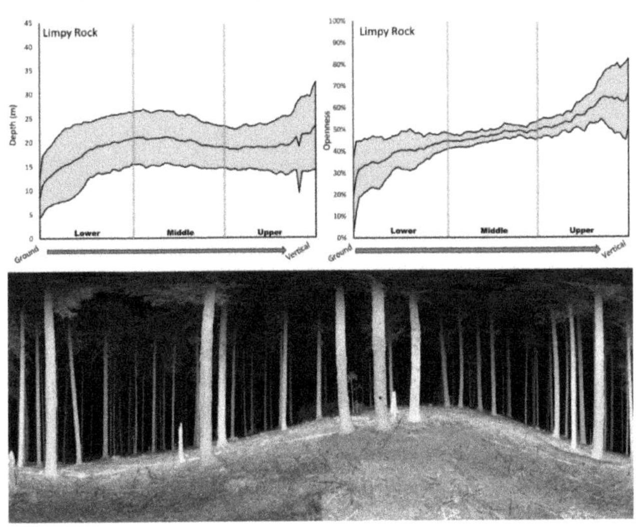

Figure 5. *Cont.*

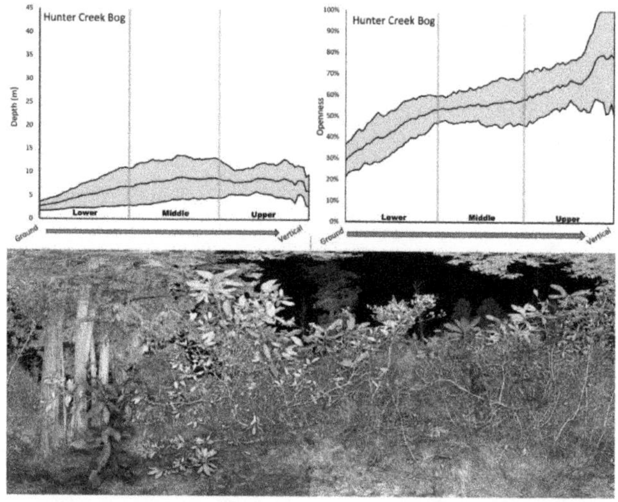

Figure 5. Structural signatures (shaded area is one standard deviation above and below mean) and representative photo of Limpy Rock RNA (**above**) and Hunter Creek Bog ACEC (**below**).

Some stands had increasing structural variation from the ground to the canopy (e.g., Little Sink; Figure 6). Other stands exhibited low variation in the lower and middle layers but abrupt increases in variation in the upper layers (e.g., Mokst Butte; Figure 6). Stands with dense understories tended to have low mean depth values (e.g., Hunter Creek Bog & Mokst Butte). Depth signature values tended to increase going from the ground to upper canopy. This increase was more rapid in stands with an open understory (e.g., Limpy Rock). Stands comprised of a mix of open and closed canopies could exhibit markedly varying depth signatures at the upper canopy layers (e.g., Mokst Butte). Stands with large canopy gaps had openness signatures that could reach 100% (e.g., Hunter Creek Bog and Mokst Butte). Stands with dense vegetation at all levels had a correspondingly low openness signature (e.g., Little Sink).

Figure 6. *Cont.*

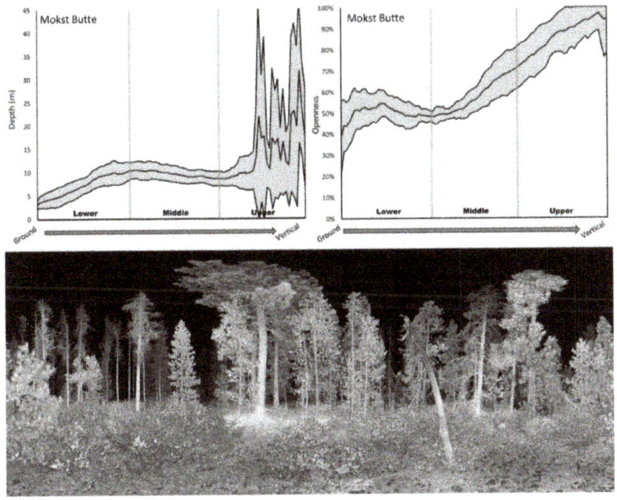

Figure 6. Structural signatures (shaded area is one standard deviation above and below mean) and representative image of Little Sink RNA (**above**) and Mokst Butte RNA (**below**).

3.2. *Depth and Openness Statistical Tests*

No correlation was found between depth and openness ($r2 < 0.01$; $p = 0.82$) suggesting the two metrics were independent.

3.2.1. Ordination

The top 5 variables that most contributed to PC 1 were the depth angular increments 55 to 59. This is just above the middle view from the scanner (Figure 3). The top 5 variables that most contributed to PC 2 were the Openness angular increments 38 to 42. Our openness metric had a negative relationship to the GNN DDI values when only usingthe angular increments that most contributed to the PCA axis and adding the GNN DDI values (Figure 7).

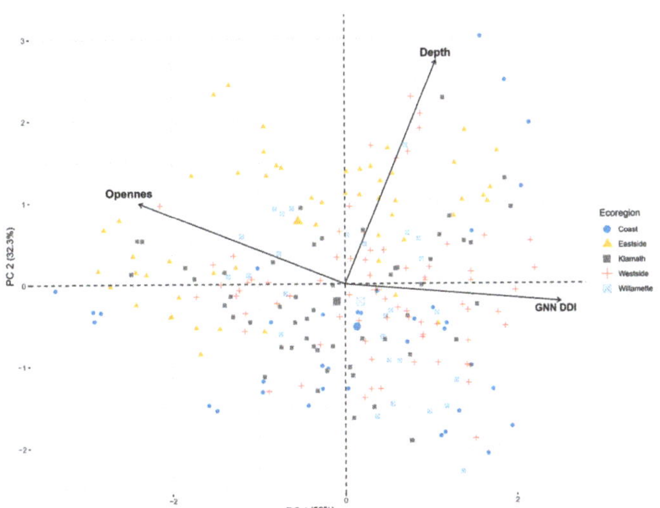

Figure 7. Ordination plot of all 243 locations sampled during the study. PC 1 and PC 2 axis are label with the amount of variation they explain. Depth, openness, and GNN DDI vectors are oriented relative to how much they contribute to PC 1 and PC 2. Our openness metric has a negative relationship with GNN DDI.

3.2.2. Depth and Openness ANOVA

There were significant differences in the depth values between stands at each of the 100 angular increments (all p values < 0.05 with the majority $< 2.2 \times 10^{-16}$). There were also significant differences in the openness values between stands at each of the 100 angular increments (all p values < 0.05 with the majority $< 2.2 \times 10^{-16}$). With p values less than 0.05 for an ANOVA test, we accept the null hypothesis that at least one of the group means was significantly different than the others.

When grouped by ecoregion, there was a significant difference between all parings of the depth and openness metrics except for the depth metric between ecoregions of Klamath mountains and the coast range, as well as Willamette valley and the westside of the cascade mountains (Figure 8). Our stands in the eastside of the cascades were most dissimilar to stands in both the Klamath and coast ecoregions. Similar trends were also seen with the Openness metric.

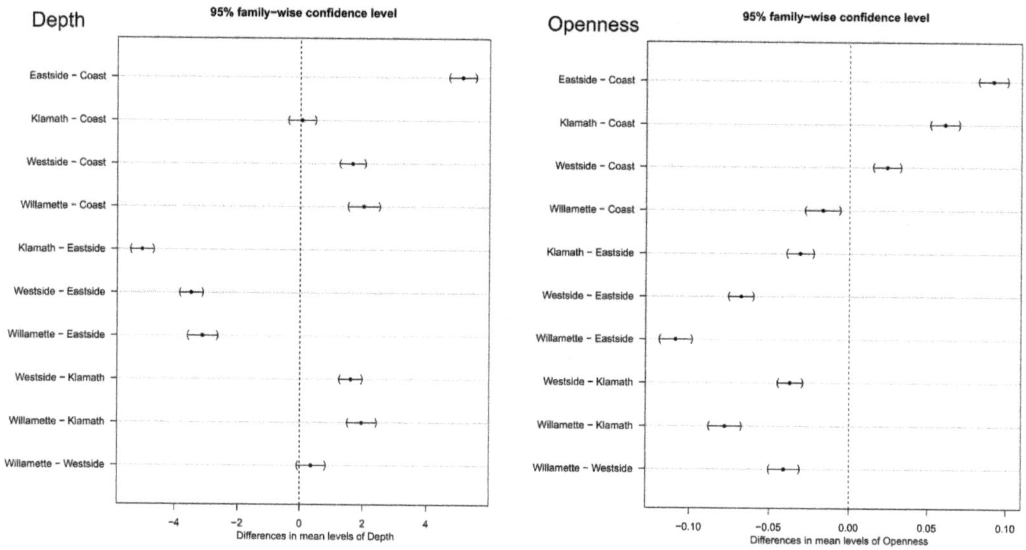

Figure 8. Tukey HSD results for the post hoc ecoregion comparison of both the depth and openness metrics. The closer to the centerline, the more similar the two regions were in their respective metric.

3.3. Isovist Statistical Tests

When grouped by stand, the isovist percentage of area visible was significantly different ($p = 4.116 \times 10^{-5}$). In general, the area visible decreased as the radius of the isovist increased (Figures 9 and 10). This decrease was typically about 40%. The percentage visible ranking changed for many stands as the isovist polygon radius increased. For example, New River (NR) had the least area visible with a 5 m radius isovist but at 25 m and beyond, Wechee butte had the least area visible. While the rate of decrease was different between stands, the spread of percentage visible was about 30–35% for the 5 m distance all the way through the 55 m distance. When grouped by ecoregion, most of the combinations of ecoregions had no significant difference in averaged percentage visible. The only ecoregion that did have a statistically significant difference was the coast range (Figure 11).

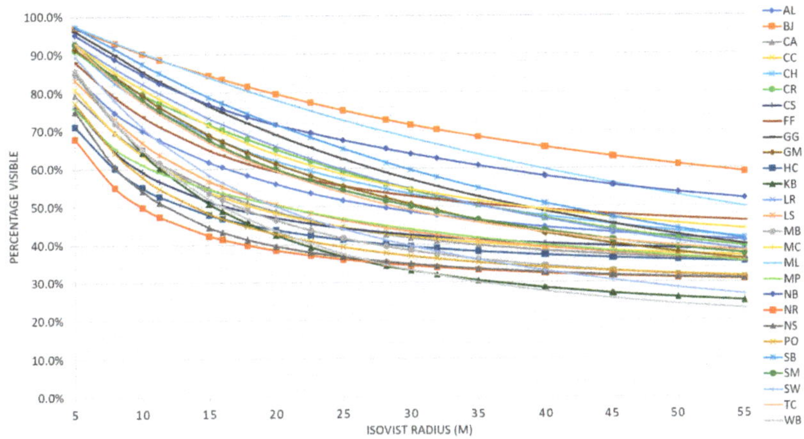

Figure 9. Average percent visible of area for each stand as measured by isovists. Area visible uniformly decreased as the radius of the isovist increased.

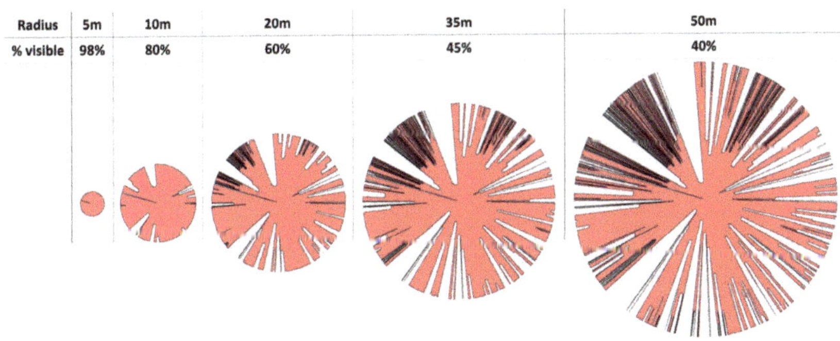

Figure 10. Example isovist polygon at differing radii. The percentage of visible area declines as the radius of the isovist is increased.

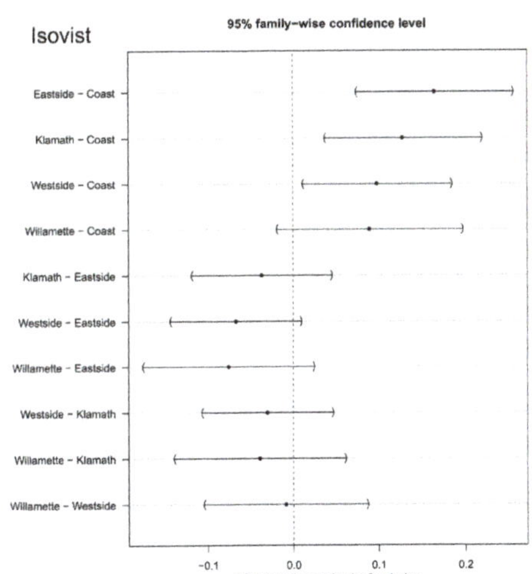

Figure 11. Tukey HSD paired comparison between isovist values grouped by ecoregion.

4. Discussion

4.1. Depth and Openness

Our results suggest that depth and openness are two distinct measures of structural complexity. Each of our 27 stands exhibited unique signatures and could be clearly distinguished from each other, suggesting that both depth and openness can capture the unique variation in structural complexity found in natural forests and at relatively fine scales. The structural signatures can be qualitatively assessed for stand openness and heterogeneity while also providing quantitative values for statistical analysis.

One strength of these metrics is that specific height strata within a forest can be extracted. For example, if only understory vegetation is of interest, then only the lowest angular increment layers can be included in an analysis. Our suggestion of 100 angular increments is a subjectively assigned value and fewer, or more increments can be used as is warranted by the research question being addressed. It is important to consider that our sampling was not a horizontal metric of canopy slices at different height strata, but rather the angular view of different height strata from a ground perspective.

We have shown that the depth and openness metrics do return significantly different values for stands at every angular increment, but more importantly, they also can be used to differentiate forest structure when aggregated to ecoregion. The Tukey HSD results show which regions have dramatically different vegetation structures and which ecoregions tend to have structurally similar forests. Although not explicitly examined here, we would expect a much higher degree of similarity among signatures in young forests managed for timber production where within-stand structure tends to be relatively homogeneous (e.g., single tree species of a similar age cohort, uniform spacing between individual trees).

4.2. Isovists

The ability to quantify the structural differences between forest types and forest of different ecoregions was less using the isovist approach compared to our depth and openness metrics. However, this does not indicate that the isovists do not quantify an important element of forest structure, as they are a direct measure of visual occlusion. As such, their utility may be better suited to ecological questions where occlusion can be an important component of habitat. Many wildlife species change behavior based on the area visible at any given location or have active preferences for certain amounts of visibility for hunting, denning, or relaxation [43,69,70]. For example, Canada lynx (*Lynx canadensis*), a species of special conservation concern, appears to require high amounts of understory cover (i.e., lower percentage of area visible) as part of quality habitat [71,72].

Our metrics are a direct measurement of understory vegetation structure and occlusion that can be used as localized assessment of habitat quality. Beyond a localized assessment, these metrics offer a high potential to be used as ground truth data for upscaling on the ground measurements of understory vegetation structure to a region wide model using airborne lidar. Such upscaling and modeling has already been done using cover boards to quantify horizontal visibility levels [73], but we believe that TLS offers advantages over the more subjective coverboard estimates in both time and effort required to obtain the data as well as providing a more robust measurement of cover.

Calculating visual occlusion using TLS point clouds is not new. With robust multi-scan TLS point clouds 3D viewshed analysis has been done for several wildlife studies [74,75] using R packages such as viewshed3D [76]. The advantage of our method is the ability to correct for slope without a ground model, but undoubtedly if full and comprehensive multi-scan TLS point clouds are available then the generation of a 3D viewshed has obvious benefits over our 2D option.

4.3. Potential Applications

We suggest our new metrics have a wide range of potential applications for both management and research. For example, a primary goal for much of the timber harvesting on federal lands in the Pacific Northwest is to move the structural trajectories of young, managed forests towards characteristics more typical of late-seral forest. TLS signatures could be used to help define and describe the structural variation found in natural (unmanaged) forests, including diverse late-seral forests as evaluated in this study. In managed forests, TLS signatures could help (1) assess pre-thinning structural condition; (2) assess immediate post-harvest condition to determine if thinning met structural goals; (3) monitor treated stands at regular intervals to follow the trajectory of structural development; and (4) establish benchmark data from relevant natural stands for desired future condition.

Biomass estimates may also be possible with our proposed metrics. Biomass is an important measurement as it relates to carbon storage, timber volume, and species habitat. If the foliage density and viewshed at different angular increments can be related to on the ground manual measurements of forest biomass, then such biomass measurements could be done at further areas with a greatly reduced labor cost. Multi position TLS sans are already widely used for biomass estimates [77–79] and biomass modeling from airborne and space borne lidar has provided valuable estimates [80–82], but there is still an excellent opportunity for future research deriving biomass estimates in highly occluded areas with a single point TLS scan.

A subset of the full signature may also help assess change and characteristics at different vertical strata within a forest. For example, upper canopy measures could be used to evaluate effects of defoliating pathogens and insects. Lower sections could be used to derive indices for understory shrub and sapling density or fire fuel loads. Similar to biomass estimates, TLS has seen considerable research to quantify fire fuel loads, but the vast majority of the work is dependent on multiple scans [38,39]. Both the point cloud data and photographs generated with each TLS scan could also be used to determine actual height layers of vegetation, and in some cases, plant species composition.

The point cloud data could be synthesized in other ways than we described here. For example, we chose to divide the vertical scan lines into 100 sections to create signatures because increasing the number of sections did not produce a visually discernible difference in the level of detail. The number of sections could be increased if a higher resolution signature was desired when more detail was necessary. There is also potential for further development of structural measures using three-dimensional isovists for measuring openness as distance increased in any direction.

4.4. Limitations

One important limitation of TLS is that the ability to detect objects diminishes with increasing distance to objects (reliability of scan returns diminished substantially > 60 m for this study). Multiple scans could be registered together to produce a more complete three-dimensional model of each plot. Not only would this allow three-dimensional modeling of individual trees, it would also allow calculation of structural signatures based on height above ground (rather than angular increment). However, labor costs (field and post-processing) to create these composite datasets would be markedly higher compared to single-point scans, especially for forests with high levels of occlusion. Additionally, initial filtering of the point clouds could have a substantial effect on our metrics. Structurally complex objects such as trees are especially susceptible to noise due to edge effects and backscattering. Overly aggressive point filtering can remove valid points and reduce point cloud accuracy [83]. We suspect our methods are likely very susceptible to errors due to noise and filtering and this is one avenue of future research. Lastly, while lidar technologies are becoming more common and ubiquitous within the field of forest ecology, the units are still expensive and the skillset to process data to final and meaningful ecological metrics requires extensive training.

For our processing we relied on manual removal of ground points within the raster depth maps and the creation of a hillslope horizon line that accounted for vegetation. This is a subjective process and needs refinement. For the purposes of accounting for slope, this process does have advantages over traditional lidar normalization for single scans, but further removing the subjective element would be ideal. There may be a way to create a hybrid approach by classifying ground points before creating raster depth maps and then removing all pixels within a set radius from a ground point. Further, this approach for accounting for slope does not account for micro topography and abrupt changes in slope.

5. Conclusions

While no single metric of structure can possibly satisfy all needs for all applications, we believe our depth and openness measures have significant advantages over many past measures for quantifying forest structure. In addition to capturing the multi-dimensional properties of forest structure, these new measures (1) are spatially and temporally scalable (a few centimeters to tens of meters) in all directions; (2) are precise with low observer bias; (3) can be rapidly collected in the field; (4) are based on point cloud data that is unlikely to become obsolete in the foreseeable future; and (5) can be further synthesized in ways to meet specific research questions and management needs. Next steps include testing the usefulness of these new measures for quantifying wildlife habitat and periodic (5 to10-yr) re-measurement of our study plots to test their sensitivity to capture structural change over time. Additionally, using these metrics as ground truth data to create regional models needs to be more fully explored. TLS metrics that are derived directly from point clouds may be better suited as model inputs to estimate forest structure using larger area data sets such as aerial lidar.

Supplementary Materials: The following are available online at https://www.mdpi.com/article/10.3390/rs15010145/s1, Supplement A—Lidar plot scan images; Supplement B—Lidar plot Signatures.

Author Contributions: Conceptualization, J.L.B., T.M.W., M.J.O., W.J.R.; methodology, J.L.B., T.M.W., M.J.O.; software, J.L.B., M.J.O.; formal analysis, J.L.B., T.M.W., M.J.O.; resources, T.M.W., W.J.R.; writing—original draft preparation, J.L.B.; writing—review and editing, T.M.W., M.J.O., W.J.R.; visualization, J.L.B.; supervision, T.M.W., M.J.O., W.J.R.; project administration, T.M.W.; funding acquisition, T.M.W., W.J.R. All authors have read and agreed to the published version of the manuscript.

Funding: Funding was provided by the USDI Bureau of Land Management, Oregon State Office, and the USDA Forest Service, Pacific Northwest Research Station.

Data Availability Statement: The data presented in this study are available on request from the corresponding author.

Acknowledgments: Luke Painter provided guidance and mentorship on manuscript preparation. We thank Leica Geosystems for providing Cyclone licenses used for exploratory analysis.

Conflicts of Interest: Michael Olsen has financial interests in the company EzDataMD LLC related to the commercialization of technology Involving point cloud analysis. The conduct, outcomes, or reporting of this research could benefit EzDataMD LLC and could potentially benefit them.

References

1. Lindenmayer, D.B.; Franklin, J.F. *Conserving Forest Biodiversity: A Comprehensive Multiscaled Approach*; Island Press: Washington, DC, USA, 2002.
2. Carey, A.B. *AIMing for Healthy Forests: Active, Intentional Management for Multiple Values*; General Technical Report PNW-GTR-721; U.S. Department of Agriculture, Forest Service, Pacific Northwest Research Station: Portland, OR, USA, 2007.
3. Shugart, H.; Saatchi, S.; Hall, F. Importance of Structure and Its Measurement in Quantifying Function of Forest Ecosystems. *J. Geophys. Res. Biogeosci.* **2010**, 115. [CrossRef]
4. Chazdon, R.L. Sunflecks and Their Importance to Forest Understorey Plants. *Adv. Ecol. Res.* **1988**, *18*, 1–63.
5. Beier, P.; Drennan, J.E. Forest Structure and Prey Abundance in Foraging Areas of Northern Goshawks. *Ecol. Appl.* **1997**, *7*, 564–571. [CrossRef]
6. Pardini, R.; de Souza, S.M.; Braga-Neto, R.; Metzger, J.P. The Role of Forest Structure, Fragment Size and Corridors in Maintaining Small Mammal Abundance and Diversity in an Atlantic Forest Landscape. *Biol. Conserv.* **2005**, *124*, 253–266. [CrossRef]
7. Musselman, K.N.; Molotch, N.P.; Margulis, S.A.; Kirchner, P.B.; Bales, R.C. Influence of Canopy Structure and Direct Beam Solar Irradiance on Snowmelt Rates in a Mixed Conifer Forest. *Agric. For. Meteorol.* **2012**, *161*, 46–56. [CrossRef]
8. Davis, C.R.; Belote, T.; Williamson, M.; Larson, A.; Esch, B. A Rapid Forest Assessment Method for Multiparty Monitoring across Landscapes. *J. For.* **2016**, *114*, 125–133. [CrossRef]

9. Everett, R.L.; Leader, A.T. *Eastside Forest Ecosystem Health Assessment*; General Technical Report PNW-GTR-330; U.S. Department of Agriculture, Forest Service, Pacific Northwest Research Station: Portland, OR, USA, 1994.
10. Massie, M. Assessment of the Vulnerability of Oregon and Washington's Natural Areas to Climate Change. Master's Thesis, Oregon State University, Portland, OR, USA, 2014.
11. Rapp, V. *Northwest Forest Plan—The First 10 Years (1994–2003): First-Decade Results of the Northwest Forest Plan*; General Technical Report PNW-GTR-720; U.S. Department of Agriculture, Forest Service, Pacific Northwest Research Station: Portland, OR, USA, 2008.
12. Acker, S.A.; Sabin, T.E.; Ganio, L.M.; McKee, W.A. Development of Old-Growth Structure and Timber Volume Growth Trends in Maturing Douglas-Fir Stands. *For. Ecol. Manag.* **1998**, *104*, 265–280. [CrossRef]
13. McComb, W.C.; McGrath, M.T.; Spies, T.A.; Vesely, D. Models for Mapping Potential Habitat at Landscape Scales: An Example Using Northern Spotted Owls. *For. Sci.* **2002**, *48*, 203–216.
14. McElhinny, C.; Gibbons, P.; Brack, C.; Bauhus, J. Forest and Woodland Stand Structural Complexity: Its Definition and Measurement. *For. Ecol. Manag.* **2005**, *218*, 1–24. [CrossRef]
15. Staudhammer, C.L.; LeMay, V.M. Introduction and Evaluation of Possible Indices of Stand Structural Diversity. *Can. J. For. Res.* **2001**, *31*, 1105–1115. [CrossRef]
16. Tuchmann, E.T.; Connaughton, K.P. *The Northwest Forest Plan: A Report to the President and Congress*; DIANE Publishing: Collingdale, PA, USA, 1998.
17. Pommerening, A. Approaches to Quantifying Forest Structures. *Forestry* **2002**, *75*, 305–324. [CrossRef]
18. Bruggisser, M.; Hollaus, M.; Kükenbrink, D.; Pfeifer, N. Comparison of Forest Structure Metrics Derived from UAV Lidar and ALS Data. *ISPRS Ann. Photogramm. Remote Sens. Spat. Inf. Sci.* **2019**, *4*, 325–332. [CrossRef]
19. Moeser, D.; Morsdorf, F.; Jonas, T. Novel Forest Structure Metrics from Airborne LiDAR Data for Improved Snow Interception Estimation. *Agric. For. Meteorol.* **2015**, *208*, 40–49. [CrossRef]
20. Uuemaa, E.; Antrop, M.; Roosaare, J.; Marja, R.; Mander, Ü. Landscape Metrics and Indices: An Overview of Their Use in Landscape Research. *Living Rev. Landsc. Res.* **2009**, *3*, 1–28. [CrossRef]
21. Frey, J.; Joa, B.; Schraml, U.; Koch, B. Same Viewpoint Different Perspectives—A Comparison of Expert Ratings with a TLS Derived Forest Stand Structural Complexity Index. *Remote Sens.* **2019**, *11*, 1137. [CrossRef]
22. Vales, D.J.; Bunnell, F.L. Comparison of Methods for Estimating Forest Overstory Cover. *I. Observer Effects. Can. J. For. Res.* **1988**, *18*, 606–609.
23. Beland, M.; Parker, G.; Sparrow, B.; Harding, D.; Chasmer, L.; Phinn, S.; Antonarakis, A.; Strahler, A. On Promoting the Use of Lidar Systems in Forest Ecosystem Research. *For. Ecol. Manag.* **2019**, *450*, 117484. [CrossRef]
24. Camarretta, N.; Harrison, P.A.; Bailey, T.; Potts, B.; Lucieer, A.; Davidson, N.; Hunt, M. Monitoring Forest Structure to Guide Adaptive Management of Forest Restoration: A Review of Remote Sensing Approaches. *New For.* **2020**, *51*, 573–596. [CrossRef]
25. Wulder, M.A.; White, J.C.; Nelson, R.F.; Næsset, E.; Ørka, H.O.; Coops, N.C.; Hilker, T.; Bater, C.W.; Gobakken, T. Lidar Sampling for Large-Area Forest Characterization: A Review. *Remote Sens. Environ.* **2012**, *121*, 196–209. [CrossRef]
26. Calders, K.; Lewis, P.; Disney, M.; Verbesselt, J.; Herold, M. Investigating Assumptions of Crown Archetypes for Modelling LiDAR Returns. *Remote Sens. Environ.* **2013**, *134*, 39–49. [CrossRef]
27. Zimble, D.A.; Evans, D.L.; Carlson, G.C.; Parker, R.C.; Grado, S.C.; Gerard, P.D. Characterizing Vertical Forest Structure Using Small-Footprint Airborne LiDAR. *Remote Sens. Environ.* **2003**, *87*, 171–182. [CrossRef]
28. Ackers, S.H.; Davis, R.J.; Olsen, K.A.; Dugger, K.M. The Evolution of Mapping Habitat for Northern Spotted Owls (Strix Occidentalis Caurina): A Comparison of Photo-Interpreted, Landsat-Based, and Lidar-Based Habitat Maps. *Remote Sens. Environ.* **2015**, *156*, 361–373. [CrossRef]
29. Coops, N.C.; Tompalski, P.; Goodbody, T.R.; Queinnec, M.; Luther, J.E.; Bolton, D.K.; White, J.C.; Wulder, M.A.; van Lier, O.R.; Hermosilla, T. Modelling Lidar-Derived Estimates of Forest Attributes over Space and Time: A Review of Approaches and Future Trends. *Remote Sens. Environ.* **2021**, *260*, 112477. [CrossRef]
30. Lister, A.J.; Andersen, H.; Frescino, T.; Gatziolis, D.; Healey, S.; Heath, L.S.; Liknes, G.C.; McRoberts, R.; Moisen, G.G.; Nelson, M. Use of Remote Sensing Data to Improve the Efficiency of National Forest Inventories: A Case Study from the United States National Forest Inventory. *Forests* **2020**, *11*, 1364. [CrossRef]
31. Tinkham, W.T.; Mahoney, P.R.; Hudak, A.T.; Domke, G.M.; Falkowski, M.J.; Woodall, C.W.; Smith, A.M. Applications of the United States Forest Inventory and Analysis Dataset: A Review and Future Directions. *Can. J. For. Res.* **2018**, *48*, 1251–1268. [CrossRef]
32. Campbell, M.J.; Dennison, P.E.; Hudak, A.T.; Parham, L.M.; Butler, B.W. Quantifying Understory Vegetation Density Using Small-Footprint Airborne Lidar. *Remote Sens. Environ.* **2018**, *215*, 330–342. [CrossRef]
33. Hilker, T.; van Leeuwen, M.; Coops, N.C.; Wulder, M.A.; Newnham, G.J.; Jupp, D.L.B.; Culvenor, D.S. Comparing Canopy Metrics Derived from Terrestrial and Airborne Laser Scanning in a Douglas-Fir Dominated Forest Stand. *Trees* **2010**, *24*, 819–832. [CrossRef]
34. Ruiz, L.Á.; Crespo-Peremarch, P.; Torralba, J. Modelling Canopy Fuel Properties and Understory Vegetation with Full-Waveform LiDAR. In Proceedings of the International Conference on Smart Geoinformatics Applications (ICSGA 2021), Phuket, Thailand, 24–25 February 2021; pp. 29–32.

35. Calders, K.; Adams, J.; Armston, J.; Bartholomeus, H.; Bauwens, S.; Bentley, L.P.; Chave, J.; Danson, F.M.; Demol, M.; Disney, M.; et al. Terrestrial Laser Scanning in Forest Ecology: Expanding the Horizon. *Remote Sens. Environ.* **2020**, *251*, 112102. [CrossRef]
36. Henning, J.G.; Radtke, P.J. Detailed Stem Measurements of Standing Trees from Ground-Based Scanning Lidar. *For. Sci.* **2006**, *52*, 67–80.
37. Palace, M.; Sullivan, F.B.; Ducey, M.; Herrick, C. Estimating Tropical Forest Structure Using a Terrestrial Lidar. *PLoS ONE* **2016**, *11*, e0154115. [CrossRef]
38. Loudermilk, E.L.; Hiers, J.K.; O'Brien, J.J.; Mitchell, R.J.; Singhania, A.; Fernandez, J.C.; Cropper, W.P.; Slatton, K.C. Ground-Based LIDAR: A Novel Approach to Quantify Fine-Scale Fuelbed Characteristics. *Int. J. Wildland Fire* **2009**, *18*, 676–685. [CrossRef]
39. Rowell, E.; Loudermilk, E.L.; Hawley, C.; Pokswinski, S.; Seielstad, C.; Queen, L.; O'Brien, J.J.; Hudak, A.T.; Goodrick, S.; Hiers, J.K. Coupling Terrestrial Laser Scanning with 3D Fuel Biomass Sampling for Advancing Wildland Fuels Characterization. *For. Ecol. Manag.* **2020**, *462*, 117945. [CrossRef]
40. Wilson, N.; Bradstock, R.; Bedward, M. Detecting the Effects of Logging and Wildfire on Forest Fuel Structure Using Terrestrial Laser Scanning (TLS). *For. Ecol. Manag.* **2021**, *488*, 119037. [CrossRef]
41. Ashcroft, M.B.; Gollan, J.R.; Ramp, D. Creating Vegetation Density Profiles for a Diverse Range of Ecological Habitats Using Terrestrial Laser Scanning. *Methods Ecol. Evol.* **2014**, *5*, 263–272. [CrossRef]
42. Kazakova, A.N. Quantifying Vertical and Horizontal Stand Structure Using Terrestrial LiDAR in Pacific Northwest Forests. Master's Thesis, University of Washington, Seattle, WA, USA, 2014.
43. Olsoy, P.J.; Forbey, J.S.; Rachlow, J.L.; Nobler, J.D.; Glenn, N.F.; Shipley, L.A. Fearscapes: Mapping Functional Properties of Cover for Prey with Terrestrial LiDAR. *BioScience* **2015**, *65*, 74–80. [CrossRef]
44. Shokirov, S.; Levick, S.R.; Jucker, T.; Yeoh, P.; Youngentob, K. Comparison of TLS and ULS Data for Wildlife Habitat Assessments in Temperate Woodlands. In Proceedings of the IGARSS 2020—2020 IEEE International Geoscience and Remote Sensing Symposium, Online, 26 September–2 October 2020; pp. 6097–6100.
45. Soma, M.; Pimont, F.; Allard, D.; Fournier, R.; Dupuy, J.-L. Mitigating Occlusion Effects in Leaf Area Density Estimates from Terrestrial LiDAR through a Specific Kriging Method. *Remote Sens. Environ.* **2020**, *245*, 111836. [CrossRef]
46. Wan, P.; Wang, T.; Zhang, W.; Liang, X.; Skidmore, A.K.; Yan, G. Quantification of Occlusions Influencing the Tree Stem Curve Retrieving from Single-Scan Terrestrial Laser Scanning Data. *For. Ecosyst.* **2019**, *6*, 43. [CrossRef]
47. Litkey, P.; Liang, X.; Kaartinen, H.; Hyyppä, J.; Kukko, A.; Holopainen, M.; Hill, R.; Rosette, J.; Suárez, J. Single-Scan TLS Methods for Forest Parameter Retrieval. In Proceedings of the SilviLaser, Edinburgh, UK, 17–19 September 2008.
48. Xia, S.; Wang, C.; Pan, F.; Xi, X.; Zeng, H.; Liu, H. Detecting Stems in Dense and Homogeneous Forest Using Single-Scan TLS. *Forests* **2015**, *6*, 3923–3945. [CrossRef]
49. Pokswinski, S.; Gallagher, M.R.; Skowronski, N.S.; Loudermilk, E.L.; Hawley, C.; Wallace, D.; Everland, A.; Wallace, J.; Hiers, J.K. A Simplified and Affordable Approach to Forest Monitoring Using Single Terrestrial Laser Scans and Transect Sampling. *MethodsX* **2021**, *8*, 101484. [CrossRef]
50. Moskal, L.M.; Zheng, G. Retrieving Forest Inventory Variables with Terrestrial Laser Scanning (TLS) in Urban Heterogeneous Forest. *Remote Sens.* **2011**, *4*, 1–20. [CrossRef]
51. Richardson, J.J.; Moskal, L.M.; Bakker, J.D. Terrestrial Laser Scanning for Vegetation Sampling. *Sensors* **2014**, *14*, 20304–20319. [CrossRef] [PubMed]
52. Kato, A.; Moskal, L.M.; Batchelor, J.L.; Thau, D.; Hudak, A.T. Relationships between Satellite-Based Spectral Burned Ratios and Terrestrial Laser Scanning. *Forests* **2019**, *10*, 444. [CrossRef]
53. Gallagher, M.R.; Maxwell, A.E.; Guillén, L.A.; Everland, A.; Loudermilk, E.L.; Skowronski, N.S. Estimation of Plot-Level Burn Severity Using Terrestrial Laser Scanning. *Remote Sens.* **2021**, *13*, 4168. [CrossRef]
54. Tan, K.; Ke, T.; Tao, P.; Liu, K.; Duan, Y.; Zhang, W.; Wu, S. Discriminating Forest Leaf and Wood Components in TLS Point Clouds at Single-Scan Level Using Derived Geometric Quantities. *IEEE Trans. Geosci. Remote Sens.* **2021**, *60*, 1–17. [CrossRef]
55. Anderson, C.T.; Dietz, S.L.; Pokswinski, S.M.; Jenkins, A.M.; Kaeser, M.J.; Hiers, J.K.; Pelc, B.D. Traditional Field Metrics and Terrestrial LiDAR Predict Plant Richness in Southern Pine Forests. *For. Ecol. Manag.* **2021**, *491*, 119118. [CrossRef]
56. Wallace, L.; Hillman, S.; Hally, B.; Taneja, R.; White, A.; McGlade, J. Terrestrial Laser Scanning: An Operational Tool for Fuel Hazard Mapping? *Fire* **2022**, *5*, 85. [CrossRef]
57. Murgoitio, J.; Shrestha, R.; Glenn, N.; Spaete, L. Airborne LiDAR and Terrestrial Laser Scanning Derived Vegetation Obstruction Factors for Visibility Models. *Trans. GIS* **2014**, *18*, 147–160. [CrossRef]
58. Omernik, J.M. Ecoregions of the Conterminous United States. *Ann. Assoc. Am. Geogr.* **1987**, *77*, 118–125. [CrossRef]
59. Wilson, T.M. Pacific Northwest Interagency Natural Areas Network. Available online: http://www.fsl.orst.edu/rna/index.html (accessed on 28 March 2015).
60. LEMMA Landscape Ecology, Modeling, Mapping & Analysis Home Page. Available online: http://lemma.forestry.oregonstate.edu/ (accessed on 30 March 2015).
61. ESRI. *ArcGIS Desktop*; Version 10; ESRI: Redlands, CA, USA, 2014.
62. FARO Scene [Computer Software]; Version 5.3; FARO: Lake Mary, FL, USA. 2014. Available online: http://www.faro.com (accessed on 31 March 2015).
63. Cao, N.; Zhu, C.; Kai, Y.; Yan, P. A Method of Background Noise Reduction in Lidar Data. *Appl. Phys. B* **2013**, *113*, 115–123. [CrossRef]

64. Stovall, A.E.L.; Atkins, J.W. Assessing Low-Cost Terrestrial Laser Scanners for Deriving Forest Structure Parameters. *Preprints* **2021**, 2021070690. [CrossRef]
65. Olsen, M.J.; Ponto, K.; Kimball, J.; Seracini, M.; Kuester, F. 2D Open-Source Editing Techniques for 3D Laser Scans. In Proceedings of the 38th Annual Conference on Computer Applications and Quantitative Methods in Archaeology, Granada, Spain, 6–9 April 2010; pp. 455–461.
66. Mathworks MATLAB R2015a [Computer Program]. 2015. Available online: HTTP://www.mathworks.Com/products/matlab/ (accessed on 30 March 2015).
67. Lê, S.; Josse, J.; Husson, F. FactoMineR: An R Package for Multivariate Analysis. *J. Stat. Softw.* **2008**, *25*, 1–18. [CrossRef]
68. Abdi, H.; Williams, L.J. Tukey's Honestly Significant Difference (HSD) Test. *Encycl. Res. Des.* **2010**, *3*, 1–5.
69. Aben, J.; Pellikka, P.; Travis, J.M. A Call for Viewshed Ecology: Advancing Our Understanding of the Ecology of Information through Viewshed Analysis. *Methods Ecol. Evol.* **2018**, *9*, 624–633. [CrossRef]
70. Davies, A.B.; Tambling, C.J.; Kerley, G.I.; Asner, G.P. Effects of Vegetation Structure on the Location of Lion Kill Sites in African Thicket. *PLoS ONE* **2016**, *11*, e0149098. [CrossRef] [PubMed]
71. Mowat, G.; Slough, B. Habitat Preference of Canada Lynx through a Cycle in Snowshoe Hare Abundance. *Can. J. Zool.* **2003**, *81*, 1736–1745. [CrossRef]
72. Poole, K.G. A Review of the Canada Lynx, Lynx Canadensis, in Canada. *Can. Field-Nat.* **2003**, *117*, 360–376. [CrossRef]
73. Fekety, P.A.; Sadak, R.B.; Sauder, J.D.; Hudak, A.T.; Falkowski, M.J. Predicting Forest Understory Habitat for Canada Lynx Using LIDAR Data. *Wildl. Soc. Bull.* **2019**, *43*, 619–629. [CrossRef]
74. Galluzzi, M.; Puletti, N.; Armanini, M.; Chirichella, R.; Mustoni, A. Mobile Laser Scanner Understory Characterization: An Exploratory Study on Hazel Grouse in Italian Alps. *bioRxiv* **2022**. [CrossRef]
75. Burgett, S.; Rachlow, J.; Stein, R. Unexpected Properties of Habitat Altered by Ecosystem Engineers: A Pygmy Rabbit Case Study. 2021. Available online: https://scholarworks.boisestate.edu/icur/2021/poster_session/12/ (accessed on 16 December 2022).
76. Lecigne, B.; Eitel, J.; Rachlow, J. Viewshed3d: An R Package for Quantifying 3D Visibility Using Terrestrial Lidar Data. *Methods Ecol. Evol.* **2020**, *11*, 733–738. [CrossRef]
77. Fan, G.; Nan, L.; Dong, Y.; Su, X.; Chen, F. AdQSM: A New Method for Estimating Above-Ground Biomass from TLS Point Clouds. *Remote Sens.* **2020**, *12*, 3089. [CrossRef]
78. Hackenberg, J.; Wassenberg, M.; Spiecker, H.; Sun, D. Non Destructive Method for Biomass Prediction Combining TLS Derived Tree Volume and Wood Density. *Forests* **2015**, *6*, 1274–1300. [CrossRef]
79. Holopainen, M.; Vastaranta, M.; Kankare, V.; Räty, M.; Vaaja, M.; Liang, X.; Yu, X.; Hyyppä, J.; Hyyppä, H.; Viitala, R. Biomass Estimation of Individual Trees Using Stem and Crown Diameter TLS Measurements. *ISPRS—Int. Arch. Photogramm. Remote Sens. Spat. Inf. Sci.* **2011**, *3812*, 91–95. [CrossRef]
80. Coomes, D.A.; Dalponte, M.; Jucker, T.; Asner, G.P.; Banin, L.F.; Burslem, D.F.R.P.; Lewis, S.L.; Nilus, R.; Phillips, O.L.; Phua, M.-H.; et al. Area-Based vs Tree-Centric Approaches to Mapping Forest Carbon in Southeast Asian Forests from Airborne Laser Scanning Data. *Remote Sens. Environ.* **2017**, *194*, 77–88. [CrossRef]
81. Duncanson, L.; Neuenschwander, A.; Hancock, S.; Thomas, N.; Fatoyinbo, T.; Simard, M.; Silva, C.A.; Armston, J.; Luthcke, S.B.; Hofton, M.; et al. Biomass Estimation from Simulated GEDI, ICESat-2 and NISAR across Environmental Gradients in Sonoma County, California. *Remote Sens. Environ.* **2020**, *242*, 111779. [CrossRef]
82. Ene, L.T.; Gobakken, T.; Andersen, H.-E.; Næsset, E.; Cook, B.D.; Morton, D.C.; Babcock, C.; Nelson, R. Large-Area Hybrid Estimation of Aboveground Biomass in Interior Alaska Using Airborne Laser Scanning Data. *Remote Sens. Environ.* **2018**, *204*, 741–755. [CrossRef]
83. Mahoney, M.J.; Johnson, L.K.; Bevilacqua, E.; Beier, C.M. Filtering Ground Noise from LiDAR Returns Produces Inferior Models of Forest Aboveground Biomass in Heterogenous Landscapes. *GISci. Remote Sens.* **2022**, *59*, 1266–1280. [CrossRef]

remote sensing

MDPI

Article

3D LoD2 and LoD3 Modeling of Buildings with Ornamental Towers and Turrets Based on LiDAR Data

Elżbieta Lewandowicz [1], Fayez Tarsha Kurdi [2,*] and Zahra Gharineiat [2]

[1] Department of Geoinformation and Cartography, Institute of Geodesy and Civil Engineering, Faculty of Geoengineering, University Warmia and Mazury in Olsztyn, 10-719 Olsztyn, Poland

[2] School of Surveying and Built Environment, Faculty of Health, Engineering and Sciences, University of Southern Queensland, Springfield Campus, Springfield, QLD 4300, Australia

* Correspondence: fayez.tarshakurdi@usq.edu.au

Abstract: This paper presents an innovative approach to the automatic modeling of buildings composed of rotational surfaces, based exclusively on airborne LiDAR point clouds. The proposed approach starts by detecting the gravity center of the building's footprint. A thin point slice parallel to one coordinate axis around the gravity center was considered, and a vertical cross-section was rotated around a vertical axis passing through the gravity center, to generate the 3D building model. The constructed model was visualized with a matrix composed of three matrices, where the same dimensions represented the X, Y, and Z Euclidean coordinates. Five tower point clouds were used to evaluate the performance of the proposed algorithm. Then, to estimate the accuracy, the point cloud was superimposed onto the constructed model, and the deviation of points describing the building model was calculated, in addition to the standard deviation. The obtained standard deviation values, which express the accuracy, were determined in the range of 0.21 m to 1.41 m. These values indicate that the accuracy of the suggested method is consistent with approaches suggested previously in the literature. In the future, the obtained model could be enhanced with the use of points that have considerable deviations. The applied matrix not only facilitates the modeling of buildings with various levels of architectural complexity, but it also allows for local enhancement of the constructed models.

Keywords: 3D modeling; buildings; LiDAR; cross-section; rotating surface

Citation: Lewandowicz, E.; Tarsha Kurdi, F.; Gharineiat, Z. 3D LoD2 and LoD3 Modeling of Buildings with Ornamental Towers and Turrets Based on LiDAR Data. *Remote Sens.* **2022**, *14*, 4687. https://doi.org/10.3390/rs14194687

Academic Editors: Jorge Delgado García and Tarig Ali

Received: 31 July 2022
Accepted: 17 September 2022
Published: 20 September 2022

Publisher's Note: MDPI stays neutral with regard to jurisdictional claims in published maps and institutional affiliations.

1. Introduction

Remote measurement systems enable the development of digital models that depict real-world objects with increasing accuracy. Light detection and ranging (LiDAR) technology, which collects point clouds using airborne laser Scanning (ALS), has particularly contributed to advancements in remote measurement. Airborne LiDAR data are described by three coordinates (attributes), which, in combination with aerial color images (red, green, and blue; RGB), have led to the development of a new functionality in 3D modeling [1–3]. LiDAR data can be labeled by automatic point classification, where thematic subsets are created based on attributes [2,4,5]. Classification is a crucial process in 3D modeling, because the represented objects are characterized by increasing complexity [6,7]. A point cloud can be classified using supervised learning methods that rely on statistical formulas [8–10] or supervised methods based on machine learning classifiers [11–13]. During the classification process, subsets must be verified, to identify and eliminate outliers. Classification is combined with segmentation, to select subsets of points that represent individual objects. The selected subsets facilitate 3D city modeling, mainly 3D buildings that represent most infrastructure objects in urban areas. The segmentation of LiDAR datasets can be simplified with the use of vector sets depicting the ground floors of buildings [14–16]. Classified and segmented point clouds from aerial images fulfil the requirements for modeling buildings at the LOD0, LOD1, and LOD2 levels of detail, as long as their visibility is not disturbed by natural and artificial curtains. Additional data (for example terrestrial scans) are needed to generate LOD3 models, because extended facades of buildings are often rendered in insufficient detail based on aerial images. LOD4 modeling is also applied to model building interiors based on indoor scans. Building information modeling (BIM) technologies are increasingly being used to create virtual 3D models of buildings with architectural details, in building design and management. The BIM dataset generated in the process of designing, modeling, and managing buildings meets the requirements for creating LOD models at different levels of detail. BMI technologies offer an alternative to the above solutions [14,17]. The construction of virtual 3D city models in the CityGML 3.0 standard [18–20] requires models of urban objects with varying levels of complexity and accuracy. The generation of vector and object data with various levels of detail, based on point clouds, poses a considerable challenge. Complex models (LOD3) should accurately depict the structural features of buildings, such as gates, balconies, stairs, towers, and turrets. Vector 3D models with varying degrees of complexity should be restricted to a single topology, which poses a difficult task for researchers. Such models should not only enable rapid visualization of 3D datasets at different scales and with varying complexity, but they should also facilitate data processing during comprehensive analysis [19,21]. In the next stage of designing a smart city, 3D datasets describing individual buildings must be linked with semantic data [22] to create thematic applications [23]. The CityGML 3.0 Transportation Model also requires highly detailed models of street spaces, in particular buildings. These models are utilized in autonomous vehicles [22,24] and other mobile mapping systems.

Various approaches to modeling buildings based on point cloud data have been developed. The proposed approaches rely on subsets of points describing buildings and, in the next step, subsets describing roof planes. These points can be identified with the use of various approaches:

- Methods based on building models that are represented in LOD0 [14,25,26].
- Methods involving algorithms that are based on triangulated irregular networks (TIN) [16].
- Methods involving point classification, filtration, selection, and segmentation [27,28].
- Methods where points are classified by machine learning [12].
- Methods where points are selected based on neighborhood attributes [29].
- Methods where point clouds are filtered based on a histogram of Z-coordinates [7,30].

The modeling process is two-fold: modeling of roof planes and modeling of building surfaces [30]. These processes are often conducted manually, based on defined reference models that are available in libraries or are generated for the needs of specific projects [31]. Other approaches involve different methods of processing subsets of LiDAR points. In the generated point clouds, subsets that represent roof planes are extracted by the developed algorithms [30,32]. In the next step, roof plane boundaries are modeled as straight-line segments, and the topological relationships between these elements are established [26,27,32–35]. Building facades are usually difficult to model, due to incomplete datasets in point clouds. Therefore, it was assumed that a building's outside walls should be reconstructed based on the roof boundary. This approach supported the development of numerous algorithms for automatically identifying and modeling buildings based on LiDAR data at the LOD2 level. The generated algorithms are based on modeled roof planes [32,35,36]. In the constructed models, the ground floor of a building is represented by the contour of the roof [34,35]. This simplified approach was adopted to establish topological relationships between geometric objects in building models [18,35,37]. The proposed methods and algorithms for modeling buildings at different levels of detail always give simplified results. Models rarely fully correspond to reality, because construction technologies allow creating complex spatial structures that are difficult to render in 2D or even 3D mathematical frameworks for virtual visualization of entire cities. The present study was undertaken to search for new solutions to this problem.

2. Research Objective

Most solutions for roof plane modeling in the literature are based on straight-line geometric elements. In practice, roofs and roof structures tend to be more complex in buildings that feature towers, turrets, or other ornamental structures, in the shape of spheres or curved planes. In the proposed models, such elements are usually simplified or even omitted. These differences become apparent when virtual LOD2 models of cities are compared with Street View visualizations. Two 3D building models from the Polish Spatial Data Infrastructure (SDI) geoportal are presented in Figure 1. The constructed models present selected buildings in the Polish city of Olsztyn. Complex building structures, including ornamental features, were not visualized because only straight-line 3D elements were used in the modeling process. This problem had been previously recognized by Huang et al. [35].

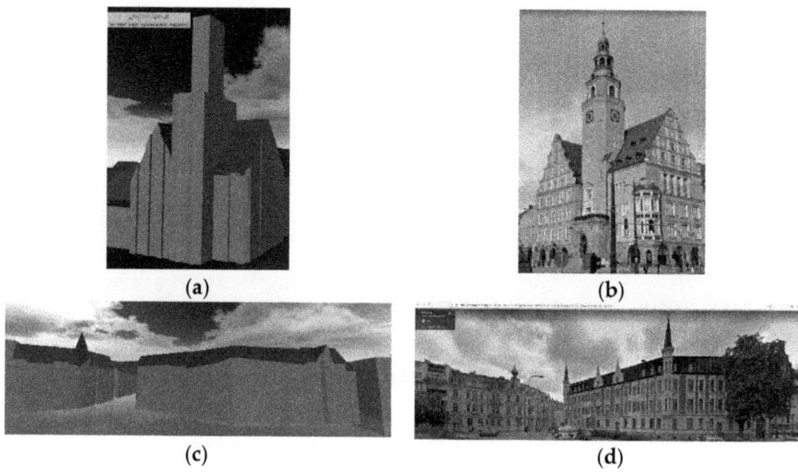

Figure 1. Selected 3D building models from the Polish SDI Geoportal, including Street View visualizations; (**a**) Model of the Olsztyn City Hall building; (**b**) Visualization of the Olsztyn City Hall building; (**c**) Models of other representative buildings and their visualizations (**d**).

These observations indicate that towers, turrets, and other ornamental features constitute structural blocks and require special modeling methods. Some of these structures can be modeled by rotating straight-line segments. New methods for the automatic generation of detailed building models are thus needed, to ensure compliance with the CityGML 3.0 standard. Therefore, the aim of this study was to develop an automated algorithm for modeling the characteristics of tall structures in buildings, represented by solids of revolution, and based on LiDAR point cloud data.

These observations indicate that:

- Towers, turrets, and other ornamental structures require special modeling methods.
- Some of these structures can be modeled by rotating straight-line segments.
- New methods for the automatic generation of detailed building models are thus needed to ensure compliance with the CityGML 3.0 standard.

Therefore, the aim of this study was to develop an algorithm for modeling the characteristics of tall structures in buildings, represented by solids of revolution, and based on point cloud data.

3. Design Concept

The expected model that will be generated by the proposed modeling algorithm should be described first. For this purpose, an experiment was designed to assess the similarity between the envisaged model and the tower point cloud, and between the envisaged model and the tower building. The test was inspired by Tarsha Kurdi and Awrangjeb [38]. A 3D point cloud distributed irregularly on the tower's outer surfaces can be measured by airborne laser scanning. On the one hand, the relatively low point density, irregular point distribution, accuracy of point location, presence of noisy points, and the geometric complexity of the scanned building decrease the similarity between the scanned tower and its point cloud. On the other hand, the generalization of the point cloud for calculating a 3D model decreases the similarity between the constructed model and the point cloud. Therefore, the fidelity of the constructed model decreases twice: during the scanning step and during the modeling step.

The experiment consisted of two stages. First, a 3D tower model was developed manually from the tower point cloud, without a reference to the original building image. In the first stage, the calculated model was named Model1, and it is shown in Figure 2b. Both the tower point cloud and the terrestrial image were used as inputs, to manually generate the 3D tower model. In the second stage, the calculated model was named Model2, and it is shown in Figure 2c. It should be noted that the model was generated with the use of the official Polish GIS model, which was imported to CAD and is presented as a skeleton in Figure 2c.

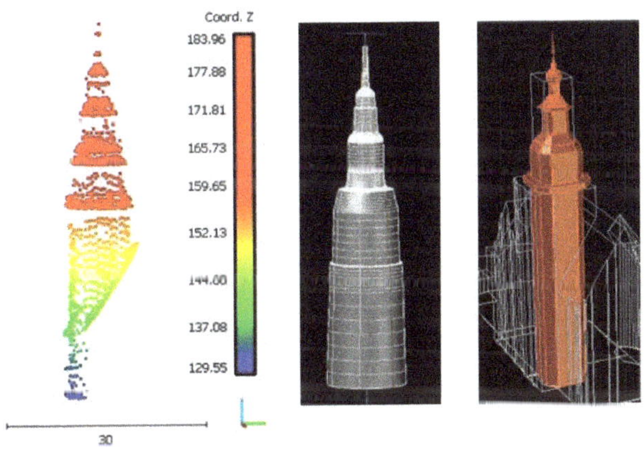

(a) (b) (c)

Figure 2. Tower of the Olsztyn City Hall building; (**a**) LiDAR point cloud; (**b**) Model 1 generated directly from a point cloud; (**c**) Model 2 generated from the point cloud and the terrestrial image shown in Figure 1b.

A comparison of the obtained tower models indicates that:

- Despite the fact that geometric details are not rendered with sufficient clarity in the point cloud, they can be identified in Model 2, but not in Model 1.
- Model 2 preserves the tower's geometric form, which can be observed in the terrestrial image.
- Some errors in the diameters of different parts of the tower body in Model 2 result from a greater focus on the image than the point cloud.
- Model 1 renders the geometric form of different tower parts with lower accuracy, but it preserves dimensions with greater accuracy.
- Model 1 represents the point cloud more accurately than Model 2, whereas Model 2 represents the original tower more accurately than Model 1.

These observations suggest that in an automatic modeling approach, based only on an airborne LiDAR point cloud, the processing parameters and measurements are applied directly to the point cloud. Thus, the expected model will more accurately represent the point cloud describing the original building. Consequently, the constructed building may be more similar to Model 1 than Model 2.

Figure 2b indicates that the tower can be regarded as a rotational surface. The tower body is composed of five vertical parts (Figure 2); therefore, five horizontal cross-sections were calculated from the tower point cloud to verify this hypothesis, as shown in Figure 3.

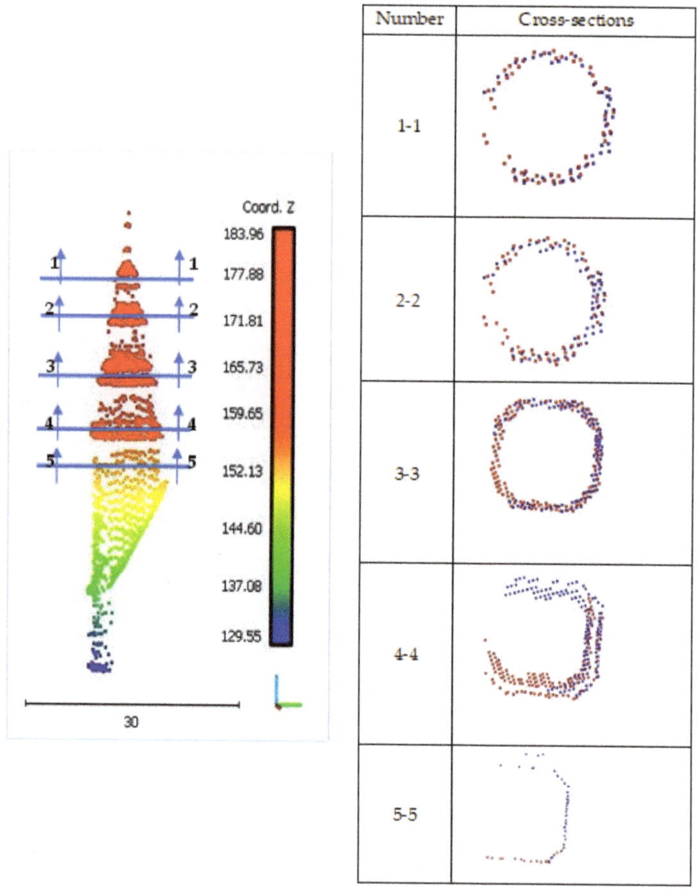

Figure 3. Five cross-sections in the tower point cloud.

The first three cross-sections are circular, but the last two are rectangular. Moreover, the point density is low in the lower part of the tower, due to airborne scanning and the presence of elements connecting the building with the tower, which is why cross-sections 4 and 5 are not complete. However, due to the tower's architectural complexity and the fact that similar towers can be presented geometrically by rotational surfaces (as discussed in Section 4), the latter hypothesis was adopted, and tower points were modeled based on rotational surfaces.

Finally, Model 1 was built with the use of the automatic modeling approach, which relies on this strategy being applied to the point cloud, and all steps of the construction process were automated.

4. Proposed Modeling Approach

The suggested modeling approach was applied to the 3D airborne LiDAR point cloud of the tower. The presented algorithm was applied to automatically generate a 3D model of the scanned tower. The proposed method consists of five consecutive steps, which are presented in Figure 4. First, to calculate the tower footprint gravity center, the tower point cloud was projected onto a horizontal plane passing through the coordinate origin.

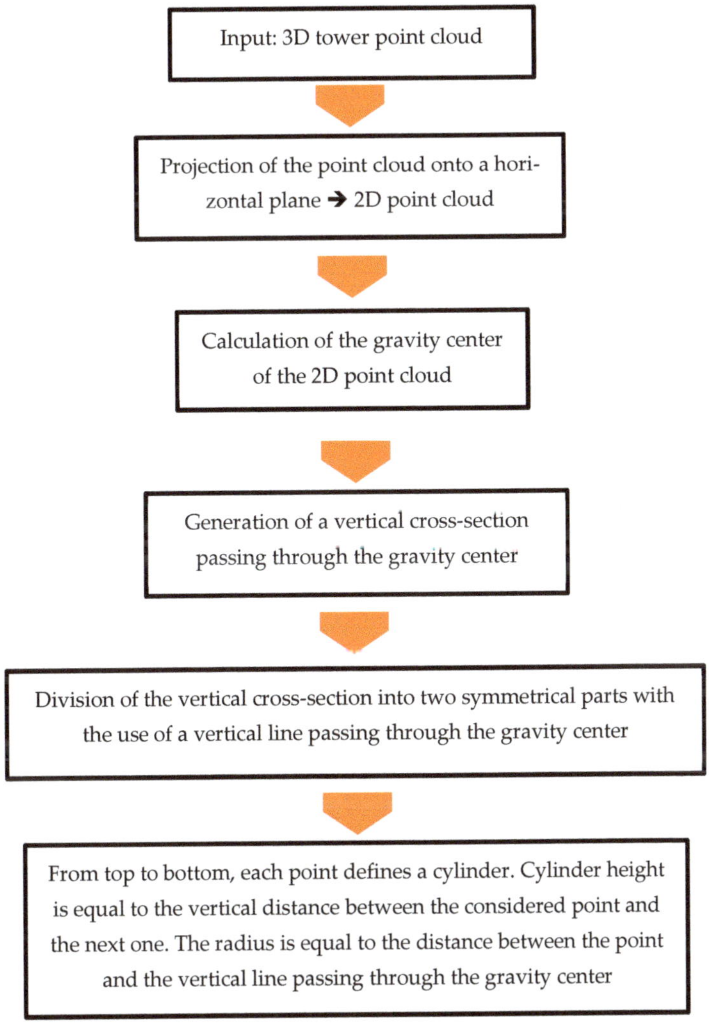

Figure 4. Workflow of a modeling algorithm for generating a building composed of rotational surfaces.

The projection on the horizontal plane OXY follows the lines parallel to the *Z* axis; therefore, the result is a 2D point cloud with only *X* and *Y* coordinates (the same *X* and *Y* coordinates as in 3D space). In other words, this operation could be realized by considering only the coordinates *X* and *Y*, to define a new 2D point cloud that represents the tower footprint. The elimination of the *Z* coordinate from the original point cloud enabled the generation of the target 2D footprint point cloud. The resulting 2D point cloud of the tower footprint is presented in Figure 5a. Due to an irregular distribution of 3D points on tower surfaces, the density of the obtained point cloud is also irregular: greater on the right side and smaller on the left side.

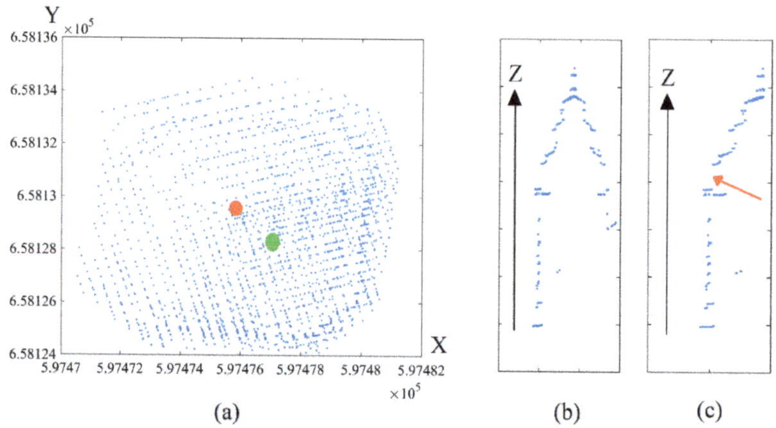

Figure 5. (**a**) Projection of a 3D tower point cloud on a horizontal plane passing through the coordinate origin; the red circle is the gravity center calculated based on extreme coordinate values; the green circle is the gravity center calculated based on static moments. (**b**) Vertical cross-section passing through the gravity center. (**c**) Semi-vertical cross-section.

In the second step, the gravity center coordinates of the projected point cloud (Figure 5a) are calculated. For this purpose, the static moments of a 2D cloud are analyzed by considering the points as infinitely small elements. Therefore, static moment equations were applied to the new point cloud, to calculate the gravity center of the tower footprint (Equation (1) [39]).

The application of the static moment principle shifted the gravity center, due to an irregular point density (see the green circle in Figure 5a). The tower footprint is symmetrical, and this problem can be resolved by calculating gravity center coordinates using extreme values of X and Y coordinates (minimum and maximum), as indicated in Equation (2) (see the red circle in Figure 5a).

$$X_g = \frac{\sum_{i=1}^{n} X_i}{n} \qquad Y_g = \frac{\sum_{i=1}^{n} Y_i}{n} \tag{1}$$

$$X_g = minX + \frac{Xmax - Xmin}{2},\ Y_g = minY + \frac{Ymax - Ymin}{2} \tag{2}$$

where X_g and Y_g are the coordinates of the gravity center; n is the number of points; Xi and Yi are point cloud abscissas and ordinates in OXY.

In the third step, a vertical cross-section passing through the gravity center was calculated, by identifying the points located in a slice with thickness ε around the considered vertical plane. The ε value was considered according to Equation (3) (Figure 5b).

$$\varepsilon = 2 \times T_d \quad T_d = \frac{1}{\sqrt{\vartheta}} \tag{3}$$

where ε is the thickness of the vertical cross-section slice; T_d is the mean horizontal distance between two neighboring points [29]; and θ is the point density.

In the fourth step, the symmetrical vertical cross-section was divided into two parts using the vertical line passing through the gravity center (Figure 5c). The obtained semi-cross-section represents the basic graph that revolves around the vertical line passing through the gravity center, to approximate the surface of revolution which represents the 3D tower model.

Mathematically, when a line segment is revolved around an axis, it draws a band. This band is actually a piece of a cone called the frustum of a cone. This cone could be a cylinder when the line segment is parallel to the rotating axis. Finally, from top to bottom, each point defines a cylinder, and the cylinder's height is equal to the vertical distance between the considered point and the next one, and the radius is equal to the distance between the point and the vertical line passing through the gravity center. The analyzed semi-cross-section (Figure 5c) is not continuous and contains gaps (see the red arrow in Figure 5c). Due to the low point density and irregular point distribution, these gaps are presented by the frustums of cones connecting the two consecutive cylinders.

The 3D tower model is calculated with the use of a matrix. Three matrices were used for this purpose: X, Y, and Z (Equations (4)–(6)). These matrices represent the coordinates of rotating surface pixels and have the same number of rows and columns. The number of rows is equal to the number of points in the semi cross-section, whereas the number of columns can be selected arbitrarily, but it must be greater than seven and multiples of four added to one. In the model presented in Figure 6, the number of columns is equal to 25.

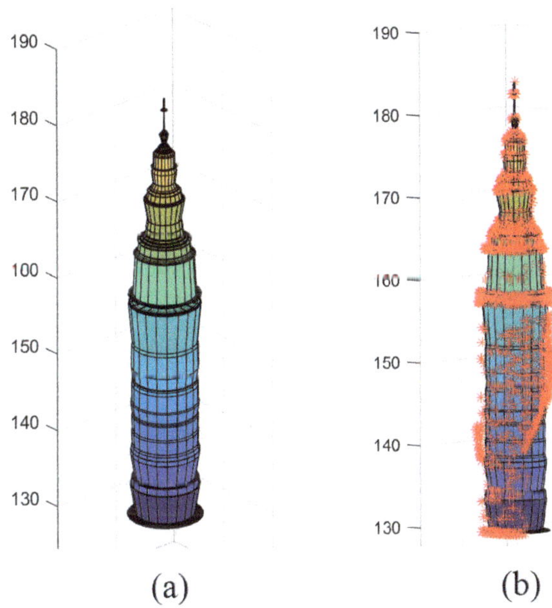

Figure 6. (**a**) Automatically generated 3D tower model; (**b**) superimposition of the tower point cloud onto the tower model.

$$X = \begin{bmatrix} X_g & X_g+\beta_{1,1} & X_g+2\beta_{1,2}\ldots\ldots X_g \\ X_g & X_g+\beta_{2,1} & X_g+2\beta_{2,2}\ldots\ldots X_g \\ X_g & X_g+\beta_{3,1} & X_g+2\beta_{3,2}\ldots\ldots X_g \\ & . & .\quad .\quad . \\ X_g & X_g+\beta_{n,1} & X_g+2\beta_{n,2}\ldots\ldots X_g \end{bmatrix} \tag{4}$$

$$Y = \begin{bmatrix} Y_1 & Y_1+\alpha_{1,1} & Y_1+2\alpha_{1,2}\ldots\ldots Y_1 \\ Y_2 & Y_2+\alpha_{2,1} & Y_2+2\alpha_{2,2}\ldots\ldots Y_2 \\ Y_3 & X_3+\alpha_{3,1} & Y_3+2\alpha_{3,2}\ldots\ldots Y_3 \\ . & . \quad . & .\quad .\quad . \\ Y_n & Y_N+\alpha_{n,1} & Y_n+2\alpha_{n,2}\ldots\ldots Y_n \end{bmatrix} \qquad Z = \begin{bmatrix} Z_1 Z_1\ldots\ldots Z_1 \\ Z_2 Z_2\ldots\ldots Z_2 \\ Z_3 Z_3\ldots\ldots Z_3 \\ .\ .\ .\ .\ . \\ .\ .\ .\ .\ . \\ Z_n Z_n\ldots\ldots Z_n \end{bmatrix} \tag{5}$$

$$\alpha_{i,j} = (Y_g - Y_i)\sin\left(\frac{2j\pi}{m} + \frac{3\pi}{2}\right),\ \beta_{i,j} = (Y_g - Y_i)\cos\left(\frac{2j\pi}{m} + \frac{3\pi}{2}\right) \tag{6}$$

where X_g and Y_g are the coordinates of the gravity center (Equation 2); X_i, Y_i, and Z_i (i = 1 to n) are the point coordinates of the semi cross-section; j = 1 to m; n is the number of points in the semi cross-section; α_i and β_i are the step values of X and Y, respectively; and m is the number of columns in matrix X.

Figure 6 presents the 3D model of the tower point cloud shown in Figure 2a. This model was constructed automatically using the described approach, based on rotational surfaces. The generated model is similar to that shown in Figure 2b, because only the tower point cloud was considered in both models. Both models represent a rotational surface and consist of five parts that are vertically superimposed. Furthermore, their dimensions are similar to the mean dimensions measured directly from the point cloud. In contrast, the model shown in Figure 2c, where both the tower point cloud and the tower image were considered, differs more considerably from the calculated model.

The suggested algorithm was implemented in MATLAB software, and the "surf (X, Y, Z)" command was used to visualize the calculated building model. However, the suggested algorithm has the following pseudocode (Algorithm 1):

Algorithm 1

Input (point cloud (X, Y, Z), m, θ)
Point cloud sorted in ascending order based on Z values
$X_g = minX + \frac{Xmax - Xmin}{2}$
$Y_g = minY + \frac{Ymax - Ymin}{2}$
$T_d = \frac{1}{\sqrt{\theta}}$
i = **find** ($X > X_g - T_d$ **and** $X < X_g + T_d$ **and** $Y \leq Y_g$)
SCS = [Y(i), Z(i)]
for i = 1 **to** length (SCS), **Step** = 1
  for j = 0 **to** m, **Step** = 1
    Z_b (i, j+1) = SCS (i, 2)
    X_b (i, j+1) = X_g + (Y_g − SCS (i, 1)) × $\cos(\frac{3\times\pi}{2} + \frac{2\times j\times\pi}{m})$
    Y_b (i, j+1) = X_g + (Y_g − SCS (i, 1)) × $\sin(\frac{3\times\pi}{2} + \frac{2\times j\times\pi}{m})$
  Next j
Next i
Surf (X, Y, Z)

Where SCS is the list of semi-cross-section points; Surf is the 3D visualization function; X_b, Y_b, and Z_b are the three matrices of the building model (Equations (4) and (5)).

In the previous pseudocode, the suggested algorithm was very short and simple. The algorithm outputs three matrices that can be exported in raster or vector format. The datasets used in the suggested approach will be presented in the next section. The remaining results will be discussed and the accuracy of the modeling process will be estimated in Section 6.

5. Datasets

The Polish SDI was developed by the Head Office of Geodesy and Cartography, and constitutes a data source that is widely used in research. ALS data from LiDAR measurements conducted in 2018 (12 point/m^2), as well as LOD2 3D building models generated in the CityGML 2.0 standard, were selected from SDI resources for the needs of the study.

The 3D building models were generated by compiling three data sources: 2D building contours from the Database of Topographic Objects in 1:10 000 scale, LiDAR data (building class), and the Digital Terrain Model (DTM) with a mesh size of 1 m. Buildings were modeled based on 2D building contours. The height of building contours was determined based on the minimum height of building contour vertices in the DTM dataset. The 3D building models were downloaded from the SDI, as individual vector files presenting roof planes, building walls, and 2D building contours.

The most characteristic buildings in Olsztyn, including the Olsztyn City Hall with an ornamental tower (Figure 1a,b), a building with a chimney (Figure 7a), and water towers (Figure 7b), were selected for the study. Two water towers with characteristic shapes, located in the cities of Bydgoszcz (Figure 7c) and Siedlce (Figure 7d), were additionally selected. ALS data were obtained from SDI resources.

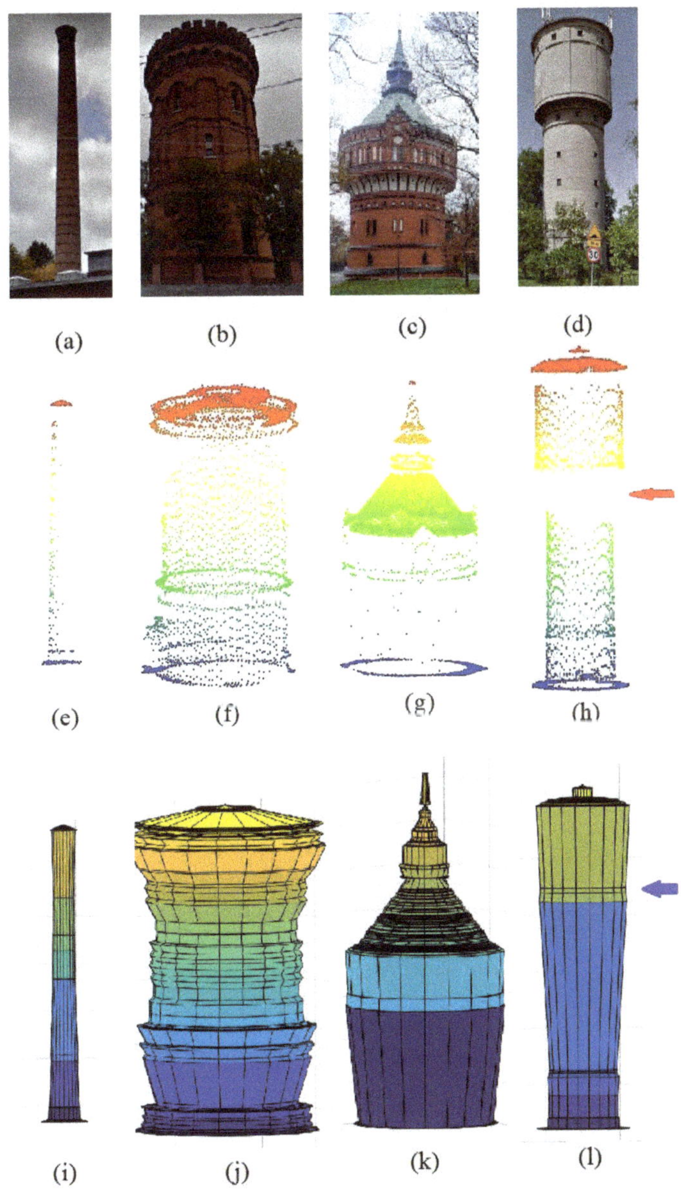

Figure 7. Modeling four tower point clouds; (**a–d**): tower images from Google Street View; (**e–h**): Tower point clouds; (**i–l**): 3D tower models.

6. Results, Accuracy Estimation, and Discussion

In the literature, the accuracy of 3D building models generated based on LiDAR data can be estimated using two approaches. In the first approach, the generated model is compared with the reference model [15,40–42]. In the second approach, a LiDAR point cloud is the reference model [15,38,43–45]. In the second approach, the accuracy is estimated by calculating the distances between the 3D model and the point cloud. In the present study, the reference model was the building point cloud. To estimate the accuracy of the generated model, the point cloud is superimposed onto the built model, and the distances between cloud points and the model are calculated. The point cloud is superimposed onto the 3D tower model in Figure 6b. A histogram of the distances between cloud points and the generated tower model (see Figure 6b) is shown in Figure 8. Figure 8 shows that a high percentage of points fit the constructed tower model accurately, and the distances between the points and the model are less than 0.30 m. A distance of less than 0.35 m is regarded as acceptable. The accuracy of altimetry measurements, the texture of building surfaces and ornaments, and the presence of noise shift cloud points around the mean building surfaces. Moreover, some parts of the building do not support the hypothesis postulating that the building is represented by a rotational surface. For example, the lower part of the tower in Figure 3 is not exactly a rotational surface.

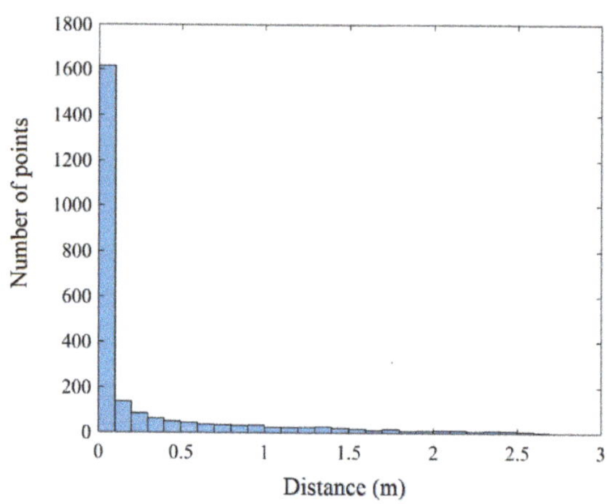

Figure 8. Histogram of deviations between cloud points and the 3D tower model.

In addition to the point cloud presented in Figure 2a, four different tower point clouds were assessed, to analyze the accuracy of the models developed with the suggested approach. The images, the point cloud, and the constructed models of four tower point clouds are presented in Figure 7. The total number of points, the number of points that deviate from the model within the interval (0, 0.3 m), the number of points with deviations greater than 0.3 m, and the standard deviation of the distances from the constructed model are presented in Table 1.

Table 1. Standard deviations and the number of points that deviate from the tower model within the specified intervals.

Tower Number Name–City	Number of Points	Number of Points		σ (m)
		Dist ϵ (0, 0.3 m)	Dist > 0.3 m	
1 Olsztyn City Hall	2330	1833	497	0.49
2 Building with a chimney in Olsztyn	330	244	86	0.9

Table 1. *Cont.*

Tower Number Name–City	Number of Points	Number of Points		σ (m)
		Dist ϵ (0, 0.3 m)	Dist > 0.3 m	
3 Water tower in Olsztyn	4974	2217	2757	0.84
4 Water tower in Bydgoszcz	5500	5246	254	0.21
5 Water tower in Siedlce	4811	3825	986	1.4

Dist: deviation between a point and the constructed model; σ: standard deviation of the distance between the cloud point and the tower model.

It should be noted that the same approach was used to calculate the values presented in Table 1 and the histogram in Figure 8. Hence, the tower point cloud was superimposed onto the calculated model, and the deviation of each point from the constructed model was calculated. This operation was performed to calculate the deviation for each point. A histogram of deviations is presented in Figure 8. The calculated deviations were analyzed, and the results are presented in Table 1.

To calculate the deviation of a given Point P (X_p, Y_p, Z_p) in Figure 9, the Z coordinate was used to determine the point's location in matrix Z in Equation 5. The point can have three locations. In the first case, if the point is located in a row where $Z_p = Z_i$ ($I = 1$ to n), the deviation can be calculated according to Equation (7) (as shown in Figure 9).

$$div = \sqrt{(X_p \quad X_g)^2 + (Y_p - Y_g)^2} - Y_g + Y_i \qquad (7)$$

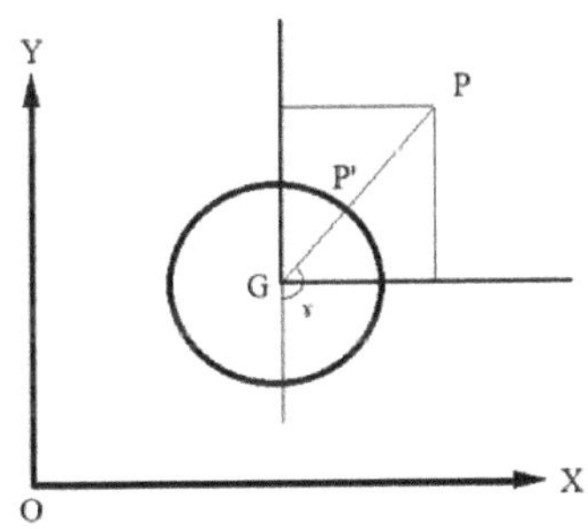

Figure 9. Deviation of Point P located in the horizontal plane, where $Z = Z_p$.

In the second case, if Point P is located between two rows, the deviation is calculated for each of the two rows, and the final deviation is estimated for the last two values. In the third case, if Point P is located outside matrix Z values (up or down), it is considered a noisy point and neglected.

In Figure 9, distance P P′ represents deviation, and distance P′G represents the radius of rotation, which equals $Y_i - Y_g$. In Equation (6), angle ɤ equals ((2.j.π)/m). It should be noted that the deviation can be negative or positive, depending on the point's location in the calculated model. The deviation is negative if the point is located inside the model, and positive if the point is located outside the calculated model. Standard deviation always has a positive value.

In Table 1, tower No. 1 is the tower shown in Figure 2, whereas towers No. 2, 3, 4, and 5 are the towers shown in Figure 7, in the same order.

In Table 1, Tower 5 has the greatest standard deviation ($\sigma = 1.4$ m). This tower consists of two main parts, separated by a step. The points located down the step are not shown (see the red arrow in Figure 7h) because the object was scanned from an aerial view, and the missed points are located within the hidden area. In fact, the suggested modeling algorithm replaces the missing points with the frustum of a cone, which is why the obtained model was deformed in the hidden area (see the blue arrow in Figure 7l). Despite a high standard deviation, most points fit the calculated model.

In building No. 3, the number of points that well fit the tower model is almost equal to the number of points with a deviation greater than 0.3 m. In Figure 7b, the deviations can be attributed to the highly ornamental building facades. From another viewpoint, the reasonable standard deviation (σ = 0.84 m) confirms this result. Tower No. 2 has a simple architectural design, and the deviation is below the minimal value of standard deviation (σ = 0.21 m).

The suggested approach has certain limitations. The proposed algorithm assumes that building facades are completely covered by LiDAR points. In fact, this hypothesis may not always be valid. Therefore, when facade points disappear for whatever reason, the analyzed building details will also disappear. Moreover, the discussed method is very sensitive to noisy points, which can substantially deform the building model. Fortunately, this issue can be resolved by considering point deviation values in addition to building symmetry. Some geometrical forms create hidden areas that cannot be accessed by laser pulses, such as the building shown in Figure 7d. These hidden areas may produce distortions in the calculated building model.

It should be noted that most algorithms for modeling buildings based on LiDAR data suggested in the literature are model-driven or data-driven approaches [39]. In these approaches, the concept of a building model relies on the assumption that the building consists of connected facets that are described by neighborhood relationships. The connections between these facets form facet borders and vertices. A comparison of the proposed modeling approach and the approaches suggested in the literature indicates that the developed algorithm does not belong to the last two modeling approaches, because the building concept differs entirely from the modeling approaches where one building is represented by three matrices that describe the building's geometric form.

However, modeling algorithms should be compared based on their performance. Therefore, three selected approaches were compared with the proposed algorithm in Table 2. In Table 2, standard deviation was used to estimate the accuracy of the generated model. Despite differences in the architectural complexity of the target buildings in the compared approaches, the accuracy of the suggested algorithm is still acceptable.

Table 2. Accuracy of the proposed approach and previous algorithms.

Approach	Standard Deviation
Proposed algorithm	0.21 m to 1.41 m
Kulawiak [3]	0.29 m to 2.36 m
Ostrowski et al. [15]	<0.3 m to >1 m
Jung and Sohn [42]	0.05 m to >3 m

To conclude, the suggested approach paves the way to developing new and general modeling methods based on a matrix representation of buildings with both simple and complex architectural features. In the future, the proposed model could be further improved by integrating point deviations and improving the model's fidelity to the original point cloud.

7. Conclusions

This article proposes a methodology for automating the modeling of buildings with ornamental turrets and towers based on LiDAR data. The proposed modeling procedure was based directly on a point cloud. A vertical axis was generated from a LIDAR data subset describing a tower. It was assumed that the tower was symmetrical about its axis. A cross-section was introduced to the point cloud, with a plane passing through the axis, which produced a vertical cross-section. The vertical cross-section was used to build a solid of revolutions, as the 3D model of the tower. The modeling algorithm relied on a matrix to generate the building model in a mathematical form.

Five tower point clouds were used to evaluate the accuracy of the suggested method. Hence, the deviation of points representing the obtained model was calculated, in addition to the standard deviation. Despite the algorithm's overall efficacy, it had three main limitations. The tower was not covered by LiDAR data in its entirety. Moreover, some geometric forms may generate hidden areas that can produce deformations in the model. Moreover, the suggested algorithm is sensitive to the presence of noisy points. However, facade ornaments, an insufficient accuracy of LiDAR data, and noisy points significantly decreased the accuracy of the generated model. In the future, the building model can be enhanced by considering points with considerable deviations. The matrix form of the proposed algorithm facilitates local enhancements. In addition to the matrix, a vertical cross-section can also be applied to develop a new approach for modeling buildings, regardless of the level of architectural complexity. Finally, additional data, such as aerial or terrestrial imagery, could be incorporated into the proposed modeling approach, to increase the model's fidelity to the original building.

Author Contributions: Conceptualization, F.T.K., E.L., and Z.G.; methodology, F.T.K., E.L.; software, F.T.K.; validation, F.T.K., formal analysis, F.T.K.; resources, E.L., data curation, E.L., F.T.K.; writing—original draft preparation, F.T.K., E.L.; writing—review and editing, Z.G.; visualization, F.T.K., E.L.; All authors have read and agreed to the published version of the manuscript.

Funding: This research was financed as part of a statutory research project of the Faculty of Geoengineering of the University of Warmia and Mazury in Olsztyn, Poland, entitled "Geoinformation from the theoretical, analytical and practical perspective" (No. 29.610.008-110_timeline: 2020–2022).

Data Availability Statement: The publication uses LAS measurement data obtained from an open Polish portal run by the Central Office of Geodesy and Cartography, and Street View images were used. Water towers were searched based on the portal https://wiezecisnien.eu/en/wieze-cisnien/ (accessed on 20 September 2022).

Acknowledgments: We would like to thank the Central Office of Geodesy and Cartography (GUGiK) in Poland for providing Lidar measurement data and data from the 3D portal.

Conflicts of Interest: There are no conflict of interest.

References

1. Dong, Z.; Liang, F.; Yang, B.; Xu, Y.; Zang, Y.; Li, J.; Stilla, U. Registration of large-scale terrestrial laser scanner point clouds: A review and benchmark. *ISPRS J. Photogramm. Remote Sens.* **2020**, *163*, 327–342. [CrossRef]
2. Cheng, L.; Chen, S.; Liu, X.; Xu, H.; Wu, Y.; Li, M.; Chen, Y. Registration of laser scanning point clouds: A review. *Sensors* **2018**, *18*, 1641. [CrossRef] [PubMed]
3. Kulawiak, M. A cost-effective method for reconstructing city-building 3D models from sparse lidar point clouds. *Remote Sens.* **2022**, *14*, 1278. [CrossRef]
4. Yan, W.Y.; Shaker, A.; El-Ashmawy, N. Urban land cover classification using airborne LiDAR data: A review. *Remote Sens. Environ.* **2015**, *158*, 295–310. [CrossRef]
5. Shan, J.; Toth, C.K. *Topographic Laser Ranging and Scanning: Principles and Processing*, 2nd ed.; Taylor & Francis Group; CRC Press: Boca Raton, FL, USA, 2018. [CrossRef]
6. Xu, Y.; Stilla, U. Towards Building and Civil Infrastructure Reconstruction from Point Clouds: A Review on Data and Key Techniques. *IEEE J. Sel. Top. Appl. Earth Obs. Remote Sens.* **2021**, *14*, 2857–2885. [CrossRef]
7. Tarsha Kurdi, F.; Awrangjeb, M.; Munir, N. Automatic filtering and 2D modeling of LiDAR building point cloud. *Trans. GIS* **2021**, *25*, 164–188. [CrossRef]
8. Amakhchan, W.; Kurdi, F.T.; Gharineiat, Z.; Boulaassal, H.; Kharki, O.E. Automatic Filtering of LiDAR Building Point Cloud Using Multilayer Perceptron Neuron Network. In Proceedings of the Conference: 3rd International Conference on Big Data and Machine Learning (BML22'), Istanbul, Turkey, 21–31 May 2022.
9. Wen, C.; Yang, L.; Li, X.; Peng, L.; Chi, T. Directionally constrained fully convolutional neural network for airborne LiDAR point cloud classification. *ISPRS J. Photogramm. Remote Sens.* **2020**, *162*, 50–62. [CrossRef]
10. Tarsha Kurdi, F.; Amakhchan, W.; Gharineiat, Z. Random Forest machine learning technique for automatic vegetation detection and modeling in LiDAR data. *Int. J. Environ. Sci. Nat. Resour.* **2021**, *28*, 556234. [CrossRef]

11. Maltezos, E.; Doulamis, A.; Doulamis, N.; Ioannidis, C. Building extraction from LiDAR data applying deep convolutional neural networks. *IEEE Geosci. Remote Sens. Lett.* **2018**, *16*, 155–159. [CrossRef]
12. Wang, X.; Luo, Y.P.; Jiang, T.; Gong, H.; Luo, S.; Zhang, X.W. A new classification method for LIDAR data based on unbalanced support vector machine. In Proceedings of the 2011 International Symposium on Image and Data Fusion, Tengchong, China, 9–11 August 2011; pp. 1–4. [CrossRef]
13. Yuan, J. Learning building extraction in aerial scenes with convolutional networks. *IEEE Trans. Pattern Anal. Mach. Intell.* **2017**, *40*, 2793–2798. [CrossRef]
14. Chio, S.H.; Lin, T.Y. The establishment of 3D LOD2 objectivization building models based on data fusion. *J. Photogramm. Remote Sens.* **2021**, *26*, 57–73. Available online: https://www.csprs.org.tw/Temp/202106-26-2-57-73.pdf$\backslash$ (accessed on 16 September 2022).
15. Ostrowski, W.; Pilarska, M.; Charyton, J.; Bakuła, K. Analysis of 3D building models accuracy based on the airborne laser scanning point clouds. In *International Archives of the Photogrammetry, Remote Sensing & Spatial Information Sciences*; ISPRS: Vienna, Austria, 2018; p. 42. Available online: https://www.int-arch-photogramm-remote-sens-spatial-inf-sci.net/XLII-2/797/2018/#:~{}:text=https%3A//doi.org/10.5194/isprs%2Darchives%2DXLII%2D2%2D797%2D2018%2C%202018 (accessed on 16 September 2022).
16. Zhang, K.; Yan, J.; Chen, S.C. A framework for automated construction of building models from airborne Lidar measurements. In *Topographic Laser Ranging and Scanning: Principles and Processing*, 2nd ed.; Shan, J., Toth, C.K., Eds.; Taylor & Francis Group; CRC Press: Boca Raton, FL, USA, 2018; pp. 563–586.
17. Van Oosferom, P.J.M.; Broekhuizen, M.; Kalogianni, E. BIM models as input for 3D land administration systems for apartment registration. In Proceedings of the 7th International FIG Workshop on 3D Cadastres, New York, NY, USA, 11–13 October 2021; International Federation of Surveyors (FIG): Copenhagen, Denmark, 2021; pp. 53–74. Available online: https://research.tudelft.nl/en/publications/bim-models-as-input-for-3d-land-administration-systems-for-apartm; https://www.proquest.com/results/883F9FCF559E442EPQ/false?accountid=14884$\backslash$ (accessed on 16 September 2022).
18. Beil, C.; Ruhdorfer, R.; Coduro, T.; Kolbe, T.H. Detailed streetspace modeling for multiple applications: Discussions on the proposed CityGML 3.0 transportation model. *ISPRS Int. J. Geo-Inf.* **2020**, *9*, 603. [CrossRef]
19. Biljecki, F.; Lim, J.; Crawford, J.; Moraru, D.; Tauscher, H.; Konde, A.; Adouane, K.; Lawrence, S.; Janssen, P.; Stouffs, R. Extending CityGML for IFC-sourced 3d city models. *Autom. Constr.* **2021**, *121*, 103440. [CrossRef]
20. Jayaraj, P.; Ramiya, A.M. 3D CityGML building modeling from lidar point cloud data. In *The International Archives of Photogrammetry, Remote Sensing and Spatial Information Sciences*; Gottingen Tom XLII-5; Copernicus GmbH: Gottingen, Germany, 2018; pp. 175–180. Available online: https://www.proquest.com/opeview/47e1bca8fac2930d4be04e70741e905f/1 (accessed on 20 September 2022). [CrossRef]
21. Lukač, N.; Žalik, B. GPU-based roofs' solar potential estimation using LiDAR data. *Comput. Geosci.* **2013**, *52*, 34–41. [CrossRef]
22. Beil, C.; Kutzner, T.; Schwab, B.; Willenborg, B.; Gawronski, A.; Kolbe, T.H. Integration of 3D point clouds with semantic 3D City Models—providing semantic information beyond classification. In *ISPRS Annals of the Photogrammetry, Remote Sensing and Spatial Information Sciences*; Gottingen Tom VIII-4/W2-2021; Copernicus GmbH: Gottingen, Germany, 2021; pp. 105–112. Available online: https://www.proquest.com/openview/6c8a777f6d7c8645edb6dad807b248aa/1?pq-origsite=gscholar&cbl=2037681 (accessed on 16 September 2022). [CrossRef]
23. Chaturvedi, K.; Matheus, A.; Nguyen Son, H.; Kolbe, H. Securing spatial data infrastructures for distributed smart city applications and services. *Future Gener. Comput. Syst.* **2019**, *101*, 723–736. [CrossRef]
24. Kutzner, T.; Chaturvedi, K.; Kolbe, T.H. CityGML 3.0: New functions open up new applications. *PFG J. Photogramm. Remote Sens. Geoinf. Sci.* **2020**, *88*, 43–61. [CrossRef]
25. Shirinyan, E.; Petrova-Antonova, D. Modeling buildings in CityGML LOD1: Building parts, terrain intersection curve, and address features. *ISPRS Int. J. Geo-Inf.* **2022**, *11*, 166. [CrossRef]
26. Biljecki, F.; Ledoux, H.; Stoter, J. Generation of multi-LOD 3D city models in CityGML with the procedural modelling engine random3dcity, ISPRS ann. Photogramm. Remote Sens. *Spatial Inf. Sci.* **2016**, *IV-4/W1*, 51–59. [CrossRef]
27. Gilani, S.A.N.; Awrangjeb, M.; Lu, G. Segmentation of airborne point cloud data for automatic building roof extraction. *GISci. Remote Sens.* **2018**, *55*, 63–89. [CrossRef]
28. Chen, D.; Zhang, L.; Mathiopoulos, P.; Huang, X. A methodology for automated segmentation and reconstruction of urban 3-D buildings from ALS point clouds. *IEEE J. Sel. Top. Appl. Earth Obs. Remote Sens.* **2014**, *7*, 4199–4217. [CrossRef]
29. Dey, E.K.; Tarsha Kurdi, F.; Awrangjeb, M.; Stantic, B. Effective selection of variable point neighbourhood for feature point extraction from aerial building point cloud data. *Remote Sens.* **2021**, *13*, 1520. [CrossRef]
30. Dong, Y.; Hou, M.; Xu, B.; Li, Y.; Ji, Y. Ming and Qing dynasty official-style architecture roof types classification based on the 3D point cloud. *ISPRS Int. J. Geo-Inf.* **2021**, *10*, 650. [CrossRef]
31. Awrangjeb, M.; Gilani, S.A.N.; Siddiqui, F.U. An effective datadriven method for 3-d building roof reconstruction and robust change detection. *Remote Sens.* **2018**, *10*, 1512. [CrossRef]
32. Tarsha Kurdi, F.; Awrangjeb, M.; Munir, N. Automatic 2D modelling of inner roof planes boundaries starting from Lidar data. In Proceedings of the 14th 3D GeoInfo 2019, Singapore, 26–27 September 2019. [CrossRef]
33. Tarsha Kurdi, F.; Gharineiat, Z.; Campbell, G.; Dey, E.K.; Awrangjeb, M. Full series algorithm of automatic building extraction and modeling from LiDAR data. In Proceedings of the Digital Image Computing: Techniques and Applications (DICTA), Gold Coast, Australia, 29 November–1 December 2021; pp. 1–8. [CrossRef]

34. Dey, E.K.; Awrangjeb, M.; Tarsha Kurdi, F.; Stantic, B. Building boundary point extraction from LiDAR point cloud data. In Proceedings of the 2021 Digital Image Computing: Techniques and Applications (DICTA), Gold Coast, Australia, 29 November–1 December 2021; pp. 1–8. [CrossRef]
35. Huang, J.; Stoter, J.; Peters, R.; Nan, L. City3D: Large-scale building reconstruction from airborne LiDAR point clouds. *Remote Sens.* **2022**, *14*, 2254. [CrossRef]
36. Borisov, M.; Radulovi, V.; Ilić, Z.; Petrovi, V.; Rakićević, N. An automated process of creating 3D city model for monitoring urban infrastructures. *J. Geogr. Res.* **2022**, *5*. Available online: https://ojs.bilpublishing.com/index.php/j (accessed on 16 September 2022). [CrossRef]
37. Tarsha Kurdi, F.; Gharineiat, Z.; Campbell, G.; Awrangjeb, M.; Dey, E.K. Automatic filtering of LiDAR building point cloud in case of trees associated to building roof. *Remote Sens.* **2022**, *14*, 430. [CrossRef]
38. Tarsha Kurdi, F.; Awrangjeb, M. Comparison of LiDAR building point cloud with reference model for deep comprehension of cloud structure. *Can. J. Remote Sens.* **2020**, *46*, 603–621. [CrossRef]
39. Tarsha Kurdi, T.; Landes, T.; Grussenmeyer, P.; Koehl, M. Model-driven and data-driven approaches using Lidar data: Analysis and comparison. In Proceedings of the ISPRS Workshop, Photogrammetric Image Analysis (PIA07), Munich, Germany, 19–21 September 2007; International Archives of Photogrammetry, Remote Sensing and Spatial Information Systems: Stuttgart, Germany, 2007; Volume XXXVI, pp. 87–92, ISSN 1682-1750.
40. Cheng, L.; Zhang, W.; Zhong, L.; Du, P.; Li, M. Framework for evaluating visual and geometric quality of three-dimensional models. *IEEE J. Sel. Top. Appl. Earth Obs. Remote Sens.* **2015**, *8*, 1281–1294. [CrossRef]
41. Rottensteiner, F.; Clode, S. Building and road extraction from Lidar data. In *Topographic Laser Ranging and Scanning: Principles and Processing*, 2nd ed.; Shan, J., Toth, C.K., Eds.; Taylor & Francis Group; CRC Press: Boca Raton, FL, USA, 2018; pp. 485–522.
42. Jung, J.; Sohn, G. Progressive modeling of 3D building rooftops from airborne Lidar and imagery. In *Topographic Laser Ranging and Scanning: Principles and Processing*, 2nd ed.; Shan, J., Toth, C.K., Eds.; Taylor & Francis Group; CRC Press: Boca Raton, FL, USA, 2018; pp. 523–562.
43. Dorninger, P.; Pfeifer, N. A comprehensive automated 3D approach for building extraction, reconstruction, and regularization from airborne laser scanning point clouds. *Sensors* **2008**, *8*, 7323–7343. [CrossRef]
44. Sampath, A.; Shan, J. Segmentation and reconstruction of polyhedral building roofs from aerial Lidar point clouds. *IEEE Trans. Geosci. Remote Sens.* **2010**, *48*, 1154–1567. [CrossRef]
45. Park, S.Y.; Lee, D.G.; Yoo, E.J.; Lee, D.C. Segmentation of Lidar data using multilevel cube code. *J. Sens.* **2019**, *2019*, 4098413. [CrossRef]

remote sensing

Article

Strategies for the Storage of Large LiDAR Datasets—A Performance Comparison

Juan A. Béjar-Martos, Antonio J. Rueda-Ruiz *, Carlos J. Ogayar-Anguita and Rafael J. Segura-Sánchez and Alfonso López-Ruiz

Cntro de Estudios Avanzados en TIC, University of Jaén, 23071 Jaén, Spain; jabm0010@red.ujaen.es (J.A.B.-M.); cogayar@ujaen.es (C.J.O.-A.); rsegura@ujaen.es (R.J.S.-S.); allopezr@ujaen.es (A.L.-R.)

* Correspondence: ajrueda@ujaen.es

Abstract: The widespread use of LiDAR technologies has led to an ever-increasing volume of captured data that pose a continuous challenge for its storage and organization, so that it can be efficiently processed and analyzed. Although the use of system files in formats such as LAS/LAZ is the most common solution for LiDAR data storage, databases are gaining in popularity due to their evident advantages: centralized and uniform access to a collection of datasets; better support for concurrent retrieval; distributed storage in database engines that allows sharding; and support for metadata or spatial queries by adequately indexing or organizing the data. The present work evaluates the performance of four popular NoSQL and relational database management systems with large LiDAR datasets: Cassandra, MongoDB, MySQL and PostgreSQL. To perform a realistic assessment, we integrate these database engines in a repository implementation with an elaborate data model that enables metadata and spatial queries and progressive/partial data retrieval. Our experimentation concludes that, as expected, NoSQL databases show a modest but significant performance difference in favor of NoSQL databases, and that Cassandra provides the best overall database solution for LiDAR data.

Keywords: LiDAR; point clouds; databases; NoSQL

Citation: Béjar-Martos, J.A.; Rueda-Ruiz, A.J.; Ogayar-Anguita, C.J.; Segura-Sánchez, R.J.; López-Ruiz, A. Strategies for the Storage of Large LiDAR Datasets—A Performance Comparison. *Remote Sens.* **2022**, *14*, 2623. https://doi.org/10.3390/rs14112623

Academic Editors: Jorge Delgado García, Fayez Tarsha Kurdi and Tarig Ali

Received: 12 April 2022
Accepted: 29 May 2022
Published: 31 May 2022

1. Introduction

LiDAR scanning has become an indisputable tool in fields such as civil engineering, surveying, archaeology, forestry or environmental engineering. The widespread use of terrestrial and airborne LiDAR, powered by the fast evolution of scanning technology, is generating an unprecedented amount of data. For instance, the scanning speeds of one million points per second with an accuracy in the range of 3–5 mm have become common today in terrestrial scanning [1]. Handling such a massive amount of data poses multiple challenges related to its storage, transmission, organization, visualization, edition and analysis, which are actually common to most Big Data applications [2].

LiDAR information has traditionally been stored and exchanged through system files usually from standard formats such as LAS/LAZ. However, the use of databases potentially has many advantages such as the centralized access, complex queries based on metadata, spatial organization of data and distributed storage in databases that support sharding. NoSQL databases are extensively used in Big Data applications because of their performance in simple retrieval operations, their flexible schema and their ability to scale horizontally.

Storing large raw LiDAR data files in a database does not provide any advantages over system files. However, as stated above, the design of an appropriate database scheme or the use of a spatial extension [3] allows organizing LiDAR data into a spatial data structure such as a regular grid, quadtree or octree. This enables spatial data queries and the selective/progressive transmission of data to clients. This organization requires splitting the original LiDAR dataset into data blocks that are associated with the corresponding tree nodes or grid cells. The size of the data block can have a remarkable impact on the performance of the database.

The main objective of the present work was to provide guidance to those responsible for the design and implementation of LIDAR information storage and processing systems on the performances of different database management systems. Four different NoSQL and relational systems (Cassandra, MongoDB, MySQL PostgreSQL) were compared in terms of the performance of their upload and retrieval operations in single and concurrent scenarios. A complex database schema was used, with two nested spatial data structures and three different data block sizes, following the conceptual model of SPSLiDAR [4].

The rest of this paper is organized as follows: Section 2 reviews some of the extensive existing literature related to the storage of LiDAR information; Section 3 describes the dataset used in the experimentation, the conceptual data model of SPSLiDAR, the features of each database management system evaluated and the implementation of the conceptual data model in them; Section 4 provides the details of the experimental comparison carried out and outlines the results that are discussed in depth in Section 5. Finally, Section 6 summarizes the conclusions and proposes some future work.

2. Previous Work

LiDAR point clouds are currently a very valuable resource for all types of decision-making processes involving spatial data from the real world. The evolution of LiDAR systems enables the acquisition of massive spatial data models [5] whose volume is steadily increasing. Current sensors allow, using various technologies, obtaining dense information at different scales [6] from small objects to large digital terrain models, integrating them into the Geospatial Big Data [7–10]. Recently, many research papers related to this matter have been published [5,10–13].

The processing of LiDAR point cloud data involves a series of steps that range from its acquisition to the extraction of relevant features, including registering, filtering, segmentation, classification and conversion to other representation schemes. Each of these steps has its own research topic, and in particular, the storage of massive point clouds is one of the most notable. In recent years, with increasing processing power, memory and communication bandwidth, the main challenge with LiDAR data is to make massive information available for use in different applications [14]. In this sense, there are different approaches focusing on mass storage in secondary memory, in a cluster of servers or in the cloud [15–19].

Organizing 3D point clouds with data structures that subdivide space is a common solution for increasing the performance of spatial queries. However, from a storage point of view, it also brings the benefit of dividing a potentially massive amount of data into smaller chunks that can be more efficiently managed, which is of key importance when transmitting data over a network. The most widely used data structure for out-of-core point cloud storage is the octree [20–23]. Other widely used structures are the Kd-tree [24], variants of the R-tree [25], a sparse voxel structure [16] and simple grids [26].

Most spatial data structures can be adapted for out-of-core storage, and in this regard, there are different variants and approaches. This is of utmost importance, since today it is common to process LiDAR datasets that do not fit into main memory. The simplest approach is direct storage in secondary memory using local files [15,19]. Another variant consists of using storage in distributed file systems [27,28] typically indexed through a spatial structure hosted in the form of a master index or in a database. LiDAR datasets usually comprise one or more point cloud files encoded in non-standard ASCII formats, the LAS format of the American Society for Photogrammetry and Remote Sensing (ASPRS), its compressed variant LAZ [29], SPD [30], PCD, HDF5 and other general 3D data formats such as OBJ or PLY. In addition, there are some proposals for specific formats, closely related to applications which involve additional point attributes and custom compression algorithms. Understandably, file formats that focus on data compression are the most useful [11,29,31].

Although file-oriented storage is the simplest and most common option, using databases is the most versatile solution. An original way of storing a point cloud model in a database is to use one row for each point in a relational database, including all its attributes. In the past, this approach was common among GIS applications where sets of points of moderate size were stored in a spatial database to support spatial queries, mainly 2D. However, this scheme is not valid for a large amount of data [32], so the next logical solution is to store a group of points in each row, as systems such as Oracle Spatial or PostGIS do. In addition to this, there are other solutions based on relational databases to work with point clouds and spatial information [8,21].

On the other hand, NoSQL databases have some advantages over relational databases in Big Data applications. Among these, document-oriented databases (MongoDB, Cassandra, Couchbase, etc.) have gained popularity due to their capability to efficiently handle large volumes of data and are scalable through the use of sharding. Because of this, they are usually the preferred option for storing point clouds with a Big Data approach for both semi-structured and unstructured datasets [26,28,32]. However, the everlasting discussion about the theoretical convenience of using the modern NoSQL approach instead of relational databases for Big Data problems should not diminish the relevance of the latter, particularly considering their maturity and widespread adoption [33]. In the case of point cloud datasets, including LiDAR, several approaches use relational database management systems (DBMSs) for processing information [21,34]. The most relevant part is the data model of a DBMS that defines the logical structure of a database and that determines how the data can be stored, organized, and retrieved. In this work, we use SPSLIDAR [4], a data model with a reference implementation for a LiDAR data repository that can be used with any spatial indexing.

3. Materials and Methods

In order to study the performance of the different database engines integrated into SPSLiDAR, various datasets belonging to the city of Pamplona (Spain) were used. We chose this particular area because of the public availability of LiDAR data acquired at different moments and densities (0.5–10 points/m^2). The areas covered by the LiDAR datasets are located in the UTM zone 30N and were captured over 6 years to include a variety of terrains (urban, rural, roads, etc.). Table 1 summarizes the characteristics of the four datasets, and Figure 1 shows a partial rendering of Dataset 1.

Table 1. Datasets characteristics.

	Dataset 1	Dataset 2	Dataset 3	Dataset 4
Name	Pamplona 2011	Pamplona 2017b (reduced version of dataset 4)	Pamplona 2017c (reduced version of dataset 4)	Pamplona 2017
Number of points	30,401,539	244,894,563	534,073,172	1,068,146,345
Density (points/m^2)	0.5	2.5	5	10
Grid size (meters)	10,000	10,000	10,000	10,000
Size (MBs)	138.29	1850.21	4085.94	8054.36
Point data record length	34	38	38	38
LAS point data record format	3	8	8	8

Figure 1. Partial view of Dataset 1 (Pamplona 2011).

3.1. SPSLiDAR

In previous work, a high-level conceptual model for a repository for LiDAR data, namely SPSLiDAR, was proposed [1]. The conceptual model of SPSLiDAR, depicted in Figure 2, comprises three entities: *workspace, dataset* and *datablock*. A workspace represents a set of related datasets, such as, for instance, those generated by the LiDAR coverage of several counties or provinces, or different scanning campaigns from related archaeological sites. A dataset comprises one or more point clouds acquired at a particular site. The points of a dataset are organized into a structure of datablocks starting at one or more root datablocks. SPSLiDAR exposes its functionalities to the clients through a restful API, allowing queries by spatial and temporal criteria, and the progressive download of partial or complete datasets.

The original paper of SPSLiDAR proposed a reference implementation that used a regular grid to organize datasets and an octree for the datablocks within a dataset. MongoDB was chosen as the underlying database engine due to its flexibility, high performance and sharding capabilities that allow handling up to petabytes of information distributed across multiple data servers. The characteristics of the database management system have a significant influence on the performance of SPSLIDAR, since each time a dataset is sent to the system, one or more octrees are generated, and each one may comprise hundreds of thousands of nodes and LAZ files that must be efficiently stored in the database—both in terms of occupied time and space. Subsequently, the SPSLiDAR API enables the navigation of the dataset octrees and the retrieval of the point clouds in the LAZ files associated with the nodes of interest. The ability of the database to fetch the desired datablocks and transmit the content of the associated point cloud files has a direct impact on latency times and system throughput, especially considering that usually several clients may interact with the system at the same time.

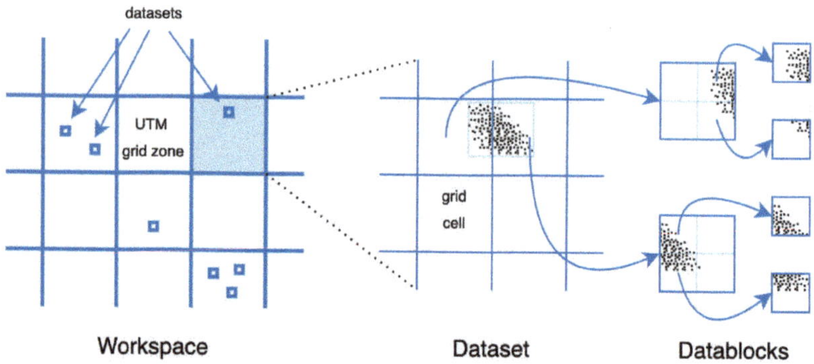

Figure 2. Spatial data structures at the workspace and dataset levels. For the sake of clarity, octrees are depicted as quadtrees at the dataset level.

3.2. *Databases*

In our experimentation, we used the original MongoDB-based implementation of the SPSLiDAR and three different adaptations corresponding to the rest of the database management systems evaluated. Even though the storage of large LiDAR datasets fits the definition of Geospatial Big Data, we did not want to restrict ourselves only to non-relational databases, usually considered the standard storage solution for applications in this domain. Relational databases have also been used for the storage of massive amounts of data [9,35] and may therefore meet the needs of our use case.

MongoDB, Cassandra, PostgreSQL and MySQL were the four technologies adopted. All of them are well established products. Among non-relational databases, MongoDB provides a document-oriented solution while Cassandra follows a wide-column approach. In the category of relational databases, PostgreSQL and MySQL are among the most widely adopted. In order to make the fairest comparison, the SPSLiDAR conceptual model was implemented in each database in the most straightforward way.

3.2.1. MongoDB

MongoDB is a non-relational document-oriented database that uses JSON to encode information. Documents in MongoDB present a flexible schema so that two documents representing a similar concept can have different fields and new attributes may be added or removed at runtime. MongoDB introduces the concept of collection as a means of grouping documents which represent the same concept or type of information.

The GridFS specification is provided by MongoDB as a means of storing binary files that surpass the 16 MB per document limit imposed by the system. GridFS uses two collections to store files. One collection stores the file chunks (*fs.chunks*) and the other stores file metadata (*fs.files*). Each file is decomposed into multiple chunks stored in the *fs.chunks* collection, each one containing the binary data of the corresponding section of the file, an order attribute that specifies its position in the sequence of chunks and a reference to the file document in the *fs.files* collection. By default, the maximum amount of data per chunk is set to 255 KB. The GridFS collections are depicted in Figure 3.

The storage of LAZ files in the database is one of the key points of the implementation as it has a direct effect on performance. From a technical point of view, we could save these files through GridFS or embedding the content in a document. We decided to use GridFS as the standard way of storing files, since this enables the construction of octrees with any arbitrary node size. A hybrid approach could also be followed to store the LAZ file through an embedded field if it does not surpass the 16 MB size limit for MongoDB documents, or via GridFS otherwise. This hybrid approach can easily be implemented in MongoDB thanks to its flexible document schema, which allows encoding the file data in a BSON

(Binary JSON) field or alternatively, to include a *DBRef* field with the *_id* of the document saved in the *fs.files* collection of GridFS.

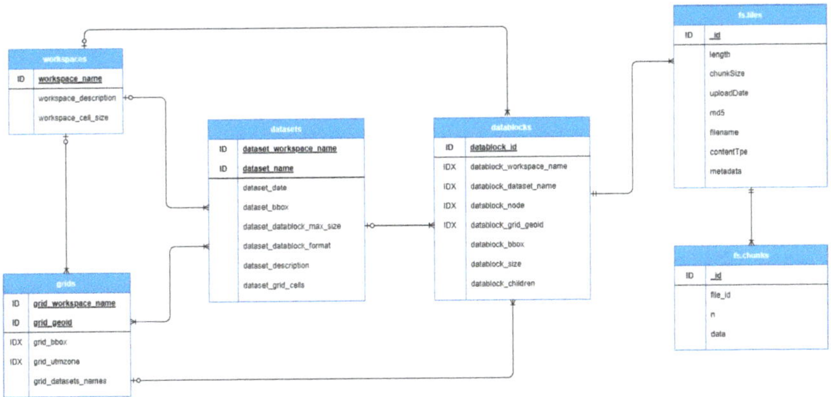

Figure 3. Database design for MongoDB. PK refers to primary key and FK refers to foreign key.

The final structure of documents and collections defined in MongoDB is depicted in Figure 3, showing the relationships among the different entities. The details of the data model, including the description of the *workspace*, *dataset* and *datablock* entities can be found in an original paper describing SPSLiDAR [4]. A mainly denormalized approach was followed: for instance, a datablock document will contain redundant information that refers to the workspace, dataset, and grid cell it belongs to. These fields were indexed in order to accelerate queries since datablock fetching is a crucial operation to access LAZ data files.

3.2.2. Cassandra

Cassandra is a non-relational database with a partitioned row model. Rows are stored in different partitions, identified by the partition key (PK) which may consist of one or more columns. A partition may contain more than one row; in these cases, a clustered key (CK) is also needed, which may also consist of more than one column and uniquely identifies each row inside a partition. Therefore, a single row is defined by a composite key formed by a partition key and optionally a clustered key.

No analogous specification to MongoDB GridFS exists in Cassandra; therefore, we implemented a custom system that follows the same basic concepts. A new entity called chunk was defined, which contains a column with binary data and is uniquely identified by a composite key formed by the partition key of the datablock it belongs to and an additional *chunk_order* attribute which identifies the position of the chunk content in the file data. As a result, all the chunks belonging to the same file are stored on the same partition and the *chunk_order* attribute facilitates a complete recovery of the chunks in ascending order, enabling a straightforward reconstruction of the file. The maximum size of content for each chunk was set to 255 KB—the same default value used by GridFS.

Figure 4 shows the tables and the columns defined for the persistence of our data model in Cassandra. As we did in MongoDB, a denormalized approach was followed. Cassandra recommends a model design based on the queries that will be performed, with data duplication encouraged to increase reading efficiency.

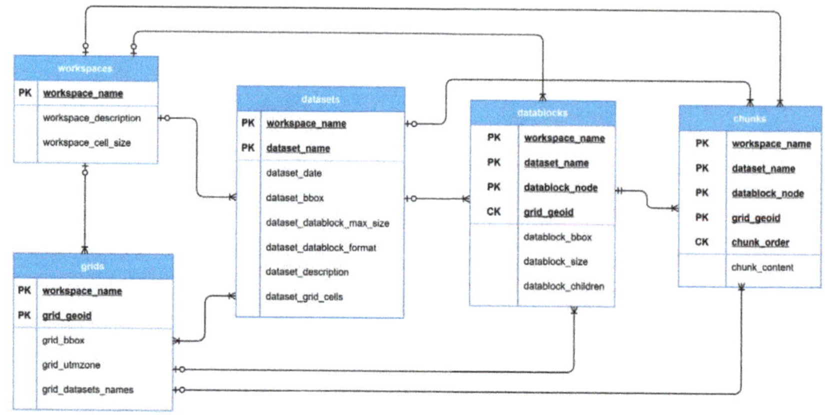

Figure 4. Database design for Cassandra implementation. PK refers to primary key and FK to foreign key.

3.2.3. Relational Databases

Regarding relational databases, we integrated two different alternatives into the system: namely PostgreSQL and MySQL. Due to their similarities, we used almost the same implementation on the persistence layer, only modifying the types so that they adjust to each database specification.

In the same way as with Cassandra, we implemented a tailored solution for the storage of files. The proposed solution defines a new entity called *chunk*, with a composite primary key formed by the primary key of its associated datablock together with an additional attribute that represents the position of the attached binary data block in the file. Normalization is usually a requirement in relational database schemas, but in order to avoid table parameters that may adversely affect performance, we denormalized both the datablocks and chunks entities. We consider that the time performance outweighs the disadvantages of a denormalized schema for these particular tables since they are expected to contain a large number of rows and support a high number of queries. In Figure 5, an entity-relational diagram of the model is shown for both relational databases.

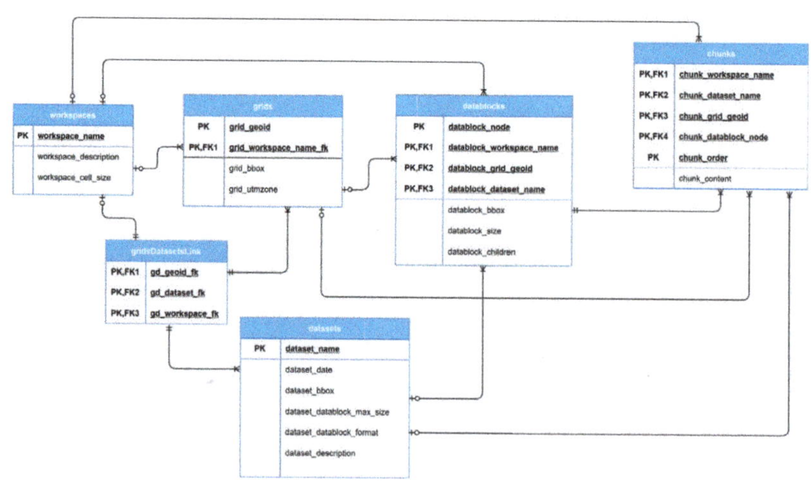

Figure 5. Database design for MySQL and PostgreSQL implementations. PK refers to the primary key and FK to foreign key.

4. Experiments and Results

To evaluate the performance of each of the databases with large LiDAR datasets, three experiments were performed. These experiments are based on those proposed in [36]. The first one (upload test) evaluates the upload time and total storage space required by the four datasets. The second test (simple requests test) measures the average download time for 1M points from a single client. For this purpose, multiple random LAZ files are requested until completing 20 M points, computing the average from the total download time. The third test (concurrent requests test) evaluates the response of the system to high workloads by measuring the average download time for multiple concurrent requests of 1M points from several clients.

The upload test and simple requests test were both implemented through a Python 3.9 script that uses the *Requests* library [37] in order to perform the petitions to the server. The concurrent requests test was developed using the *LoadTest* JavaScript package for Node.js [38]. All operations were performed through the SPSLiDAR API. The tested data correspond to the four datasets described in Table 1. For each of the datasets, three different maximum datablock sizes were tested: 10,000, 100,000, and 1,000,000 points. These values define the maximum number of points of the LAZ file stored at each node of the octree, and consequently determine its complexity since a lower or higher number of levels would be required to store the dataset.

All experiments were carried out by running the different implementations of the SPSLiDAR repository on a server with Intel i7-10700 Octa-Core at 2.9GHz, 16 GB RAM, and 1 TB SSD drive, running Windows 10 Enterprise 64. All databases were deployed through Docker using images with MongoDB 4.2.10, Cassandra 3.11.0, MySQL 8.0.24, and PostgreSQL 13.2. The client computer was a Mackbook Pro laptop with an Intel Core i7-4770HQ Quad-Core at 2.2 GHz with 16 GB RAM and a 256 GB SSD Drive, running macOS Big Sur. The connection between the server and client was made through a 1 GB/s Ethernet cable network.

4.1. Upload Test

A dataset upload comprises two stages. First, the original dataset is preprocessed and subdivided into a set of small LAZ files whose size is limited by the maximum datablock size. These files are indexed through octrees that represent the original point cloud. Second, the files and the octree metadata are stored in the database. The first stage is highly computationally demanding and consumes most of the time. Therefore, the influence of the database performance on the overall time is limited. The preprocessing stage is deterministic and is external to the persistence layer, resulting in the insertion of the same number of files in each database. Furthermore, note that, in practice, this upload operation is only to be carried out once per dataset. For this reason, we consider that the results of this test should have the least weight in the choice of the database.

Table 2 shows the average upload times for each combination of dataset and the maximum datablock size through three executions. The biggest differences among databases occurred in Dataset 4 as the octree structure generated is the most complex, with the highest number of LAZ files to store. Figure 6 shows a relative comparison of the upload times of Cassandra, PostgreSQL and MySQL with respect to MongoDB. We compared them against MongoDB because this was used in the reference implementation of SPSLiDAR.

Table 2. Dataset upload times (in seconds) for the implementations based on MongoDB, Cassandra, PostgreSQL and MySQL.

Dataset	Maximum Datablock Size	MongoDB	Cassandra	PostgreSQL	MySQL
Dataset 1	1,000,000	67	68	69	70
	100,000	218	211	224	213
	10,000	1444	1373	1553	1411
Dataset 2	1,000,000	746	793	824	727
	100,000	2089	2027	2016	1993
	10,000	13,962	12,791	12,650	12,942
Dataset 3	1,000,000	1945	1806	1899	2191
	100,000	3906	3843	3897	3962
	10,000	25,176	24,528	26,823	23,307
Dataset 4	1,000,000	4131	4220	4150	4816
	100,000	8460	8258	8887	8575
	10,000	45,035	41,290	49,714	42,018

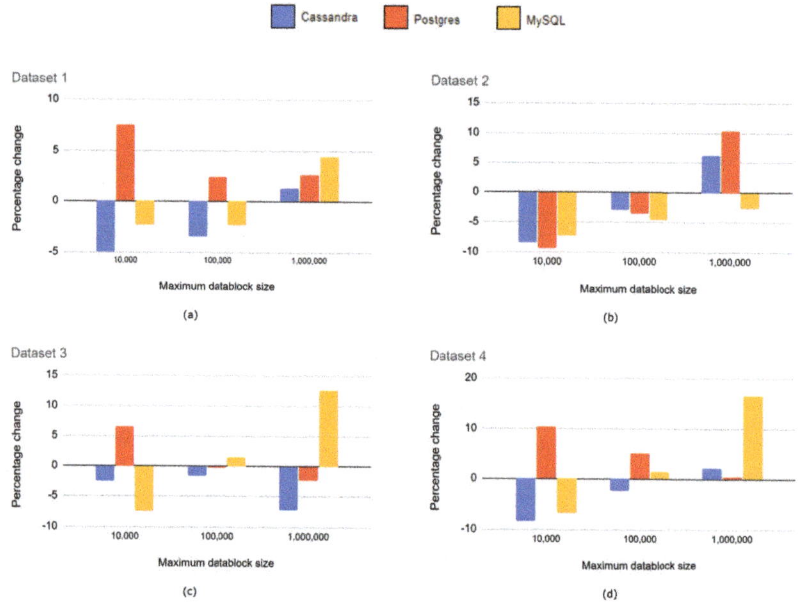

Figure 6. Relative comparison of upload times. Results for (**a**) Dataset 1; (**b**) Dataset 2; (**c**) Dataset 3; and (**d**) Dataset 4.

Table 3 shows the final storage space required at the database level by the datasets after the upload tests. This includes the LAZ files generated in the process in order for all the entities necessary for the system to work correctly. The results shown are the average of three upload operations, although we observed low variance among the results. Figure 7 depicts the information of Table 3 as a relative comparison in storage size with respect to MongoDB.

Table 3. Storage required (in MB) by the implementations based on MongoDB, Cassandra PostgreSQL and MySQL.

Dataset	Maximum Datablock Size	MongoDB	Cassandra	PostgreSQL	MySQL
Dataset 1	1,000,000	198	197	214	224
	100,000	202	200	217	220
	10,000	235	224	247	262
Dataset 2	1,000,000	1875	1882	1960	2118
	100,000	1893	1884	1970	2109
	10,000	2162	2056	2262	2603
Dataset 3	1,000.000	4233	4243	4413	4337
	100,000	4313	4308	4486	4365
	10,000	5062	4753	5377	6130
Dataset 4	1,000.000	8265	8291	8609	9116
	100,000	8388	8478	8726	9397
	10,000	9893	9160	10,270	12,406

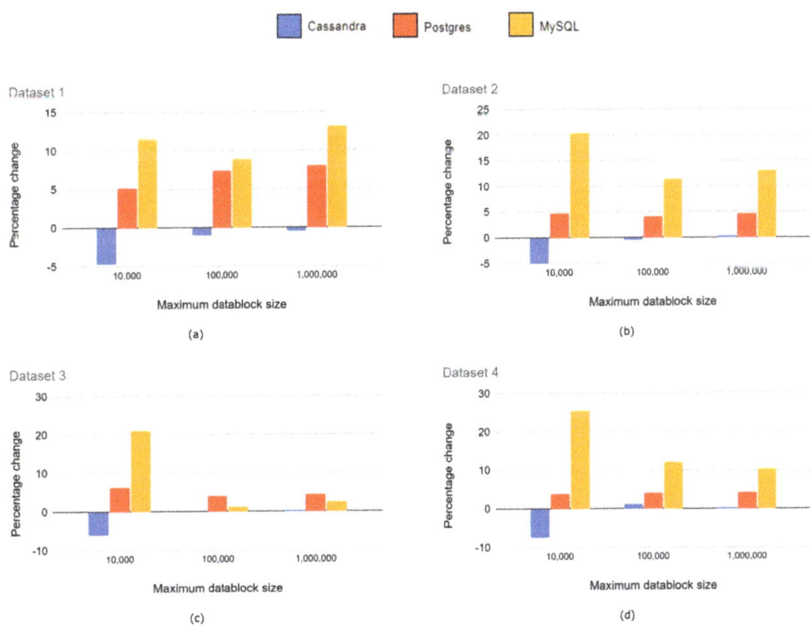

Figure 7. Relative comparison of the required storage. Results for (**a**) Dataset 1; (**b**) Dataset 2; (**c**) Dataset 3; and (**d**) Dataset 4.

4.2. Simple Requests Test

Table 4 shows a relative comparison of the average times required for reading 1M points from a single client in the different implementations tested. The results are also graphically summarized in Figure 8.

Table 4. Average times (in seconds) of a request operation of 1M points in the implementations based on MongoDB, Cassandra, PostgreSQL and MySQL.

Dataset	Maximum Datablock Size	MongoDB	Cassandra	Postgres	MySQL
Dataset 1	1,000,000	0.107	0.11	0.264	0.268
	100,000	0.333	0.303	0.965	0.425
	10,000	2.547	1.824	5.997	1.72
Dataset 2	1,000,000	0.149	0.222	0.308	0.324
	100,000	0.701	0.438	0.992	0.4
	10,000	3.718	3.036	9.48	2.792
Dataset 3	1,000,000	0.17	0.237	0.304	0.469
	100,000	0.429	0.332	1.094	0.551
	10,000	3.559	2.453	8.991	2.223
Dataset 4	1,000,000	0.348	0.188	0.304	0.572
	100,000	0.452	0.329	1.042	0.641
	10,000	3.663	2.701	6.859	2.894

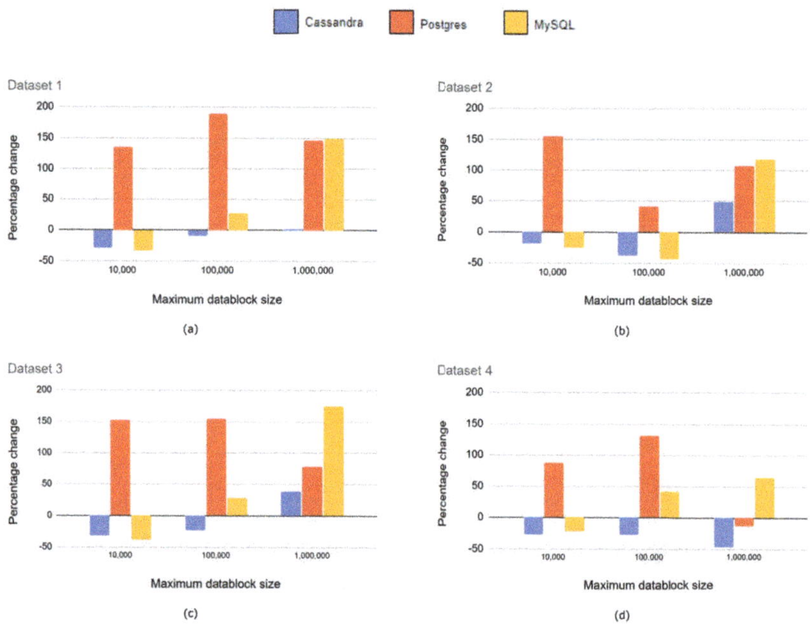

Figure 8. Relative comparison of the results of the simple requests test with (**a**) Dataset 1; (**b**) Dataset 2; (**c**) Dataset 3; and (**d**) Dataset 4.

4.3. Concurrent Requests Test

The third experiment evaluates the response time for several concurrent requests of 1M points from different clients. Tables 5–8 show the results obtained for each implementation with 10 and 100 concurrent clients, respectively. For the sake of simplicity, requests are only performed on Dataset 1. Figure 9 graphically compares the throughput of the different implementations.

Table 5. Results of the concurrent requests test in Cassandra.

Concurrent Users	Maximum Datablock Size	Average Request Time (s)	Maximum Request Time (s)	Throughput (Requests per Second)	Total Time (s)
10	1,000,000	1.085	1.512	7	1.532
	100,000	0.174	0.23	55	1.831
	10,000	0.025	0.074	386	2.59
100	1,000,000	13.36	18.572	5	18.61
	100,000	1.715	2.589	56	17.73
	10,000	0.237	0.45	419	23.86

Table 6. Results of the concurrent requests test in MongoDB.

Concurrent Users	Maximum Datablock Size	Average Request Time (s)	Maximum Request Time (s)	Throughput (Requests per Second)	Total Time (s)
10	1,000,000	1.653	1.68	6	1.691
	100,000	0.252	0.42	39	2.5705
	10,000	0.038	0.076	259	3.8628
100	1,000,000	1.614	1.657	6	16.164
	100,000	2.079	3.823	47	21.4625
	10,000	0.269	0.527	370	27.011

Table 7. Results of the concurrent requests test in PostgreSQL.

Concurrent Users	Maximum Datablock Size	Average Request Time (s)	Maximum Request Time (s)	Throughput (Requests per Second)	Total Time (s)
10	1,000,000	2.386	3.002	3	3.024
	100,000	0.224	0.503	42	2.352
	10,000	0.035	0.4	274	3.646
100	1,000,000	23.3	38.57	3	38.62
	100,000	3.763	16.51	26	38.691
	10,000	0.279	0.597	356	28.071

Table 8. Results of the concurrent requests test in MySQL.

Concurrent Users	Maximum Datablock Size	Average Request Time (s)	Maximum Request Time (s)	Throughput (Requests per second)	Total Time (s)
10	1,000,000	1.547	1.579	6	1.604
	100,000	0.232	0.296	42	2.36
	10,000	0.044	0.113	223	4.493
100	1,000,000	25.422	34.841	3	34.886
	100,000	9.316	23.556	10	95.749
	10,000	0.416	0.633	239	41.86

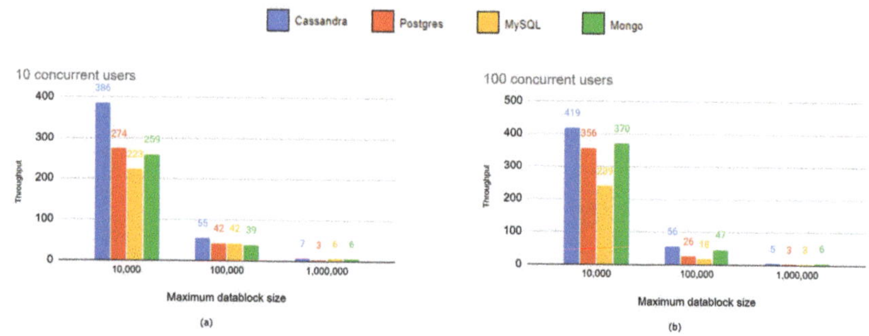

Figure 9. Comparison of the throughput of the different implementations with (**a**) 10 concurrent users and (**b**) 100 concurrent users.

5. Discussion

NoSQL databases are the primary storage option in Big Data applications. Our experiments with LiDAR datasets confirm in general terms the advantage of using NoSQL databases over relational ones, although the performance of MySQL is close or even superior in certain situations. Nevertheless, Cassandra is the clear winner, only beaten by MongoDB when the spatial data structure was organized in large datablocks (1 M). One possible explanation for this fact may be the superior performance of GridFS when handling larger LAZ files. Among all the experiments, the upload tests show less conclusive results, but in general terms, Cassandra and MySQL perform better with small or medium datablocks, and therefore, larger octrees, while MongoDB beats the rest of databases with large datablocks (i.e., smaller octrees). Regarding the storage space required, the two relational databases clearly perform the worst; most notably MySQL requires up to 20% more space than MongoDB when using small datablock sizes. This may be due to a more complex internal organization of the information or a wider use of indexes, which is necessary to be able to respond to more complex queries in the relational model.

In the simple requests test, Cassandra and MySQL showed the lowest latency with smaller and medium datablocks up to 25% better than MongoDB, although the latter was still overall the fastest database with a 1M datablock size. PostgreSQL is clearly the worst option—up to twice as slow as Cassandra in some experiments.

In the third series of experiments related to concurrent requests, Cassandra showed the highest throughput followed by MongoDB. Surprisingly, the MySQL performance was much less satisfactory in situations with concurrent requests than with single requests, coming in last position—even behind PostgreSQL. This is an important weakness of MySQL that may discourage its use in applications which need to support high concurrent workloads.

In summary, our recommendation for any project that requires storage in a database of large LiDAR datasets is Cassandra. Its performance is excellent in all operations. MongoDB can be an interesting alternative when Big Data chunks have to be stored in the database, thanks to the excellent performance of GridFS. Finally, if using relational databases is a requirement of the project, MySQL is the option of choice, although its performance under high concurrent workloads should be carefully observed.

6. Conclusions and Future Work

Databases are a versatile and robust option when storing LiDAR data. In this paper, we tried to shed some light on the most appropriate choice of database management system for this type of data. To this end, we compared the performance of four of the most popular NoSQL and relational database management systems in several areas. Our conclusion is that although NoSQL databases perform better than relational systems, the gap, particularly with MySQL, is narrow. Overall, Cassandra shines in all areas, only lagging behind MongoDB when datasets are split into large blocks.

Many approaches use a database only for storing dataset metadata or spatial indexes, keeping LiDAR data in external files. We store LiDAR data in the database because of its clear advantages: simplicity, guaranteed consistency and distributed storage when using database engines that support sharding. However, a further comparison of the performance of the in-database vs. file system storage would provide useful information for decision making when designing systems working with LiDAR data. The storage and processing of massive volumes of LiDAR data usually requires a distributed architecture. For this reason, a new experimentation similar to that carried out in this paper but using a cluster of database nodes would be relevant. However, the relational databases management systems analyzed are centralized, therefore additional solutions such as MySQL Cluster CGE would be necessary. The use of this extra infrastructure and the decisions made during the installation, configuration and tuning of the cluster and database nodes may influence the experimentation and limit the validity of the results.

Author Contributions: Conceptualization, A.J.R.-R., R.J.S.-S., C.J.O.-A.; methodology, A.J.R.-R., R.J.S.-S., C.J.O.-A.; software, J.A.B.-M.; validation, A.J.R.-R., R.J.S.-S., C.J.O.-A., A.L.-R.; formal analysis, C.J.O.-A., A.L.-R.; investigation, A.J.R.-R., R.J.S.-S., C.J.O.-A.; resources, J.A.B.-M.; data curation, J.A.B.-M., C.J.O.-A.; writing—original draft preparation, J.A.B.-M., R.J.S.-S., C.J.O.-A.; writing—review and editing, A.J.R.-R.; visualization, R.J.S.-S., C.J.O.-A.; supervision, A.J.R.-R., R.J.S.-S., C.J.O.-A.; project administration, R.J.S.-S.; funding acquisition, R.J.S.-S. All authors have read and agreed to the published version of the manuscript.

Funding: This result is part of the research project RTI2018-099638-B-I00 funded by MCIN/ AEI/ 10.13039/501100011033/ and ERDF funds "A way of doing Europe". Also, the work has been funded by the University of Jaén (via ERDF funds) through the research project 1265116/2020.

Data Availability Statement: The datasets used for the experimentation are publicly available from SITNA at the following URL: https://filescartografia.navarra.es/, accessed on 11 April 2022.

Conflicts of Interest: The authors declare no conflict of interest.

References

1. Leica Geosystems. ScanStation P40, P30 and P16 Comparison Chart. Available online: https://leica-geosystems.com/en-us/products/laser-scanners/ /media/00ac56bc2a93476b8fe3d7c1795040ac.ashx (accessed on 22 May 2022).
2. Li, S.; Dragicevic, S.; Anton, F.; Sester, M.; Winter, S.; Coltekin, A.; Pettit, C.; Jiang, B.; Haworth, J.; Stein, A.; et al. Geospatial Big Data handling theory and methods: A review and research challenges. *ISPRS J. Photogramm. Remote Sens.* **2015**, *115*, 119–133. [CrossRef]
3. Chen, R.; Xie, J. *Open Source Databases and their Spatial Extensions*; Springer: Berlin/Heidelberg, Germany, 2008; Volume 2, pp. 105–129. [CrossRef]
4. Rueda-Ruiz, A.J.; Ogáyar-Anguita, C.J.; Segura-Sánchez, R.J.; Béjar-Martos, J.A.; Delgado-Garcia, J. SPSLiDAR: Towards a multi-purpose repository for large scale LiDAR datasets. *Int. J. Geogr. Inf. Sci.* **2022**, *36*, 1–20. [CrossRef]
5. Poux, Florent. The Smart Point Cloud: Structuring 3D Intelligent Point Data. Ph.D. Thesis, Universite de Liege, Liege, Belgique, 2019. [CrossRef]
6. Bräunl, T. Lidar sensors. In *Robot Adventures in Python and C*; Springer: Cham, Switzerland, 2020; pp. 47–51. [CrossRef]
7. Evans, M.R.; Oliver, D.; Zhou, X.; Shekhar, S. Spatial Big Data. Case studies on volume, velocity, and varitety. In *Big Data: Techniques and Technologies in Geoinformatics*; CRC Press: Boca Raton, FL, USA, 2014.
8. Lee, J.G.; Kang, M. Geospatial Big Data: Challenges and opportunities. *Big Data Res.* **2015**, *2*, 74–81. [CrossRef]
9. Pääkkönen, P.; Pakkala, D. Reference architecture and classification of technologies, products and services for Big Data systems. *Big Data Res.* **2015**, *2*, 166–186. [CrossRef]
10. Deng, X.; Liu, P.; Liu, X.; Wang, R.; Zhang, Y.; He, J.; Yao, Y. Geospatial Big Data: Ew paradigm of remote sensing applications. *IEEE J. Sel. Top. Appl. Earth Obs. Remote Sens.* **2019**, *12*, 3841–3851. [CrossRef]
11. Sugimoto, K.; Cohen, R.A.; Tian, D.; Vetro, A. Trends in efficient representation of 3D point clouds. In Proceedings of the 2017 Asia-Pacific Signal and Information Processing Association Annual Summit and Conference (APSIPA ASC), Kuala Lumpur, Malaysia, 12–15 December 2017; pp. 364–369. [CrossRef]
12. Ullrich, A.; Pfennigbauer, M. Advances in LiDAR point cloud processing. In *Laser Radar Technology and Applications XXIV*; Turner, M.D., Kamerman, G.W., Eds.; SPIE: Baltimore, MD, USA, 2019; p. 19. [CrossRef]
13. Ma, X.; Liu, S.; Xia, Z.; Zhang, H.; Zeng, X.; Ouyang, W. Rethinking pseudo-LiDAR representation. In *Computer Vision—ECCV 2020*; Vedaldi, A., Bischof, H., Brox, T., Frahm, J.M., Eds.; Springer International Publishing: Cham, Switzerland, 2020; Volume 12358, pp. 311–327. [CrossRef]

14. Poux, F.; Billen, R. A Smart Point Cloud Infrastructure for intelligent environments. In *Laser Scanning: An Emerging Technology in Structural Engineering*; CRC Press: Boca Raton, FL, USA, 2019.
15. Scheiblauer, C.; Wimmer, M. Out-of-core selection and editing of huge point clouds. *Comput. Graph.* **2011**, *35*, 342–351. [CrossRef]
16. Baert, J.; Lagae, A.; Dutré, P. Out-of-core construction of sparse voxel octrees. *Comput. Graph. Forum* **2014**, *33*, 220–227. [CrossRef]
17. Richter, R.; Discher, S.; Döllner, J. Out-of-core visualization of classified 3D point clouds. In *3D Geoinformation Science*; Breunig, M., Al-Doori, M., Butwilowski, E., Kuper, P.V., Benner, J., Haefele, K.H., Eds.; Springer International Publishing: Cham, Switzerland, 2015; pp. 227–242. [CrossRef]
18. Deibe, D.; Amor, M.; Doallo, R. Supporting multi-resolution out-of-core rendering of massive LiDAR point clouds through non-redundant data structures. *Int. J. Geogr. Inf. Sci.* **2019**, *33*, 593–617. [CrossRef]
19. Schütz, M.; Ohrhallinger, S.; Wimmer, M. Fast out-of-core octree generation for massive point clouds. *Comput. Graph. Forum* **2020**, *39*, 155–167. [CrossRef]
20. Elseberg, J.; Borrmann, D.; Nüchter, A. One billion points in the cloud—An octree for efficient processing of 3D laser scans. *ISPRS J. Photogramm. Remote Sens.* **2013**, *76*, 76–88. [CrossRef]
21. Schön, B.; Mosa, A.S.M.; Laefer, D.F.; Bertolotto, M. Octree-based indexing for 3D point clouds within an Oracle Spatial DBMS. *Comput. Geosci.* **2013**, *51*, 430–438. [CrossRef]
22. Lu, B.; Wang, Q.; Li, A. Massive point cloud space management method based on octree-like encoding. *Arab. J. Sci. Eng.* **2019**, *44*, 9397–9411. [CrossRef]
23. Huang, L.; Wang, S.; Wong, K.; Liu, J.; Urtasun, R. OctSqueeze: Octree-structured entropy model for LiDAR compression. *arXiv* **2020**, arXiv: 2005.07178.
24. Goswami, P.; Erol, F.; Mukhi, R.; Pajarola, R.; Gobbetti, E. An efficient multi-resolution framework for high quality interactive rendering of massive point clouds using multi-way kd-trees. *Vis. Comput.* **2013**, *29*, 69–83. [CrossRef]
25. Gong, J.; Zhu, Q.; Zhong, R.; Zhang, Y.; Xie, X. An efficient point cloud management method based on a 3D R-tree. *Photogramm. Eng. Remote Sens.* **2012**, *78*, 373–381. [CrossRef]
26. Hongchao, M.; Wang, Z. Distributed data organization and parallel data retrieval methods for huge laser scanner point clouds. *Comput. Geosci.* **2011**, *37*, 193–201. [CrossRef]
27. Krämer, M.; Senner, I. A modular software architecture for processing of Big Geospatial Data in the cloud. *Comput. Graph.* **2015**, *49*, 69–81. [CrossRef]
28. Deibe, D.; Amor, M.; Doallo, R. Big Data storage technologies: A case study for web-based LiDAR visualization. In Proceedings of the 2018 IEEE International Conference on Big Data (Big Data), Seattle, WA, USA, 10–13 December 2018; pp. 3831–3840. [CrossRef]
29. Isenburg, M. LASzip. *Photogramm. Eng. Remote. Sens.* **2013**, *79*, 209–217. [CrossRef]
30. Bunting, P.; Armston, J.; Lucas, R.M.; Clewley, D. Sorted pulse data (SPD) library. Part I: A generic file format for LiDAR data from pulsed laser systems in terrestrial environments. *Comput. Geosci.* **2013**, *56*, 197–206. [CrossRef]
31. Cao, C.; Preda, M.; Zaharia, T. 3D point cloud compression: A survey. In Proceedings of the 24th International Conference on 3D Web Technology, Los Angeles, CA, USA, 26–28 July 2019; pp. 1–9. [CrossRef]
32. Boehm, J. File-centric organization of large LiDAR point clouds in a Big Data context. In Proceedings of the Workshop on Processing Large Geospatial Data, Dallas, TX, USA, 4 November 2014.
33. Pandey, R. *Performance Benchmarking and Comparison of Cloud-Based Databases MongoDB (NoSQL) vs. MySQL (Relational) Using YCSB*; National College of Ireland: Dublin, Ireland, 2020. [CrossRef]
34. Cura, R.; Perret, J.; Paparoditis, N. Point Cloud Server (PCS): Point clouds in-base management and processing. *ISPRS Ann. Photogramm. Remote Sens. Spat. Inf. Sci.* **2015**, *II-3/W5*, 531–539. [CrossRef]
35. Migrating Facebook to MySQL 8.0. Available online: https://engineering.fb.com/2021/07/22/data-infrastructure/mysql (accessed on 22 May 2022).
36. van Oosterom, P.; Martinez-Rubi, O.; Ivanova, M.; Horhammer, M.; Geringer, D.; Ravada, S.; Tijssen, T.; Kodde, M.; Gonçalves, R. Massive point cloud data management: Design, implementation and execution of a point cloud benchmark. *Comput. Graph.* **2015**, *49*, 92–125. [CrossRef]
37. Reitz, K. HTTP for Humans(TM)—Requests 2.25.1 Documentation. Available online: https://docs.python-requests.org/ (accessed on 9 March 2021).
38. Fernández, A. Loadtest 5.1.2. Available online: https://www.npmjs.com/package/loadtest (accessed on 9 March 2021).

remote sensing

Article

Comparative Approach of Unmanned Aerial Vehicle Restrictions in Controlled Airspaces

Stephen John McTegg, Fayez Tarsha Kurdi *, Shane Simmons and Zahra Gharineiat

School of Surveying and Built Environment, Faculty of Health, Engineering and Sciences, University of Southern Queensland, Springfield, QLD 4350, Australia; steve_mctegg@bigpond.com (S.J.M.); shane.simmons@usq.edu.au (S.S.); Zahra.gharineiat@usq.edu.au (Z.G.)

* Correspondence: fayez.tarshakurdi@usq.edu.au

Abstract: Recent public discourse regarding unmanned aerial vehicle (UAV) usage and regulation is focused around public privacy and safety. Most authorities have employed key guidelines and licensing procedures for piloting UAVs, however there is marginal consensus amongst regulators and a limited view towards unified procedures. This paper aims to analyze the key challenges that affect the use of UAVs and to determine if the current rules address those challenges. For this purpose: privacy, safety, security, public nuisance and trespass are tested. A set of criteria are developed to perform a comparative analysis against the existing UAV regulations to determine how they are meeting the specified criteria. Within this framework, five countries are selected: Australia, Canada, European Union (EU), United Kingdom (UK) and the United States of America (USA), with usage data and length of time between regulatory reviews ensuring any analysis is realized on updated protocols. The regulations of each country are then compared against the developed criteria. The findings show there are shortfalls with the majority of regulations failing to meet some criteria and the results confirm that key issues fail to be addressed. Finally, recommendations are suggested for filling the gaps in the regulations.

Keywords: unmanned aerial vehicle (UAV); drone; regulations; restrictions; privacy; safety

Citation: McTegg, S.J.; Tarsha Kurdi, F.; Simmons, S.; Gharineiat, Z. Comparative Approach of Unmanned Aerial Vehicle Restrictions in Controlled Airspaces. *Remote Sens.* **2022**, *14*, 822. https://doi.org/10.3390/rs14040822

Academic Editor: Eben Broadbent

Received: 27 December 2021
Accepted: 7 February 2022
Published: 9 February 2022

1. Introduction

The use of commercial and private unmanned aerial vehicles (UAVs) is becoming increasingly ubiquitous. In 2018, the number of remote pilot licenses in Australia had increased by 53% over the previous year [1]. In 2021, approximately 22,000 remote pilot licenses had been issued, averaging 300 new licenses per month [1].

With the increase in the number of UAVs, there comes a greater need for regulation to ensure the safety and privacy of the public. On 19 December 2018, two drones were flown close to the Gatwick Airport perimeter resulting in the airport being shut down for 33 h and causing over 140,000 travelers to be affected [2]. The incident was the first time a major airport was shut down due to a drone incursion and demonstrated how it could disrupt the operations of a major airport and air traffic routes. The case remains unsolved, but the incident focused political attention on drone operations.

As politicians and regulators aim to catch up with the increasing number of UAVs, researchers urge caution should be applied as regulation can both promote and suppress innovation [3], as UAV use and development needs surety to continue to be practicable. Within the European market, the two most prevalent issues regarding regulation are security and safety considerations [4]. However, as Europe contains multiple independent countries with non-uniform regulations, over-regulation can be the principal hurdle to surmount. In certain countries, the regulatory limitations imposed on the commercialization of the technology were "notorious" [5].

Public discourse regarding UAV usage and regulation centers around public privacy and safety, which regulations aim to address [3,6,7]. Whilst most jurisdictions have implemented key guidelines and licensing procedures for piloting UAVs, there is marginal consensus amongst regulatory bodies and a limited view towards a universal standard [8]. Identical UAVs operating in identical flight circumstances are bound by different guidelines and regulations depending on the country of operation. These regulations can differ significantly for flying height, flying proximity to people/buildings and weight classes. These regulatory differences add additional complexities for a person to navigate when planning and flying UAVs, whether it's for commercial or private use. Whilst UAV operators in Australia are bound by a single regulator [9], a person in the European Union (EU) could have to contend with and be well versed in a multitude of regulations encompassing numerous jurisdictions [10].

This paper aims to compare existing UAV regulations across a number of diverse jurisdictions to determine the key problems and issues that arise from the use of UAVs and develop a set of criteria to attempt to resolve problems identified and determine if existing regulations address these issues. By identifying if the key issues, such as privacy, safety and security are being addressed, recommendations will be made to mitigate any discrepancies found in the regulations. To conclude, the objectives of this research are as follows:

Determine the similarities and differences in existing regulations;
Define the key problems that arise from the use of UAVs;
Develop a set of criteria to resolve the key problems;
Using the criteria, examine if present restrictions address the key issues;
Provide recommendations addressing any shortfalls in the regulations.

Although this project seeks to research UAV regulations across a variety of jurisdictions, due to time and scope constraints only a selection of regulations have been analyzed. These are chosen based on UAV usage numbers, existing established UAV regulations and length of time between regulatory reviews. Countries with limited regulations or limited usage data available are excluded from the analysis. Other limitations are that although there is significant data available relating to the employment of UAVs by the military, the primary emphasis of this paper is on the commercial and private/recreational use of UAVs. Numerous countries are investigating the need for changes to their UAV guidelines with several jurisdictions in the process of transitioning to new regulations.

2. Literature Review

In this section, five points are investigated among the UAV guidelines which are privacy, safety, security, public nuisance and trespass.

2.1. Privacy

The primary concern with the increased use and development of UAVs is what Finn and Wright [11] describe as the potential for UAVs to infringe on the ethical and privacy rights of people as a threat to civil liberties. Nelson et al. [12] use Floridi's notions of privacy to ascribe tangibility to privacy and define private space as areas "free from sensory interference or intrusion, void of unwarranted interruptions", such as sound, touch and vision. Clarke [13] links the issue of privacy to UAVs and surveillance through his five dimensions of privacy (person, data, behavior, communication, personal experience), of which he attributes behavioral privacy and privacy of personal experience as being the most important as they incorporate "the interests that are most directly impinged upon by drone-base surveillance". Finn et al. [14] further defines Clarkes behavioral privacy as protection against disclosure of sensitive information, such as religious practices, sexual practices or political activities.

Australia's privacy regime is often out of date with current technology [15]. In this context, the question of the ability for UAVs to enter private property, travel unnoticed and record or live stream images and sounds creates significant opportunities for privacy breaches to enter the public domain. The privacy question comes down to how it will be regulated and finding the line between regulation that achieves the technological benefits without removing the right of recourse and remedy if an individuals' privacy has been invaded [15]. However, existing laws are either outdated or are lacking in definition and provide remedy only in limited circumstances [13].

2.2. Safety

Sanz et al. [7] define safety as the state in which the system is not in danger or at risk, free of injuries or losses. Clothier and Walker [16] state that for routine UAV procedures to be combined into the civilian airspace, UAV developers, operators and regulators must prove that the safety of UAVs is at a minimum, equal to conventionally piloted aircraft. In this context, Clarke and Moses [17] observe that there are challenges to achieving this standard due to low costs, limited safety features and the growing volume of UAVs that there will unavoidably be low standards of pilot performance and the usual high costs in detection, investigation and defining responsibility. There are three main factors that play a major role in UAV safety as follows:

Collision–Aircraft: Prior to existing regulations, both commercial and private UAV flights were able to fly close to airports, secure facilities and overpopulated areas [18]. This initial freedom posed a serious threat to safety and security. An increasing number of International Civil Aviation Organization (ICAO) members were becoming concerned with UAVs flying within close proximity to commercial aircraft which was proving hazardous to commercial aircraft [19]. Over half of all incidences concerning UAVs reported to ATSB were close encounters involving manned aircraft and half of those again involved high-capacity air transport aircraft [20]. ATSB [20] outlines a near encounter as when a UAV interrupts or is seen in the vicinity of another aircraft. Furthermore, they found that in 8% of cases, engine ingestion can be expected, and it is likely that when compared with bird ingestion, engine damage and engine shutdown is expected to be higher. This increase in damage was explained by UAVs being heavier and more rigid than the skeletal structure of a bird. At this stage, IATA [21] emphasizes that manned aircraft encounters with UAVs are the top safety risk to the aviation industry;

Collision–Terrain: Terrain collisions accounted for 26% of all reported UAV accidents between 2012 and 2017, with almost half of these events occurring from loss of control of the aircraft [20]. The collision with the terrain had the second highest incident occurrence out of all incidents. Of these accidents, 84% resulted in the UAV being significantly damaged or destroyed [20]. A terrain collision can occur when the operator fails to maintain a visual line of sight (VLOS) with the UAV and is unable to account for changes in elevation in the terrain or obstacles, such as trees and buildings;

Impact: The issue of direct physical UAV impact gives rise to potential harm to public safety. There are also indirect threats where impact can lead to fire or explosion [17]. Magister [22] links UAV design shape and injury biomechanics relating to the blunt ballistic impact of UAVs and found that the severity of injury in small UAVs (<15 kg) when operating at minimal airspeed was "less than serious" when contacted with a blunt section of the UAV. However, CASA [23] modeled human injury potential from impacts of small unmanned aircraft and determined that an abbreviated injury scale (AIS) of 3 as the highest acceptable injury allowed. They found UAV mass and velocity as well as the diameter of the UAV determined how severe the injury would be. They noted that for a 2 kg UAV travelling at 10 m/s for a head impact, it would cause a fractured skull when impacting with the flat side of the UAV. In a total loss of control scenario where the UAV falls, reaching its terminal velocity, any impact at such high speeds (> 30 m/s) would cause unacceptably severe injuries regardless of the weight of the UAV [23]. Consequently, certain design and operational criteria may reduce the risk of injury.

2.3. Security

The question of security relating to UAVs can be summarized into two major points:

Loss of Control: CASA [24] states that loss of control of a UAV may be sudden and recovery very difficult even for experienced remote pilots. Loss of control may happen through several means: hijacking, global navigation satellite system (GNSS) jamming and spoofing, hardware/software malfunction, electromagnetic interference, exceeding UAV limitations, malicious software and user error. In this context, several examples can be cited regarding loss of control. Shepard et al. [25] refer to an incident when a Central Intelligence Agency (CIA) surveillance drone was captured by Iranian forces by jamming the UAVs communication link forcing the UAV into autopilot mode using a predetermined global positioning system (GPS) guidance to return to its base in Afghanistan. The UAV was able to be commandeered by spoofing the UAV with new GPS coordinates causing it to land in Iranian territory. Commercial UAVs on the market today are open to hijacking or hacking by electronic interference. Skyjack is a UAV created by Samy Kamkar that will find vulnerable UAVs in the air, discover an open network and change its service set identifier (SSID), which eliminates any connected users [26]. Zhi et al. [27] discuss a hijack method that involves GPS spoofing by broadcasting a false location and time with a universal software radio peripheral (USRP), which leads to control over the target UAV. GPS signals are susceptible to interference (both intentional and unintentional) due to the low power they have at the earth's surface [26]. Arteaga et al. [26] note that radio-frequency communication can be exploited. They discuss a process called flooding where an adversary floods the wireless fidelity (WIFI) channel with information interrupting the communication on the channel. This is commonly referred to as denial of service (DoS). Electromagnetic interference may cause a loss of connection. In 2018, a UAV collided with a cruise ship near Fort Hill Wharf, Northern Territory, Australia, when the UAV lost signal. The UAV operator started return home procedures when the UAV deviated from its path and collided with the ship and was destroyed [20]. Finally, Maldrone is software that produces a backdoor in the targeted UAV software and waits for a reverse TCP connection, which once received allows an attacker to gain control over the now infected UAV [26];

Anonymity: UAVs can be purchased and flown anonymously. This creates security problems as it is difficult to track and identify an offending UAV pilot that breaks the law. Therefore, automatic dependent surveillance broadcast (ADS-B) transponders are used on aircraft and obtain their position via GNSS. This position is then broadcast to control towers and other aircraft which helps maintain safe self-separation while in the air. A leading drone manufacturer is installing ADS-B receivers in all drones above 250 g [28]. Another parts manufacturer has developed an ADS-B beacon measuring 50 mm × 50 mm and weighing 50 g. This allows it to be placed on smaller UAVs allowing their position to be tracked.

2.4. Public Nuisance

Noise and nuisance were found to be the primary issues regarding public nuisance and UAVs:

Noise: Noise causing environmental harm/nuisance negatively affects human health and wellbeing by interfering with recreational activities, sleep and relaxation [29]. UAVs are not quiet in their operation. Christian and Cabell [30] found that participants subjected to psychoacoustic testing had a systemic variance between the annoyance response produced by the noise of the UAVs and that of the road vehicles. Intaratep et al. [31] analyzed the acoustic properties of a popular brand of quadcopter UAV and found that the sound intensity at maximum thrust (80 dB) is equivalent to a freight train passing 15 m away. At this level, it may cause annoyance in populated areas subject to high drone usage. Christiansen et al. [32] measured the noise level of two multi-rotor UAVs at varying heights above and below the water to determine the negative impact on UAVs flying over marine animals. A mean value measured at a height of 10 m for each UAV was 80 and 81 dB re 20 μPa for the in-air recordings and 95 and 101 dB re 1 μPa for the underwater recordings. However, decibels measured in the air do not directly link to decibels measured underwater. Christiansen et al. [32] observed that the in-air dB values of UAVs flying at low altitudes will most likely have a disturbing effect on sea otters and pinnipeds when they were on land or with their heads protruded from the water;

Nuisance: During the 2017 fire season in Victoria, Australia, there were four instances where amateur UAV operators had engaged in nuisance behavior during fire suppression activities and standard operating procedure when unauthorized UAVs are flying over emergency areas is to ground the aerial fleet [19]. Pomeroy et al. [33] confirm that flights over both Gray and Harbor seals were variable, with individual variation amongst seals of the same species. Gray seals generally changed behavior from alert to moving at altitudes from 10–50 m and a lateral distance of 15 to 210 m whilst Harbor seals showed little reaction at a 30 m altitude. Nelson et al. [12] posit that as UAVs become more ubiquitous and people become more familiar with them, a decrease in concern for privacy will occur and the issues of UAV-bystander interaction will more likely shift towards nuisance rather than privacy.

2.5. Trespass

In Australia, New South Wales farmers had limited legal remedy when they've caught UAVs trespassing on their property [19]. Farmers suggest punitive monetary penalties for the lack of due diligence and potential harm caused by UAV operators. Additionally discussed by the committee is the use of geofencing to prevent access to private property by limiting the distance the UAV could travel from its pilot thus reducing the risk of trespass and privacy infringement. New South Wales, South Australian, Tasmanian, Victorian and Western Australian legislation states that there is no trespass or nuisance by aircraft flying over the property at a reasonable height as long as air navigation regulations are adhered to [34].

3. Methodology

To achieve our stated objectives, we pursued the following approach: Determine the countries for analysis, extract the UAV regulations from the selected countries, develop criteria from the identified issues found in the regulations and the examination of existing regulations compared against the prescribed issues using the developed criteria. The first three steps are detailed in this Section with the remaining step applied in Section 4.

3.1. Country Selection

To select five representative countries to analyze the UAV regulations, two criteria are used. The first one is the number of users and ascertains if there is a strong user base. The premise behind this question is that if a country has a large number of users, then the regulations will be robust as there would be a greater opportunity for issues to present themselves and be addressed. The second criterion determines if there has been recent UAV regulatory reform. This criterion ensures that any comparison performed will be realized on regulations that have been updated in parallel with any technological developments. Consequently, greater consideration is given to countries that are in the process of amending or transitioning to new regulations. In this context, three research segments were identified to direct how the user base will be determined which are:

Countries that are acknowledged in prior research and market analysis;
Estimated numbers of UAVs;
Estimated users including reported licensed and/or certified operators.

Table 1 sorts in descending order the country's most often identified in the literature [18,35,36].

From Table 1, it can be seen that Australia, Canada and USA are mentioned across the three references, with China and UK mentioned twice. The USA have the largest market size, is the most active country online and have the most documented flight operations per country. China and France follow in second and third place, respectively [36].

Table 1. Countries Most Mentioned in the Literature.

Choi-Fitzpatrick et al. [35]	Vela et al. [18]	GUAA [36]
USA	USA	USA
UK	Brazil	China
Australia	Australia	France
India	UK	Germany
Canada	Indonesia	Great Britain
China	Mexico	Australia
	South Africa	Japan
	Chile	Canada
	Colombia	Switzerland
		Korea

Concerning the estimated numbers of UAVs, Figure 1 illustrates the estimated numbers of UAVs of counties where UAVs are widespread [4]. The EU, USA, Germany, Canada, China and Australia have the largest documented numbers of UAVs reported.

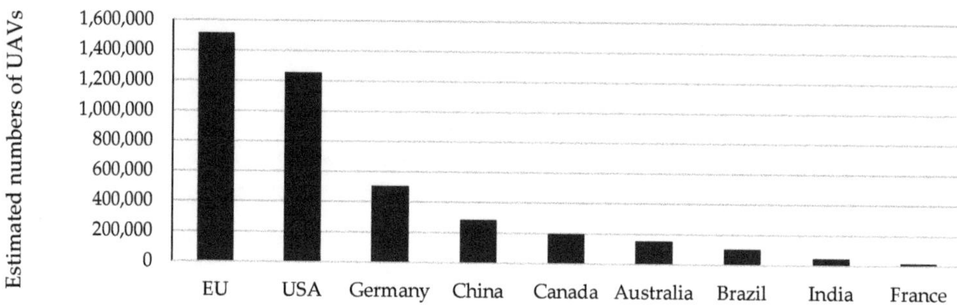

Figure 1. Estimated numbers of UAVs per county.

An overview of the estimated numbers of UAVs and recreational/commercial users from the most commonly mentioned countries from the previous studies is detailed below:

Australia—estimated 100,000 and 150,000 UAVs in the country and estimate of 50,000 recreational users and 1720 commercial users [9];

Brazil—34,000 civil UAS registered with 65% being for recreational purposes and 35% for commercial, estimated actual total numbers 100,000 [37];

Canada—337,468 UAVs in Canada 74% recreational and 26% commercial. 12% is the ratio between manned aircraft pilots between the US and Canada. The same ratio was used to compare the number of UAVs in the US to estimate the number of UAVs in Canada [38]. However, this number was revised down in 2018 to 193,500 estimated UAVs being flown in Canada by 140,800 operators [39];

China—24,407 certificates to fly were administered at the end on 2017, however certification is only needed if the UAV weighs over 7 kg [40];

European Union (macro-class)—1–1.5 million leisure UAVs and 10,000 commercial UAVs [41];

France—7471 referenced operators and 13,647 referenced UAVs in December 2018 [42];

Germany—500,000 drones in Germany. 455,000 are for private use and 19,000 for commercial use. Over 10,000 employed in the drone industry [36];

India—A rough calculation of 40,000 UAVs, predominately civilian, but including military and law enforcement UAVs as well [43];

United Kingdom—CAA estimated that during 2018 there would be 170,000 registered drone operators. These estimates were based on registrations in other countries such as the US and Ireland [9]. There were 5383 registered commercial UAV operators as of September 2019 [9];

United States—900,000 registered owners and 1.25 million estimated UAVs [44].

Finally, the estimated numbers of licensed operators are shown in Figure 2. The USA has the highest number of users by far, followed by the UK and Canada of the countries mentioned in Figure 2. The sources of statistics of Figure 2 are presented in Table 2. At this stage, the regulations that have not been amended or revised in the last three years will be excluded from the analysis. This ensures that comparisons are performed against regulations that are attempting to address recent developments in technology and user requirements.

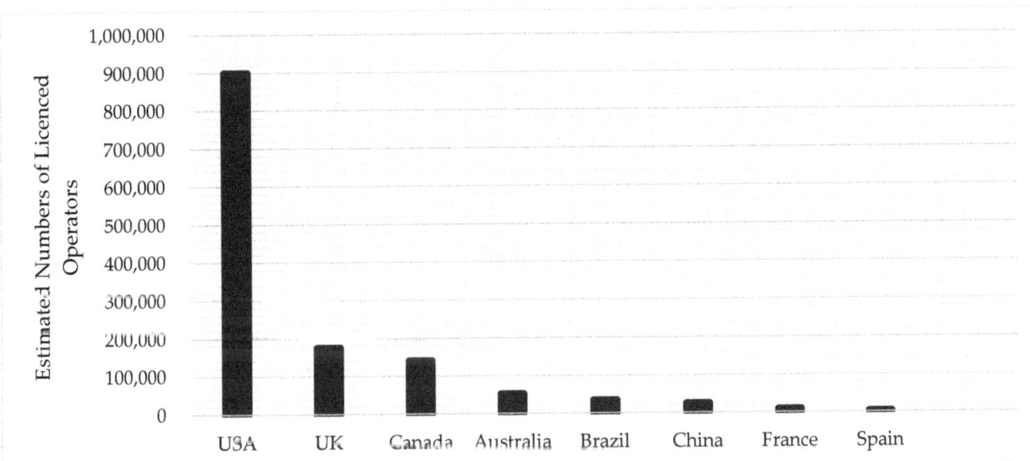

Figure 2. Estimated numbers of licensed operators per county.

Table 2. Sources of statistics in Figure 2.

USA	UK	Canada	Australia	Brazil	China	France	Spain
FAA [44]	CAA [9]	Canada Gazette [39]	CASA [1]	Unmanned Airspace [37]	Wangshu [40]	French Civil Aviation Authority [42]	Molina and Campos [4]

For further analysis and by intersecting the data shown in Table 1, Figures 1 and 2, the following five countries are selected:

- Australia—Large numbers of UAVs with a large user base and regulations recently updated;
- Canada—Large numbers of UAVs with a large user base and regulations recently updated;
- European Union—Currently in regulatory transition, encompasses a large population and numerous jurisdictions;
- United Kingdom—Large numbers of UAVs with a large user base and regulations recently updated;
- United States of America—The largest market of UAV operators with regulations recently updated.

3.2. UAV Regulations in Selected Countries

UAV regulations for the five chosen countries are summarized (see Appendix A). The primary source of information for the regulations is the respective government bodies which are available online. These regulations are collated by year of inception. This ensures that comparisons are performed against regulations that are attempting to address recent developments in technology and user requirements. The new European regulations will be included in the comparison because no other regulation will encompass such a wide variety of jurisdictions and a large population base.

At this stage, it is important to note that there is considerable variety of UAV restrictions between the selected countries, e.g., in the context of UAV security regulations in controlled airspaces, the regulations have stipulated guidelines regarding UAV usage near controlled or restricted airspaces. The shape of the restricted zones generally varies significantly across jurisdictions from radial to shapes incorporating overlays of circles, stadiums (oval shaped) and polygons with straight edges to encompass flight paths. For more details see Appendix A.

3.3. Comparison Criteria Development

Section 2 identified privacy, safety, security, public nuisance and trespass as the overarching challenges with UAVs use. Privacy proved to be one of the most researched issues with the primary concern being the ability for UAVs to infringe on people's privacy rights. Although no author outlined guidelines allowing for the practical avoidance of privacy infringement, details could be inferred from the inherent nature for UAVs to "enter private property, travel unnoticed and record or live stream images and sounds create significant opportunities for privacy breaches" [15].

Research surrounding UAV safety of course proved to be munificent in highlighting numerous challenges to UAV flight. Lack of operator knowledge was demonstrated to be the primary cause of most reported near encounters with the risk of collision being the consequence of this lack of knowledge with over half of near encounters involving UAVs. Researchers established that visual line of sight was difficult to quantify and demonstrated a formula enabling maximum VLOS to be determined using the width of the UAV whilst being predicated on perfect vision in perfect visibility. Impact could be tied in with collision but was separated as research was primarily focused on UAV design and specifications, such as weight and flight speed instead of operator interference.

Security issues were at the forefront of the research with terrorism and loss of control being of primary concerns. There are numerous ways an operator can lose control of a UAV with user error and pushing the aircraft beyond its operational limits being most likely, whilst others can be put down to a technologically sophisticated adversary or factors beyond control of the operator. Anonymity was discussed with the main risk being UAVs purchased and flown without the need for registration. However, major manufacturers are stepping in with hardware developments, such as ADS-B beacons being installed on UAVs above a certain weight class and now most manufacturers require registration prior to first flight.

Regarding public nuisance, UAV noise was a cause for concern with research stating that decibel readings taken close to the UAV were equivalent to a freight train passing 15 m away. NASA concluded that UAVs were between slightly and moderately annoying even up to 100 m away. Nuisance behavior created the greatest problems causing the cessation of emergency operations when UAV operators were flying within proximity to water bombers during firefighting operations or near rescue helicopters hindering rescues.

The following criteria were developed from the issues identified during the literature review (Section 2). The purpose of the criteria below is to interrogate the regulations and determine if and how they are addressing those issues:

- **Privacy**: The regulations shall address the privacy concerns of the general public or provide guidelines to reduce the risk of infringing on a person's right to privacy;
- **Safety**: Four major points are considered regarding the issue of safety. First, operators shall possess a minimum level of knowledge to operate the UAV safely to reduce the risk of injury to people or property. Knowledge could be demonstrated in the form of an online examination, accreditation or a pilot's certificate. Second, the regulations will contain guidelines reducing the risk of collision including onboard collision sensors and alarms. Third, maximum VLOS shall be defined as visual confirmation of the

UAV with the naked eye but not further than 400 m. This value can be calculated from Equation (1) [18].

$$D_{max} = \frac{W}{2.\arcsin\left(\frac{3\theta_{acc}}{2}\right)} \quad (1)$$

where W is the UAV width of 350 mm and θ_{acc} is the visual acuity of the pilot of 0.0167°.

Fourth, the regulations will contain limitations on the design of the UAV and the flight parameters reducing the risk of harm to people from direct impact. Weight classes and flying speed limitations shall be implemented:

- **Security**: The regulations will stipulate guidelines including limitations reducing the risk of losing control of the UAV. Furthermore, regulations have ensured that UAVs and operators are identifiable and have limited the ability for UAVs to be flown anonymously;
- **Public Nuisance**: The regulations will implement guidelines addressing the issue of noise pollution from the use of UAVs, such as a maximum decibel value from an environmental agency or placing limitations on flying times and distances. Moreover, the regulations will implement guidelines reducing the risk of nuisance behavior and specifically mention avoiding emergency personnel or placing limitations on flying times and distances;
- **Trespass**: The regulations will implement limitations and guidelines preventing the act of trespass;

In the next section, the regulations of the selected countries will be directly compared with the developed criteria by simple comparative analysis. The regulation relevant to each subheading of the designed criteria will be assessed against said criteria.

4. Results

This section aims to present the results of the comparison of the criteria formed from the issues identified in Section 3.3 and the regulations of the selected countries in Section 3.1. For each one of the selected countries, the five identified criteria will be analyzed. At this stage, in the context of safety criterion, four points will be emphasized which are a minimum level of knowledge, collision avoidance, VLOS and impact, whereas in the security criterion, two points will be emphasized which are the loss of control and anonymity. Finally, in the public nuisance criterion, noise and nuisance behavior will be investigated. Consequently, it will be determined if the regulations are meeting the developed criteria, and if so, it will be highlighted how by focusing on the differences and similarities between the regulations and the criteria.

4.1. Australia

Privacy: CASA [24] stipulates that it does not consider privacy concerns when issuing approvals. However, it does recommend that operators include privacy provisions in their operations manuals. Recreational users are advised to respect personal privacy and not record or photograph people without their consent. There are no guidelines within [41] Part 101 Unmanned Aircraft and Rockets detailing how pilots can reduce the risk of privacy violation;

Safety: Drone accreditation is needed if pilots are flying for fun (recreational users), flying over their own land or flying a UAV weighing less than 2 kg. Commercial users need a remotely piloted aircraft operator's certificate (ReOC) as well as a remote pilot's license (RePL) if flying outside the drone safety regulations or fly UAVs weighing more than 2 kg. There are no guidelines within [45] Part 101 Unmanned Aircraft and Rockets stipulating the need for collision sensors or alarms onboard UAVs. There are defined operational limits, such as maximum flying heights and distances to aerodromes. However, collision is only mentioned within the definition of operated within VLOS, which states that "a person has to be able to see the aircraft to uphold the operator's separation and collision avoidance responsibilities". Advisory circulars released by CASA guide how the regulations should be interpreted, e.g., all UAVs operating in the non-excluded (included) category should be painted for maximum visibility including collision avoidance lights such as strobe lights (101-01 v2.1 [46]). If ADS-B transponders are fitted, they should meet the required standards and be turned on.

Concerning the VLOS, subpart 101.074 (3) [46] of the regulations provides this definition of VLOS: "An unmanned aircraft is being operated within the VLOS of the person operating the aircraft if the person can continually see, orient and navigate the aircraft to meet the person's separation and collision dodge responsibilities, with or without corrective lenses, but without the use of binoculars, a telescope or other similar device" [46]. The definition provided by CASA gives guidance on how an operator would define the VLOS to their aircraft. It places the onus on the operator to determine the maximum VLOS without providing a quantitative value which would reduce the risk of operators flying outside their visual limits.

Finally, recreational users are limited to UAVs that weigh up to 2 kg. Commercial operators are bound by the same weight restrictions unless they hold a remote pilot's license and operate under a remotely piloted aircraft operators' certificate. The regulations provide no speed limitations on UAV flights and there is no mention of design parameters that would reduce the risk of harm from direct impact;

Security: Concerning the loss of control, there are no guidelines specifying procedures on mitigating the risk of losing control of a UAV ([47] Part 101 Unmanned aircraft and rockets). However, a brief overview of procedures for loss of control in an extended VLOS (EVLOS) flight is given [47]. The operator's procedures should ensure that the pilot can re-establish control or end the flight without causing a hazard to life or property [47]. Ref. [47] (AC 101-01 v2.1) discusses how losing control can be harmful to the public. It also details how mission plans should have procedures relating to how emergencies, such as loss of control would be handled by the operator and provide examples of how fail-safe devices would reduce the injury to bystanders. CASA also notes that data links between the UAV and the operator should be monitored in real-time with warnings given in case of failure.

Concerning anonymity, CASA is introducing registration for all UAVs flown for both commercial and recreational purposes. The primary purpose of the registration initiative is to ensure that people fly their UAVs responsibly and safely [1]. Apart from the ADS-B requirements for larger UAVs, no other de-anonymizing requirement was found in the regulations;

Public Nuisance: There are no guidelines within [45] specifying procedures on reducing the issue of noise pollution generated by UAVs. Both 101-01 v2.1 and 101-10 v1.3 in [46] and confirm that UAV operators are bound by local noise abatement laws. These include restrictions on altitude, flight path and the time of day of flights. Additionally, the regulations state that unless additional flight permissions are obtained, the operational hours for flying a UAV are during daylight hours as flights must be within VLOS.

Ref. [45] Part 101 unmanned aircraft and rockets specify that UAVs must not be flown over an emergency operation or public safety event without the approval of the person in charge. As noted above, the risk of nuisance behavior at night is reduced as most UAV operators can only fly during daylight hours as flights must remain within VLOS;

Trespass: There are no guidelines within [45] Part 101 Unmanned Aircraft and Rockets addressing the issue of trespass.

4.2. *Canada*

Privacy: Although not preserved within the Canadian Aviation Regulations [48], Transport Canada [49] provides clear privacy guidelines for both recreational and commercial UAV users. Operators are directed to a dedicated online resource specifically detailing the privacy guidelines for flying UAVs. This resource discusses that although Canadas's privacy regulations do not specifically reference UAVs, the privacy laws in Canada do apply to information that could be collected, such as video and pictures. Recreational users must abide by the previously mentioned five privacy principles, while commercial operators are bound by the Personal Information Protection and Electronic Documents Act (PIPEDA). PIPEDA applies to all businesses within Canada and ensures that consent is obtained when collecting personal information and that the information is handled with a high degree of professionalism;

Safety: To be allowed to fly a UAV in Canada with a MTOW between 250 g and 25 kg, recreational and commercial operators must obtain a pilot certificate with either basic or advanced endorsement. Concerning collision, general operating and flight rules specifically mentions the risk of collision and provides procedures and guidelines to reduce its risk. Pilots should not increase the risk of collisions by flying close to other aircraft (901.18 [48]). Take-off and landing sites must be suitable for the operation and that there is no risk of collision with aircraft, bystanders or obstacles (901.33 [48]) and if the risk of collision becomes too great then operators must cease any flights immediately (901.49 [48]).

Concerning the VLOS, (900.01 [48]) provides the following definition under the regulation for VLOS means unaided visual contact at all times with a remotely piloted aircraft that is adequate to be able to preserve control of the aircraft, know its location and be able to scan the airspace in which it is functioning to perform the detect and avoid functions in regard to other aircraft or objects". No pre-defined distance is given regarding the maximum visual line of sight. The responsibility is placed on the pilot to ensure that they believe they have control of the UAV at all times.

Finally, the main operating categories being basic and advanced have weight limitations on the types of UAVs that can be flown. UAVs weighing up to 25 kg can be flown providing the operators adhere to the regulations under each of those categories. Drones weighing over 25 kg can be flown but need special permission from Transport Canada [49] in the form of a Special Flight Operations Certificate. The regulations provide no speed limitations on UAV flights and there is no mention of design parameters that would reduce the risk of harm from direct impact;

Security: Loss of control is termed a "fly-away" within the regulations and defined as a loss of the command-and-control link between the operator and the UAV in which the operator is unable to control the UAV. An operator should not pilot a UAV unless they have procedures to handle emergencies, such as loss of the command-and-control link or a fly-away (901.23 (1) [48]). Pilots should cease operations if the UAV becomes uncontrollable (901.49 (1)(f) [48]). Minimal advice can be found within the regulations regarding how an operator would limit the risk of a fly-away.

Concerning anonymity, all UAVs weighing from 250 g and up to and including 25 kg need to be registered with Transport Canada. The operator's name and address, date of birth, purchase date, make, model, serial number, weight and type of UAV are recorded against a registration number. This registration number must be marked on the UAV with a permanent marker, permanent label or engraving. There are no other real-time identification requirements within the regulations;

Public Nuisance: First, there are no guidelines within [48] addressing the issue of noise pollution from the use of UAVs. Compounding the issue, night flights are permitted as well providing the UAV has position lights enabling it to be seen during the night either with or without night vision goggles worn by the operator (901.39 (1) [48]).

Second, A UAV must not be flown over or within an emergency security perimeter established by a public authority responding to an emergency (901.12 (1) [48]). As noted above, nuisance behavior is not limited to daylight hours as the regulations allow for UAV flights at night as VLOS can be established using positioning lights;

Trespass: There are no guidelines within [48] addressing the issue of trespass.

4.3. European Union

Privacy: Regulations have been established to address the risk of privacy infringement [50]. One of the main aims of the regulations is to mitigate the hazards around the protection of personal data and privacy [50]. If a UAV is capable of capturing personal data via a sensor (e.g., camera) attached to the UAV then the operator must be registered within a Member State (Article 14 (5)(a)(ii) [51]). A competent authority is tasked with maintaining a registration system for operators whose operations present a risk to privacy and personal data protection (Article 18 (m) [51]). A statement is required from the operator confirming that the planned operation will comply with all EU and national rules, in particular how it will address privacy and data protection (Article 12 (1)(c) [51]). Privacy is defined as one of the deciding factors when determining geographical areas where UAVs can operate (Article 15 (1) [51]). Knowledge regarding privacy and data protection is to be demonstrated via an examination delivered by a competent authority (UAS.OPEN.020 (4)(b) [51]). These regulations are aimed towards operator awareness and education rather than hard guidelines. Operators must demonstrate knowledge regarding privacy issues and the regulations have been constructed to reduce the risk of privacy breaches.

Safety: All operators wishing to operate UAVs with CE markings from C1 to C4 must pass an online exam. If operators want to fly close to bystanders, an additional theoretical exam must be taken and delivered by a recognized entity. Concerning collision, the only guideline in place within the regulations for reducing the risk of collision is (UAS.OPEN.060 (2) [50]). During the flight, the remote pilot shall keep the UAV in VLOS to reduce the risk of collision with manned aircraft with the flight being discontinued if there is an increase risk to aircraft, people, animals, environment or property. Any UAV operations in the certified category must abide by (no 1332/2011 [51]), which define operating procedures for airborne collision avoidance. These regulations include the need for airborne collision avoidance systems (AUR.ACAS.1005 (3) [51]) and what to do when they sound a collision alarm (AUR.ACAS.2005 (1) (2) [50]). These regulations, however, only apply to the certified category of flight operations which are deemed high risk and form a small portion of total flights. The European Union has defined five classes of UAVs based upon their weight (C0, C1, C2, C3, C4) with the C1, C2 and C3 class UAVs to be equipped with a geo-awareness system that allows for uploading and updating airspace limitations, warning alerts to the remote pilot of imminent airspace breach detections and also be able to alert the operator if the geo-awareness system is not functioning properly (Part 2–5 [52]). The C1, C2 and C3 UAVs are to be equipped with lights with the stated purpose being to aid in controlling the UAV and for increased visibility at night to able people to distinguish between manned aircraft and the UAV.

Concerning the VLOS, it is described as: "a type of UAS operation in which, the remote pilot is able to maintain continuous unaided visual contact with the unmanned aircraft, allowing the remote pilot to control the flight path of the unmanned aircraft in relation to other aircraft, people and obstacles for the purpose of avoiding collisions" (article 2 [51]). No pre-defined distance is given regarding the maximum VLOS. During a flight, the pilot shall maintain VLOS and consistently scan the surrounding airspace to avoid the risk of collision (UAS.OPEN.060 (2) [51]).

The C0 class UAV, the maximum take-off mass (MTOM) is less than 250 g and a maximum speed of 19 m/s (part 1 [52]). A C0 class UAV has to be designed to minimize damage to people from impact, such as no sharp corners and design limits on propellers to reduce injury from the fan blades. A C1 class UAV is to be made from materials and have physical and performance specifications that if an impact at its maximum speed (terminal velocity) to the human head exerts less than 80 J or has an MTOM of less than 900 g and a maximum speed limit of 19 m/s (Part 2 (1) [52]). A C2 class UAV must weigh less than 4 kg and by the nature of its design, limit the injury caused to people from impact by avoiding sharp edges and reducing the damage that can be caused by fan blades (Part 3 [49]). A C3 class UAV must be under 25 kg with Part 4 limiting the physical dimensions of the UAV to less than 3 m (Part 4 and 5 [52]);

Security: Concerning loss of control, guidelines are provided for how UAV manufacturers can provide solutions to mitigating the risk of losing control of the UAV unexpectedly (parts 1–5 [49]). The regulations stipulate that for classes C0–C4, the manufacturer of the UAV must place on the market a manual of operations that states how the UAV will behave during a loss of data link. There is also a requirement under each of the classes that when a loss of data link has occurred, there will be a reliable and predictable way to recover the data link or the flight will be terminated.

Concerning anonymity, the Commission Delegated Regulations [52] have provided clear procedures to reduce the possibility of operators flying UAVs anonymously. All UAVs with an MTOM of more than 250 g must have a direct remote identification system equipped on the UAV. Direct remote identification is defined as a system that broadcasts information locally about the UAV, such as the operator details and UAV specifications, without the need to physically access aircraft (Article 2 [51]). A UAV shall have a direct remote identification system that will periodically broadcast from the UAV, in real-time during the flight, on an open transmission protocol that can be received by existing mobile devices in broadcast range, the following information: operator registration number, unique physical serial number of the UAV, the geographical position, height above the surface and take-off point, the route that was taken by the UAV and the geographical position of the UAV [52];

Public Nuisance: The Commission Delegated Regulations state it is important to limit the noise emissions generated by UAVs to the greatest possible extent in order to provide the highest level of environmental protection [52]. The Noise Test Code establishes procedures for manufacturers to measure the noise generated from their UAVs, including microphone placement and operating conditions during the test (Part 13 [52]). It details the maximum sound power levels per class of UAV with those levels reducing over the course of two and four years to give manufacturers a grace period to adjust their UAV designs (Part 15 [52]). All operators are required to plan flights so as to minimize nuisances, including noise to people and animals (UAS.SPEC.050 (1)(a)(v) [51]). Whilst it is specified that operators should choose a UAV for the operation it is designed for to minimize noise and other emissions (UAS.SPEC.050 (1)(i)(iii) [51]).

UAV operators must not fly close to areas where an emergency situation is ongoing unless they have permission from the emergency services (UAS.SPEC.060 (3) [51]). Nuisance is mentioned throughout the regulations and is generally coupled with noise emissions requiring operators to plan flights to minimize nuisances (UAS.SPEC.050 (1)(a)(v) [51]);

Trespass: Neither the Commission Implementing Regulations [51] nor the Commission Delegated Regulations [52] provide guidelines regarding UAVs and trespass.

4.4. United Kingdom

Privacy: The Civil Aviation Authority (CAA) [9] states its duty is limited to safety and ensuring pilots are operating within the confines of their granted permissions. Their responsibility does not include concerns over privacy and directs people to the Information Commissioners Office as any privacy issues will not be dealt with by the UK CAA [9]. However, Air Navigation Order 2016 (ANO2016) [53] offers rules for UAV operators to avoid privacy issues by clearly delineating between UAVs that have surveillance capabilities and UAVs that do not. In this context, small unmanned surveillance aircraft are UAVs that can achieve surveillance and data collection [53]. These aircraft have greater flight restrictions placed on them with the regulations defining guidelines, such as keeping greater distances from people during take-off or landing and increasing the distance when flights are planned over or within congested areas and open-air assemblies. It can be inferred that the purpose for defining UAVs with surveillance capabilities and placing tighter flight restrictions on those aircraft, is to reduce the risk of the general public being surveilled and having their right to privacy invaded;

Safety: Currently, recreational users are not required to demonstrate a minimum level of knowledge and can fly UAVs legally following the regulations. However, from November 2019, recreational users will have to pass a "drone test" and register with the CAA. Guidelines are provided for collision avoidance with manned aircraft by restricting flights in certain airspaces involving the aerodrome traffic, proximity to aerodrome boundaries and restricting the flying height of the UAV (article 94(4) [53]). The UAV must be in constant unaided visual contact with operators to maintain their collision avoidance responsibilities (article 94(3) [53]). Sensors and collision alarms forming part of an onboard collision avoidance system of technical ability at least equivalent to manned aircraft specifications are only required on beyond VLOS (BVLOS) (see Appendix A) flights and are not required for recreational and most commercial applications [9].

Concerning VLOS, the UAV must be in constant unaided visual contact VLOS with the operator to maintain their collision avoidance responsibilities [53]. Extended VLOS operations are defined as flights that are performed beyond 500 m [54]. Therefore, it can be construed from the regulations that the maximum VLOS before entering into extended VLOS operations is up to 500 m.

Finally, two main operating categories are defined in the regulations, small unmanned aerial vehicles and unmanned aerial vehicles. Small unmanned aerial vehicles are defined as any unmanned aircraft weighing not more than 20 kg without its fuel [53]. Small UAVs are bound by the operating guidelines in [53], whereas UAVs weighing more than 20 kg around are subject to the entire UK aviation regulations. No speed guidelines are found in the regulations or design limitations to reduce the severity of the injury to a person from an impact with a UAV;

Security: Concerning loss of control, UAV operations in both segregated and non-segregated airspace must have procedures in place for emergency recovery after a loss of control data link [14]. Standard operating procedures should contain guidelines for loss of data link and abort procedures after a critical system failure. These recommendations are aimed at commercial operators flying in an air traffic service area. No loss of control procedures are found in the regulations for recreational users.

No procedures were found in the regulations to ensure UAVs and their operators are identifiable and unable to fly UAVs anonymously. CAA [9] proposes to introduce a registration scheme by November 2019 for all UAVs weighing over 250 g to be required to be registered with the CAA. Furthermore, operators will be registered instead of drones with the registration number of the operator to be applied to all UAVs flown by that operator [9];

Public Nuisance: No guidelines addressing the issue of noise pollution regarding the use of UAVs were found in ANO2016. It does not make decisions regarding whether an amount of noise would be annoying or damaging to people [9].

Additionally, no information was found within ANO2016 regarding nuisance flying, such as avoiding emergency personnel. Limitations on how close operators can fly to people are clearly defined in the regulations. Night-time VLOS flights are permitted provided the guidelines for VLOS are adhered to. This criterion may be satisfied by an alternative act outside the scope of this project [54];

Trespass: Operators must be aware of relevant trespass laws when conducting a flight and to obtain permission before entering or operating from private property [54].

4.5. United States of America

Privacy: No mention of guidelines addressing privacy concerns is found in Part 107 of the USA Federal Aviation Administration (FAA) regulations [44]. Section 357 of FAA Reauthorisation Act 2018 [55] states that it is the policy of the United States that UAVs shall be operated in a matter that respects and protects personal privacy in line with Federal, State and local law. Consideration was given to include privacy provisions [44]; however, given the FAA's longstanding mission and authority as a safety agency, it would be overreaching for the FAA to enact regulations concerning privacy rights [44];

Safety: No minimum level of knowledge is required under the regulations for recreation users to fly UAVs; however, the FAA is implementing an aeronautical knowledge and safety test and is currently developing a training module and exam in consultation with the industry. Commercial operators must pass a knowledge test as regulated under Part 107 which includes a multitude of topics regarding collision, UAVs are prohibited to fly at night and can only fly at civil twilight if fitted with anti-collision lights which are visible for 3 statute miles (part 107.29 [44]). All small unmanned aircraft are required to yield to all aircraft and that no one should create a collision hazard by flying a UAV too close to another aircraft (107.37 [44]). UAVs must not interfere with the operations of any airports with (107.41 [44]) declaring classes of prohibited airspace (107.43 [44]). No reference has been referred in the regulations necessitating collision sensors or collision alarms to be equipped for UAVs.

Concerning the VLOS, it is defined as with vision unaided by any device other than corrective lenses the operator of the UAV must be able to see the UAV throughout the entire flight (107.31 [44]). No pre-defined distance is assumed regarding the maximum VLOS.

Finally, the ground speed is limited of small UAVs (55 lbs/25 kg) to 44 m/s (107.51 [44]). No design limitations to reduce the severity of the injury to a person from an impact with a UAV were present in the regulations;

Security: Concerning the loss of control, no loss of control guidelines are found in the regulations. During an in-flight emergency, the operator can deviate from any rule necessary to meet the emergency (107.21 [44]). Second, all UAVs flown either recreationally or commercially have to be registered and the UAV noted with the registration number by engraving, permanent label or permanent marker. There are no other real-time, in-flight identification requirements within the regulations;

Public Nuisance: First, no guidelines addressing the issue of noise pollution were found in the regulations. Second, operators are advised not to fly near emergencies, such as accident response, firefighting and hurricane recovery;

Trespass: There are no guidelines within Part 107 or FAA Reauthorization Act 2018 addressing the issue of UAVs and trespass.

The comparative analysis performed above provides a detailed insight into how current regulations are addressing the privacy, safety and security concerns of the general public. As was expected and noted during the research phase of this project, not all criteria were met by the regulations. This is discussed at length in the following section.

5. Discussion

This section discusses whether the regulations have met the prescribed criteria developed in Section 3.3. The differences and similarities between the regulations and criteria are highlighted.

5.1. Privacy

Whether the regulations addressed the privacy concerns of the general public proved to be a complex question. There is no common theme running through each of the regulations tying into what the general public will see as a simple remit, protect a person's right to privacy. The regulations traverse the full breadth of the issue, from deflecting the issue of privacy to another authority, to regulations whose primary aim is to address privacy issues. Only Canada and the EU have attempted to guide operators to uphold privacy rights. From the regulations of the five countries that were analyzed, all five were administered by safety authorities. Three of the regulators state that privacy is not part of their responsibilities. However, two of these regulators include privacy provisions in their regulations with one delineating between surveillance and non-surveillance UAVs and the other providing a privacy policy. One of the countries provide robust privacy guidelines external to the regulations with detailed information regarding operator responsibilities and applicable acts. Two countries place restrictions on UAVs that are capable of surveillance with only one stating these restrictions are for privacy purposes.

Australian regulations provide limited guidance on reducing the risk of infringing on people's privacy, e.g., there are no additional limitations placed on UAVs equipped with surveillance equipment, no requirement for real-time inflight identification of the operator and UAV and no requirement for operators to demonstrate privacy knowledge. Operators look to the regulator seeking guidance on how to navigate the privacy issues but find limited information and are advised external to the regulations to respect personal privacy and not record or photograph people without their consent. Newly designed regulations have provided an opportune time to address the concerns of the general public. Through consultation with the Office of the Australian Information Commissioner [56], privacy guidelines could have been developed that would put them in line with their equivalent safety authorities in other countries that have been successful in addressing privacy concerns.

Canada also does not provide privacy guidelines within their regulations. However, Transport Canada [50] has given guidance to recreational users by providing an online reference that explains how users can apply privacy rules to the flying of UAVs. Commercial users are directed to the relevant privacy information and given a brief description of how they can protect people's privacy. The EU is implementing the most robust and progressive privacy guidelines of any regulator. Operator registration is needed for flying UAVs capable of performing surveillance, flight planning must take into account privacy concerns and operators must demonstrate knowledge of privacy and data protection. Additionally, operators will be able to be identified in real-time during flights, reducing the ability for operators to infringe on privacy rights anonymously. Although in the UK, CAA considers that its responsibility does not concern privacy, it has segregated UAVs that can conduct surveillance from those that cannot and included greater flight restrictions distancing UAVs from bystanders. However, distance does not negate the ability of sensors to capture and store private information and data. CAA underlines that privacy issues should be directed to the Information Commissioners Office or local police.

The USA provides a privacy policy within its regulations stating that UAV flights should protect and respect personal privacy consistent with law, but provides no guidelines directing operators on how those flights can stay within those legal boundaries. UAVs and privacy ultimately are a multifaceted complex issue that is far beyond the scope of this paper. Although the use of UAVs is regulated primarily by safety authorities, no other organization is better positioned to provide guidelines for the protection of privacy as these safety authorities have extensive knowledge of the minutia of UAVs and their applications.

5.2. Safety

All regulations surveyed require operators to have some form of prerequisite knowledge demonstrated before operating UAVs. The depth of this knowledge varied greatly, however, with some regulators only requiring a basic understanding of general safety, while other regulators are expecting broader knowledge on topics, such as meteorology, navigation and air law. All regulations in the comparison have limitations in place on how close UAVs can be flown to aerodromes to reduce the risk of collision. Equally, height restrictions are in place limiting the risk of UAV incursion into regulated airspace and therefore collision as well. Regarding the implementation of collision sensors and alarms, including broadcast beacons, this is only a requirement on larger UAVs above certain weight/size class or operations performed BVLOS for all regulations.

All regulations analyzed include a clear definition of VLOS based on continuous unaided visual contact with the UAV at all times. However, within their definitions, no regulator has set a maximum distance that operations can be performed and still be within VLOS. The regulators have put the onus onto the operators to use their best judgement without providing a best-case scenario distance limitation. The UK regulations come close to defining a limit stating that EVLOS is either within or beyond 500 m. Considering that within most regulations the transition of the line of sight is VLOS–EVLOS–BVLOS, then under the UK regulations it will then be assumed that the boundary between VLOS and EVLOS is 500 m. All regulations included in the analysis have weight class restrictions. Once again, these weight classes vary throughout the regulations with some regulators limiting users to UAVs with a maximum weight of 2 kg, whilst a majority of the regulations allowed up to 20–25 kg. Australia and the EU focus on keeping UAV weight classes low, with Australia limiting UAVs to 2 kg without the need for a license and certification whilst the EU requires UAVs weighing up to 2 kg to keep 50 m clear of people and a 150 m for

UAVs weighing more up to 25 kg. The regulations of the remaining countries allow UAVs up to 20–25 kg which rise the risk of injury and falls outside the recommended weights (see Section 4).

UAV speed is only regulated in two countries, the USA and the EU. The USA regulations have limited all small UAVs (<25 kg) to 44 m/s and the EU have stated that C0 class under 250 g and C1 class under 900 g must be kept under 19 m/s. These speeds are outside the recommendations put forth in the literature. The EU regulators have set precise design constraints for each class of UAV. The design specifications cover maximum energy levels when impacting the human body and travelling at maximum speed. No sharp edges are permitted on the UAV and propeller blades must be designed to limit injury. No other regulations analyzed have applied limitations to UAV design apart from weight restrictions.

5.3. Security

No analyzed jurisdiction provides guidelines on preventing the loss of control of the UAV during a flight within their regulations. The responsibility is predominantly placed on the operator to develop detailed protocols to be included in their standard operating procedures. All regulators require either the operator or the UAV to be registered with a competent authority allowing for the possible identification of the operator or UAV. The EU goes even further and requires all UAVs weighing greater than 250 g to have a direct remote identification system onboard that will allow the operators details, geographical location of the UAV and operator and the route taken during the flight, to be accessible by a mobile device without the need to access the UAV. This system is a considerable step forward in eliminating the possibility of UAVs being flown anonymously and could solve numerous other issues, such as invasion of privacy, ensuring safety regulations are being adhered to and determining if a UAV has flown over private property without permission.

Finally, all analyzed jurisdictions have controlled or restricted airspace zone shapes that may be either radial or a stadium shape and can include straight edge protrusions to protect runway flight paths. Appendix A shows the various configurations for the combination shape, Table 3 outlines the general provision for the shape criteria of each jurisdiction.

Table 3. General class for controlled or restricted airspace zone shape criteria of each jurisdiction (Author note: USA controlled airspace zones are predominantly radial and can be individually tailored polygons surrounding an airport).

	Australia	UK	USA	Canada	EU
Radial	X	X	X	✓	X
Combination–Radial/Stadium curved shapes with Straight edge protected flight paths	✓	✓	X	X	X
Variable–radial and/or combination and/or polygonal shapes depending upon airport or jurisdiction	X	X	✓	X	✓

✓: Criteria is considered; X: Criteria is not considered.

5.4. Public Nuisance

Only the EU regulations provide guidelines addressing the issue of noise pollution. The regulators state that noise emissions generated by UAVs must be limited to the greatest possible extent and provide a noise test code within the regulations. The noise test code aims to direct manufacturers on how to measure the noise generated by their UAVs to meet the maximum sound power levels laid out in the regulations. Operators are also required to plan their flights to minimize nuisance from noise pollution. No other regulations analyzed included noise pollution reduction guidelines. Australian regulators directed the operators to local noise reduction laws, whilst Canada and the UK go the other way and allow night flights provided safety regulations are adhered to. These night flights can increase the risk of nuisance behavior through the interruption of peoples sleep. All restrictions provide relief to emergency personnel from the incursion of UAVs into airspace near emergency procedures or public safety events by prohibiting flights unless authorization is granted from the person in charge. Throughout the regulations, nuisance behavior is also tied to noise pollution requiring operators to plan flights to reduce nuisances.

5.5. Trespass

No analyzed jurisdiction tries to address rules regarding the issue of trespass. In the next section, recommendations are suggested to address the regulatory shortfalls and current gaps for meeting the concerns of stakeholders.

6. Recommendations

From the comparative analysis, it is clear that deficiencies have been identified in the regulations. Existing investigations aim to identify and address these deficiencies. However, there still appears to be marginal consensus which is evidenced when the regulations are positioned against the prescribed criteria. What is apparent is regulators within their distinct regulatory environments have attempted to amalgamate the necessary guidelines and relevant acts making it easier for operators to identify information relevant to their area of interest. Some regulations have still proven to be either ambiguous in their guidelines placing the burden on the operator to decide on the most appropriate course of action, or have become overly complex when trying to address the plethora of issues that present when dealing with what is ultimately an extremely complicated matter. It is necessary to provide a clear mandate including procedures and guidelines on how to mitigate the risk of unmanned aerial vehicles infringing on the privacy rights of people. The following recommendations aim to address the shortfall within the regulations and provide a basis for future investigations:

- Specify guidelines and procedures centered around sound privacy principles informing unmanned aerial vehicle operators of their obligations to protect a persons' right to privacy;
- Ensure all operators attain a minimum level of knowledge regardless of maximum take-off mass (MTOM) and usage;
- Existing requirements for demonstrating a minimum level of knowledge shall include topics beyond the scope of safety and shall include a broader level of aeronautical knowledge;
- Set a maximum VLOS for a UAV under 350 mm in diameter to no greater than 400 m in accordance with existing visual acuity research to reduce the risk of collision and losing control of the aircraft;
- Limit aircraft MTOM to no greater than 2 kg and limit airspeed to 7.5 m/s in accordance with existing research thus reducing the potential for impact injury;
- Place design restraints on UAVs limiting sharp edges, increasing large curves and implementing frangible parts to absorb impact loads;
- Mandate clear loss of control protocols and procedures by incorporating manufacturing and design input and provide guidance on how to regain control of the UAV including reference to how interference can affect flight control;
- Require all aircraft regardless of MTOM and usage to incorporate direct remote identification allowing real-time identification of the operator and UAV during flights reducing the risk of privacy, safety and security infringements including trespass;
- Specify an upper limit on the noise generated by UAVs in accordance with existing environmental protection guidelines and reduce noise pollution that would otherwise cause harm/nuisance and negatively affect human health.

7. Conclusions

This paper aimed to examine how the current unmanned aerial vehicle regulations are addressing the challenges and issues affecting the use of UAVs. From these issues, criteria were created that enabled a comparative analysis between the criteria and the regulations to gain insight into whether the regulations were addressing the problems identified. Initial research aimed to clarify what those challenges were and found that the primary issues were centered on privacy, safety and security. There was little consensus amongst regulatory bodies with the regulations differing greatly between countries. Privacy issues were a primary concern and although it has been extensively investigated, UAVs and their privacy implications were tethered only lightly to privacy law. Concerning safety, a lack of operator knowledge was the main driver in the increase of near encounters as people are unaware of the regulations or unfamiliar with the safety risks. Loss of control and anonymity were the primary impediments found in the security analysis. Usage statistics, existing regulations and length of time between regulatory reviews determined Australia, Canada, the EU, the UK and the USA were chosen for analyses. Criteria were then developed from the issues found in the literature review which included five themes: privacy, safety, security, public nuisance and trespass. The regulations of the five countries were analyzed and compared against the developed criteria which demonstrated a shortfall in the regulations with all regulations failing to meet some of the criteria. These results confirm that although there have been new regulations developed, privacy, safety and security are still issues needing attention. Trespass proved to be a complex issue that ultimately fell outside the scope of this paper and is included for investigation's sake. Further investigations are requested to define if and how UAVs can trespass. We have not conducted a statistical analysis of measures regulated by relevant authorities and all the information and data provided in the manuscript is directly sourced from the regulatory bodies of the selected countries and is outside the scope of this paper. Hence, quantitative data and statistical analysis of measures would be an excellent subject for a future paper and would represent a further contribution to knowledge. Finally, the recommendations were suggested to act as a guide for filling the gaps found in UAV regulations and should not be used as a reference for training of operators or issuing licenses.

Author Contributions: Conceptualization: S.J.M., S.S.; Methodology: S.J.M., S.S.; Software: Z.G., F.T.K.; Validation: S.J.M., S.S., F.T.K., Z.G.; Formal analysis: S.J.M., S.S., F.T.K., Z.G.; Investigation: S.J.M., S.S., F.T.K., Z.G.; Resources: S.J.M., S.S., F.T.K., Z.G.; Data curation: S.J.M., S.S., F.T.K., Z.G.; Writing—original draft preparation: F.T.K.; Writing—review and editing: S.J.M., S.S., F.T.K., Z.G.; Visualization: S.J.M., S.S., F.T.K., Z.G.; Supervision: S.S., Z.G. All authors have read and agreed to the published version of the manuscript.

Funding: This research received no external funding.

Institutional Review Board Statement: Not applicable.

Informed Consent Statement: Not applicable.

Data Availability Statement: The sources of data supporting reported results are cited accurately and can be found in the reference section.

Conflicts of Interest: The authors declare no conflict of interest.

Appendix A

UAV Regulations in five countries

This appendix summarizes the UAV regulations in five countries which are: Australia, Canada, European Union (EU), United Kingdom (UK) and United States (USA).

Appendix A.1. Australia

The CASA is the principal government body charged with regulating the flying of UAVs in Australia. The legislative instrument used to regulate the flying of UAVs is the Civil Aviation Safety Regulations Part 101 (Unmanned Aircraft and Rockets) Manual of Standards 2019. The information below provides a brief summary of the regulations applicable to recreation and commercial users [47].

Recreational users follow the standard operating conditions, which are designed to protect the operator and the people around them. They include only flying one UAV at a time and only flying within VLOS. Flights must remain under 120 m (400 ft) above ground level and no closer than 30 m to people and not over or above people at any time. Operators must not fly near emergency situations, is prohibited or restricted airspace and no closer than 5.5 km to a controlled aerodrome or airfield. Recreational UAVs are limited to a maximum weight of 2 kg.

Commercial users operating UAVs under 2 kg can operate under an excluded category provided they adhere to the standard operating conditions, apply for an aviation reference number and notify CASA prior to completing the flight. Commercial users operating UAVs over 2 kg must have a Remote Pilot License (RePL) and a Remotely piloted aircraft Operator's Certificate (ReOC) or be working for a ReOC holder. ReOC's and RePL's enable pilots to fly outside the standard operating conditions, such as follows:

Flying closer than 30 m to people: Closer than 30 m but not closer than 15 m providing that the UAV has dual parallel redundant battery system with duplicated battery mounting and able to fly with one motor inoperative at the MTOW. Return home functions must be operational with at least 7 GNSS satellites. A risk assessment must be performed with all identified risks appropriately mitigated and consent from all people located within 30 m of the UAV. Written consent is preferred but not mandatory;

Area Approvals and Permissions: Flying 120 m above ground level in or within 5.5 km of a controlled and non-controlled airspace. Moreover, flying over or within 5.5 km of a controlled and non-controlled aerodrome or movement area;

Extended VLOS (EVLOS): A risk assessment must be performed with all identified risks appropriately mitigated prior to application. All areas of the operational area must always be under supervision from an observer. Either the pilot or an observer must always have direct VLOS to the UAV. Both pilot and observer need CASA approval to conduct EVLOS;

BVLOS: A risk assessment must be performed with all identified risks appropriately mitigated prior to application. All flights must be conducted to the same level of safety as manned flights focusing on aircraft controllability, fail-safe mechanisms, collision avoidance and navigational and height accuracy. The UAV must be equipped with position lights, anti-collision/strobe lights and landing lights, transponders, such as an ADS-B unit, navigation equipment and aeronautical radio.

Flight restrictions around aerodromes: According to the Australian Government, Federal Register of Legislation Civil Aviation Safety Regulations 1998 [45], Part 101 (Unmanned Aircraft and Rockets) Manual of Standards 2019 [47] (as amended): No-fly zone of a controlled aerodrome means any areas and airspace that are below 400 ft and:

Subject to this section, a person must not: (a) conduct Remotely Piloted Aircraft (RPA) operations; or (b) Fly an RPA; in the no-fly zone of a controlled aerodrome.

A person who is: (a) A certified RPA operator, or (b) The remote pilot of a certified RPA operator; may conduct, or fly as the remote pilot in, an RPA operation in the no-fly zone of a controlled aerodrome if the operation is a tethered operation in accordance with Section 4.4.

Approach and departure paths—controlled aerodromes:

1. Figure A1 shows the approach and departure paths of a controlled aerodrome.

   As shown in Figure A1, the approach and departure path are up to 400 ft, as follows:

(a) Anywhere on or from the ground upwards in the area that is the runway or the runway strip;
(b) Anywhere in the following areas which are the approach and departure paths for the controlled aerodrome: (i) Subject to subparagraph; (ii) On or from the ground upwards in the area that is shaded black; (A) To a distance of 7 km from the end of the runway strip; (B) To a width that is initially 1 km until the splay exceeds 1 km, and then to the width of the splay up to 3.85 km; (iii) anywhere from 300 ft (90 m) above the ground (referenced to the aerodrome elevation) in the area that is between 7 km

and 8.5 km from the end of the runway strip, with an initial splay width of 3.85 km and a final splay width of 4.65 km (the area that is crosshatched);

(c) Anywhere from 150 ft (45 m) above the ground (referenced to the aerodrome elevation) in the area that is shaded grey.

The area that is shaded black, which shows the approach and departure paths and the ground below them, is described as follows:

(a) Symmetrical trapezoids with the shorter side coincident with the ends of a nominal 100 m wide runway strip and extending out at an angle of 15 degrees on either side to a distance of 8.5 km, the width of the splay at that distance being no greater than 3.85 km;

(b) A rectangle extending 500 m on either side of the runway centerline and overlying the runway strip until it intersects the trapezoids at a distance of approximately 1.68 km from the end of the runway strip.

The area that is shaded grey is described as the racetrack shape comprised of two semi-circles each: (a) with a radius of 4 km from the point on the centerline at each end of the runway in the direction of the closest threshold (point 1); (b) ending at the point that is perpendicular to point 1; and (c) extending in lines parallel to the centerline until the lines extended from one semi-circle meet the lines extended from the other semi-circle.

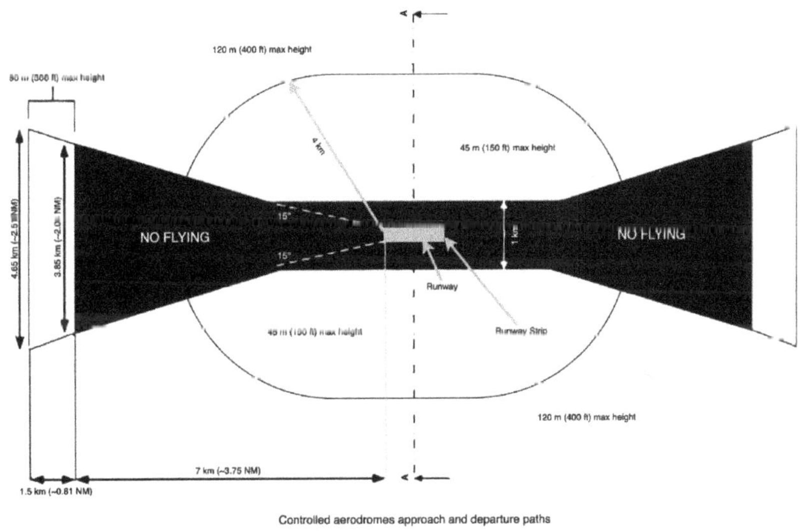

Figure A1. Controlled aerodromes approach and departure paths, Figure reproduced courtesy of the copyright holder, Australian Government 2021.

Appendix A.2. Canada

Transport Canada is responsible for the development of aviation regulations in Canada. In 2019, Transport Canada published updated regulations for flying UAVs with the new regulations apply to UAVs that weigh from 0.25 kg to 25 kg [49]. All UAVs must be registered and marked with a registration number and all pilots are required to pass an online exam to be awarded a pilot certificate in their operational category of choice. Two operational categories were introduced: basic and advanced. Commercial and recreational pilots are not treated differently and what defines which category is weight, distance from bystanders and the airspace rules for the area the UAV will be flown. The information below provides a brief summary of the regulations applicable to recreation and commercial users.

Basic operational category:

- Must be over the age of 14 or under supervision of a person who is;
- Holds a "basic" pilot certificate;
- Fly in uncontrolled airspace;
- Fly more than 30 m horizontally from bystanders;
- Don't fly over bystanders.

Advanced operational category:

- Must be over 16 years of age or under supervision of a person who is;
- Holds an "advanced" pilot certificate;
- Fly within controlled airspace;
- Fly over bystanders;
- Fly closer than 30 m to bystanders but not less than 5 m.

Any flights that deviate from the basic category automatically fall under the advanced category. Additional rules apply to the advanced category depending on which condition will not be met, such as seeking permission from air traffic control to fly in controlled. Standard guidelines, such as flying within VLOS, below 122 m and no closer than 30 m apply to all flights, as well as avoiding emergency operations (bushfires) and advertised events (parades), 5.6 km from airports and 1.9 km from heliports, as well as far away from other aircraft, such as other UAVs, helicopters and aeroplanes. UAVs that weigh under 0.25 kg do not fall under the new guidelines however they must fly within VLOS and always fly responsibly. UAVs that weigh more than 25 kg must obtain special permission from Transport Canada and apply for a Special Flight Operations Certificate (SFOC). A SFOC is also required for:

- Flights BVLOS;
- Flights by a foreign operator or a pilot authorized to fly UAVs by a foreign state;
- Flights above 122 m;
- Operating more than five UAVs from a single control station;
- Flying over an advertised event;
- Transporting payloads.

Transport Canada [50] provides privacy guidelines for both recreational and commercial UAV users. Recreational users are advised to follow five privacy principles:

- Be accountable: the pilot is responsible for all personal information collected during flights;
- Limit collection: take steps to avoid blanket collection of information and only record data that is needed. Anonymizing data, such as blurring faces and number plates is suggested;
- Obtain consent: yake all reasonable steps to obtain consent from people who will be incorporated into the capture area;
- Store information securely: prohibit access to data that may contain personal information;
- Be open and responsive about your activities: respect the rights of others especially if people complain that flights are infringing on their privacy.

Commercial users, like all other businesses in Canada, are bound by the PIPEDA. Consent must be obtained when collecting, using and sharing personal information and for that consent to be valid people must comprehend the consequences of consenting to the collection of personal information. The Privacy Commissioner for Canada outlines an additional five steps on top of the principles guiding recreational users [57] as follows:

- Identify the purpose of the data collection;
- Data collected must only be used for the purpose it was collected;
- Personal information collected must be accurate, complete and up to date;
- Individuals must be informed of the use and disclosure of their personal information and have the right to access the information;
- The privacy principles of the organization can be challenged by an individual ensuring compliance.

Flight restrictions around aerodromes: According to the Transport Canada [49], a pilot may not operate an RPA in controlled airspace unless he or she has received a written RPAs Flight Authorization from the Air Navigation Service Provider (ANSP) (901.71(1) [48]). Advanced Operations are for those intending to operate an RPA (901.62 [48]):

- Controlled airspace;
- Near people (horizontally less than 30 m, up to 5 m);
- Over people (horizontally less than 5 m over people);
- Within 3 NM from the center of an airport or a military aerodrome; (5.556 km);
- Within 1 NM from the center of a heliport (not runway based–where the center of the airport is, does not include flight paths).

Appendix A.3. European Union

In June 2019, the European Union Aviation Safety Agency [50] published preliminary UAV regulations that are applicable across all member states of the European Union. At the time of writing, two primary acts contain the regulations: The Implementing Act and The Delegated Act. These acts were in force at the end of June 2019 but will only be applicable from June 2020. Recreational and commercial users are now viewed similarly with the main intention being a risk-based approach. Three new categories are introduced: open (low risk), specific (medium risk) and certified (high risk). The certified category relates to larger UAVs that fly in controlled airspace and will require the pilot to be licensed, the UAV to have an airworthy certificate and safety controlled by the National Aviation Authority. The open category is considered to cover 90% of recreational and commercial flights. It has three subcategories, A1, A2 and A3, and depending on the operational limits, requirements of the pilot and technical requirements for the UAV, these will decide which subcategory the operation falls under. The open category does not need prior authorization for flights and there is no requirement for a pilot's license.

The general provisions for the open category are flights cannot go above 120 m and when flying within a horizontal distance of 50 m from a manmade structure greater than 105 m in height then the maximum height can be increased by a further 15 m at the request of the person responsible for the structure. The minimum age for the open and specific categories is 16 years; however, there is no age restriction if flying a toy UAV with a MTOW less than 250 g and operating under the supervision of a remote pilot. Additionally, from 2022, UAVs will be required to have product regulation markings/labelling (CE) which will detail the technical specifications of the aircraft. The labels will include information, such as MTOW, maximum speed, maximum height attainable, electrical voltage and geo awareness systems allowing airspace and altitude restrictions to be uploaded with appropriate warning systems in place alerting when nearing those restrictions.

There are four categories of CE markings (C0–C4) with each marking indicating a larger, heavier and more technical UAV. The purpose of these markings is to help identify which subcategory the operator will fly in. The information below provides a brief summary of the regulations applicable to both recreational and commercial users [50]:

Open–A1:

- Never fly over groups of people;
- Flying over bystanders is allowed if flying C0 rated UAV;
- Flying over bystanders is only allowed if flying C1 rated UAV and fly over time is as short as possible;
- Flying a C1 rated UAV requires online training and examination.

Open–A2 (Only applies when operating a C2 rated UAV):

- Don't fly over people or crowds;
- Don't fly closer than 5 m to people and only if active low speed function is activated, otherwise stay back 30 m;
- Pilot must hold a certificate of remote pilot competency, completed online training course, self-practical training and pass theoretical exam.

Open–A3 (Only applies when operating a C2, C3 or C4 rated UAV or a self-built UAV):

- Only fly where the operator reasonably expects no bystander will be put in danger during the flight;
- Don't fly closer than 150 m from residential, commercial, industrial or recreational areas;
- Same pilot competencies regarding certificate and training as Open–A2.

Specific Category: All flights operating outside the general provisions of the open category fall under the specific category as they are deemed to have an increased risk. If a standard scenario is not available for the intended operation, then a risk assessment needs to be provided to authorities using a specified methodology. Standard scenarios relate to particular flight operations and have had predetermined safety objectives and mitigation steps established by EASA. The advantage of these pre-packaged risk assessments is that it reduces the burden on the operator and the official assessing the application. An example of a standard scenario would be flying BVLOS above 120 m over sparsely populated areas.

Flight restrictions around aerodromes: According to European Union Aviation Safety Authority [50], each EASA member state will determine drone geographical zones, which are areas where drones may not fly (e.g., national parks, city centers or near airports) or may fly only under certain conditions, or where they need a flight authorization. Therefore, it is important for you to consult your National Aviation Authority to check where you can and cannot fly your drone (FAQ n.116463 [50]), e.g., in Portugal, stay a minimum of 8 km away from airports.

Under new EASA regulations introduced 1.1.2021, including detailed regulations and provisions for the operation of unmanned aircraft systems and remote pilot operators where rules and procedures are established for operators and flight risk level criteria are used to establish three categories of operations: 'open', 'specific' and 'certified' categories. The UK has similar categories of operation subject to European Union exit transition changes in the future (Sep 2021 Easy Access Rules for Unmanned Aircraft Systems [50]).

Flights within a controlled airspace could undertake operations in either the open or the specific category depending upon the limitations placed in the open category (regardless of category a permit would be required from the nearest airport to fly within their controlled airspace). If a flight exceeds any limitations of the open category, then UAS operations for the 'specific' category shall require an operational authorization issued by the competent authority pursuant to Article 12 or an authorization received in accordance with Article 16, or, under circumstances defined in Article 5(5), a declaration to be made by a UAS operator (Sep 2021 Easy Access Rules for Unmanned Aircraft Systems [50]).

Specific Category: Considering the moderate level of risk involved, flights in this category require authorization before the operation. The permission is given considering the mitigation measures identified in an operational risk assessment, except in specific standard scenarios where an operator declaration is sufficient [50].

Appendix A.4. United Kingdom

The Civil Aviation Act and ANO2016 [50] are the principal pieces of legislation governing the use of UAVs in the UK. The regulations have been updated to allow for a simpler set of rules to apply to all UAVs that weigh 20 kg or less, which under the legislation are viewed as small unmanned aircraft (SUA). UAVs that have a mass greater than 20 kg are not deemed to be SUA and must comply with all the requirements of the ANO, such as licensed flight crew, airworthiness certificates and permits to fly. Recreational and commercial operators follow different exemptions and permissions. Exemptions allow for exceptions to the law, whilst permissions are used when the law prevents the activity but has enabled the possibility within the law for that activity to take place. The information below provides a brief summary of the regulations applicable to recreation and commercial users [9].

Recreational Operators:

- Responsibility falls on the pilot to ensure all flights are performed in a safe and responsible manner;

- Flying UAVs over 0.25 kg require passing a test and registering the UAV with the CAA;
- The UAV must always be within VLOS;
- Flights must remain under 400 ft (122 m);
- Additional restrictions apply if flying a UAV weighing more than 7 kg in certain types of airspace.

If the UAV is fitted with equipment capable of performing surveillance or capturing data, such as a camera, then it is classed as a small unmanned surveillance aircraft (SUSA) and must adhere to additional regulations as follows:

- Flights must remain 150 ft (50 m) from people, vessels, vehicles and structures;
- Flights must remain 500 ft (150 m) from congested areas (areas that are used for residential, commercial, industrial or recreational and built-up areas);
- Flights must remain 500 ft (150 m) from open-air assemblies of more than 1000 from any person;
- During take-off and landing the SUSA must remain 30 m from any person.

To operate beyond the regulations the CAA allows for permissions and exemptions to be granted. There is a general exemption in place to allow for first person view flying (FPV). Ordinarily, the headset would obstruct the operator's field of view to the UAV and therefore violate the VLOS rule. However, CAA [9] has given an exemption to this rule providing the operator follows the remaining rules and is accompanied by a competent observer.

Commercial Operators:

- Any commercial operation using UAVs must have a permission issued by the CAA;
- Operators must have appropriate insurance coverage which is a condition of all exemptions and permissions granted by CAA;
- Commercial operators must adhere to the same regulations as recreational operators;
- Permissions are required from the CAA to fly outside the standard;
- A 'standard permission' enables commercial flight over or within 150 m of congested areas provided the pilot submits an operation manual, evidence of competency and proof of insurance cover;
- Reduced distance permissions allow UAVs to fly within 50 m of people within a congested area and less than 150 m from open air assemblies;
- Flights above 400 ft require operators to submit a risk assessment demonstrating the flight will be performed safely;
- BVLOS / EVLOS and UAVs over 20 kg require exemptions and pilots must submit a safety case with risk assessment proving flights will be conducted safely;
- BVLOS flights require the aircraft to have onboard collision avoidance equivalent to manned aircraft, such as a detect and avoid capability and a block of airspace enabling the UAV to be segregated from other aircraft;
- The collection of images of recognizable people are subject to the general data protection regulation and the data protection.

Flight restrictions around aerodromes: This section is taken from CAA 2020, Unmanned Aircraft System Operations in UK Airspace–Guidance [51]. Flights of unmanned aircraft around aerodromes that are designated as "protected aerodromes" are restricted. Unmanned aircraft of any size must not be flown within the flight restriction zone (FRZ) of a protected aerodrome, without appropriate permission. The flight restriction zone consists of the following three elements (Figure A2):

- A zone with the same dimensions as the aerodrome traffic zone: a 2 nm (3.7 km) or 2.5 nm (4.63 km) radius "cylinder" around the aerodrome, extending 2000 ft above ground level, centered on the longest runway;
- Runway protection zones: a rectangle extending 5 km from the threshold of each runway away from the aerodrome, along the extended runway centerline and 500 m either side, also to a height of 2000 ft above ground level;
- Additional zones: in the case where a line that is drawn 1 km beyond the boundary of an aerodrome extends beyond the aerodrome traffic zone, and so would not be

protected by it, the flight restriction zone will include a "bump" (the airfield boundary + 1 KM) to protect this part of the aerodrome.

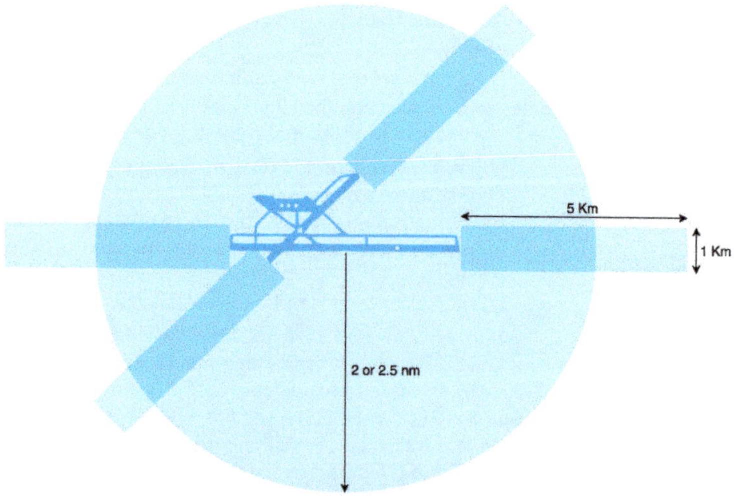

Figure A2. Flight Restriction Zone, adapted from [9].

Drones and airports: This section is taken from (gatwickairport) website [58]. Anyone flying a drone must stay well away from aircraft, airports and airfields. There is a 5 km (author note-larger than 4.63 km) flight restriction zone around Gatwick (Figure A3) and it is illegal to fly any unauthorized drones within this area. Drones must not fly above 400 ft (approx. 120 m) at any time. It is a criminal act to break the no-fly zone and the operator could put lives at risk and go to prison for up to five years.

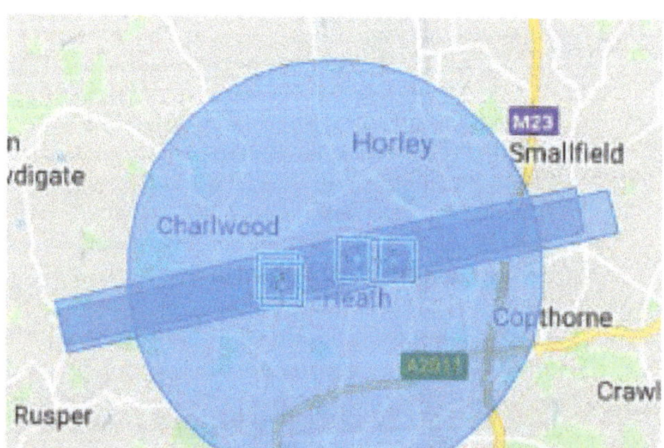

Figure A3. Gatwick airport (FRZ) UK [32].

Appendix A.5. United States

The FAA [44] is an agency of the USA department of transportation which is charged with the regulation of all civil aviation. The FAA reauthorization act signed in October 2018 and establishes changes for the safe integration and use of UAVs, including new rules regarding recreational use of UAVs. Section 349 of said act lays out the regulations pertaining to recreational operators as follows:

- The aircraft must be flown for recreation purposes only;
- The aircraft must weigh less than 55 pounds (25 kg);
- The aircraft is operated within the safety guidelines of a community-based organization (CBO) which were codeveloped with the FAA;
- VLOS must be maintained at all times;
- Flights must not enter prohibited airspace or fly near other aircraft;
- Operators must not fly over groups of people, public events or stadiums;
- Flights must not fly near emergencies, such as brushfires and law enforcement activities;
- Flights close or within airspace at or near airports must comply with airspace restrictions and prohibitions and must have prior authorization from an administrator;
- In uncontrolled airspace flights must be below 400 ft;
- The aircraft must be registered with registration number marked on the outside of the aircraft by engraving, permanent label or permanent marker;
- Operators must pass an aeronautical knowledge and safety test with proof of passing carried during flights.

Commercial Operators can operate UAVs weighing less than 55 pounds under part 107 of the federal aviation regulations. A brief overview of the rules are as follows:

- Aircraft must weigh under 55 pounds;
- Operator must hold a remote pilot airman certificate with a small UAV rating or be supervised by someone who has one;
- UAVs must be registered and marked as per recreational rules;
- VLOS maintained at all times either by the pilot or by an observer and the aircraft must remain close enough to the pilot/operator to be seen with the naked eye;
- Aircraft must not operate over bystanders;
- Daylight operations only;
- Maximum ground speed of 100 mph (160 km/h);
- FPV can be used if see and avoid requirements are met in other ways;
- Maximum altitude of 400 ft above ground level;
- External payloads are allowed provided it is securely attached and aircraft airworthiness is not unfavorably affected.

Waivers are documents issued by the FAA [44], which allows flight operations that deviate from the standard regulations. Waivers provide an opportunity for pilots to, for example, fly at night, fly over people and fly BVLOS.

Operation in certain airspace: This section is taken from the Department of Transportation, FAA Operation and Certification of Small Unmanned Aircraft Systems. No person may operate a small unmanned aircraft in Class B, Class C or Class D airspace or within the lateral boundaries of the surface area of Class E airspace designated for an airport unless that person has prior authorization from Air Traffic Control (ATC).

Rules for recreational flyers: Fly at or below 400′ in controlled airspace (Class B, C, D, and E) only with prior authorization by using LAANC or DroneZone, e.g., controlled airspace Class C, although the configuration of each Class C area is individually tailored, the airspace usually consists of a surface area with a five NM radius (USA Department of Transportation, FAA 2016 Pilot's Handbook of Aeronautical Knowledge Chapter 15 Airspace [44]). Fly at or below 400 feet in Class G (uncontrolled) airspace.

Note: Flying drones in certain airspace is not allowed. Classes of airspace and flying restrictions can be found on our B4UFLY app or the UAS Facility Maps webpage.

References

1. Australian Civil Aviation Safety Authority (CASA). Available online: https://www.casa.gov.au/ (accessed on 20 December 2021).
2. Shackle, S. The Mystery of the Gatwick Drone. The Guardian. Available online: https://www.theguardian.com/uk-news/2020/dec/01/the-mystery-of-the-gatwick-drone (accessed on 20 October 2021).
3. Nakamura, H.; Kajikawa, Y. Regulation and Innovation: How Should Small Unmanned Aerial Vehicles be Regulated? *Technol. Forecast. Soc. Chang.* **2017**, *128*, 262–274. [CrossRef]
4. Molina, M.; Santamarina-Campos, V. *Ethics and Civil Drones-European Policies and Proposals for the Industry*; Springer Nature: Berlin, Germany, 2018. [CrossRef]
5. Morales, A.C.; Paez, D.; Arango, C. Multi-Criteria Analysis of UAVs Regulations in 6 Countries Using the Analytical Hierarchical Process and Expert Knowledge. *ISPRS-Int. Arch. Photogramm. Remote Sens. Spat. Inf. Sci.* **2015**, *XL-1/W4*, 175–181. [CrossRef]
6. Luppicini, R.; So, A. A Technoethical Review of Commercial Drone Use in the Context of Governance, Ethics, and Privacy. *Technol. Soc.* **2016**, *46*, 109–119. [CrossRef]
7. Sanz, D.; Valente, J.; Del Cerro, J.; Colorado, J.; Barrientos, A. Safe Operation of Mini UAVs: A Review of Regulation and Best Practices. *Adv. Robot.* **2015**, *29*, 1221–1233. [CrossRef]
8. Stöcker, C.; Bennett, R.; Nex, F.; Gerke, M.; Zevenbergen, J. Review of the Current State of UAV Regulations. *Remote Sens.* **2017**, *9*, 459. [CrossRef]
9. Civil Aviation Authority (CAA). *2019 Drone Registration Scheme: Charge Proposal Consultation Document*; Civil Aviation Authority: West Sussex, UK, 2019. Available online: https://consultations.caa.co.uk/finance/drone-registration/user_uploads/cap1775droneregistrationchargeconsultationdocument.pdf (accessed on 20 December 2021).
10. Herrmann, M. A Comparison of Unmanned Aerial Vehicle Regulations in the United States and Europe. In Proceedings of the 53rd ASC Annual International Conference Proceedings, Seattle, WA, USA, 12 April 2017; pp. 299–306.
11. Finn, R.L.; Wright, D. Unmanned aircraft systems: Surveillance, ethics and privacy in civil applications. *Comput. Law Secur. Rev.* **2012**, *28*, 184–194. [CrossRef]
12. Nelson, J.R.; Grubesic, T.H.; Wallace, D.; Chamberlain, A.W. The View from Above: A Survey of the Public's Perception of Unmanned Aerial Vehicles and Privacy. *J. Urban Technol.* **2019**, *26*, 83–105. [CrossRef]
13. Clarke, R. The regulation of civilian drones' impacts on behavioural privacy. *Comput. Law Secur. Rev.* **2014**, *30*, 286–305. [CrossRef]
14. Finn, R.L.; Wright, D.; Friedewald, M. Seven Types of Privacy. In *European Data Protection: Coming of Age. Gutwirth, S.*; Leenes, R., De Hert, P., Poullet, Y., Eds.; Springer: Dordrecht, The Netherlands, 2013. [CrossRef]
15. Australian Law Reform Commission. *Serious Invasions of Privacy in the Digital Era*; Discussion Paper; Australian Government: Canberra, Australia, 2014. Available online: https://www.alrc.gov.au/publication/serious-invasions-of-privacy-in-the-digital-era-dp-80/ (accessed on 20 December 2021).
16. Clothier, R.; Walker, R. Determination and Evaluation of UAV safety Objectives. In Proceedings of the 21st International Unmanned Air Vehicle Systems Conference, Bristol, UK, 12 May 2006; pp. 18.1–18.16.
17. Clarke, R.; Moses, L. The regulation of civilian drones' impacts on public safety. *Comput. Law Secur. Rev.* **2014**, *30*, 263–285. [CrossRef]
18. Vela, A.E.; Ferreira, L.; Babin, T. A Safety Analysis of UAV Mapping Operations. In *2018 IEEE/AIAA 37th Digital Avionics Systems Conference (DASC)*; IEEE: Piscataway, NJ, USA, 2018; pp. 1–8. [CrossRef]
19. Rural and Regional Affairs and Transport References Committee. *Current and Future Regulatory Requirements That Impact on the Safe Commercial and Recreational Use of Remotely Piloted Aircraft Systems (RPAS)*; Unmanned Aerial Systems (UAS) and associated systems; The Australian Senate: Canberra, Australia, 2018. Available online: https://www.aph.gov.au/Parliamentary_Business/Committees/Senate/Rural_and_Regional_Affairs_and_Transport/Drones/Report (accessed on 20 December 2021).
20. Australian Transport Safety Bureau. *A Safety Analysis of Remotely Piloted Aircraft Systems*; Australian Government: Canberra, Australia, 2017. Available online: https://www.atsb.gov.au/publications/2017/ar-2017-016/ (accessed on 20 December 2021).
21. International Air Transport Association. *Safety Report 2018*; International Air Transport Association: Quebec, QC, Canada, 2018. Available online: https://libraryonline.erau.edu/online-full-text/iata-safety-reports/IATA-Safety-Report-2018.pdf (accessed on 20 December 2021).
22. Magister, T. The small unmanned aircraft blunt criterion based injury potential estimation. *Saf. Sci.* **2010**, *48*, 1313–1320. [CrossRef]
23. Civil Aviation Safety Authority. *Human Injury Model for Small Unmanned Aircraft Impacts*; Civil Aviation Safety Authority: Clayton, Australia, 2013. Available online: https://www.casa.gov.au/files/human-injury-model-small-unmanned-aircraft-impactspdf (accessed on 20 December 2021).
24. Civil Aviation Safety Authority. *Review of Aviation Safety Regulations of Remotely Piloted Aircraft Systems*; Australian Government: Canberra, Australia, 2018. Available online: https://libraryonline.erau.edu/online-full-text/iata-safety-reports/IATA-Safety-Report-2018.pdf (accessed on 20 December 2021).
25. Shepard, D.P.; Bhatti, J.A.; Humphreys, T.E. Drone Hack: Spoofing Attack Demonstration on a Civilian Unmanned Aerial Vehicle. *GPS World* **2012**, *23*, 30–33.
26. Arteaga, S.P.; Hernández, L.A.M.; Pérez, G.S.; Orozco, A.L.S.; Villalba, L.J.G. Analysis of the GPS Spoofing Vulnerability in the Drone 3DR Solo. *IEEE Access* **2019**, *7*, 51782–51789. [CrossRef]
27. Zhi, Y.; Fu, Z.; Sun, X.; Yu, J. Security and Privacy Issues of UAV: A Survey. *Mob. Netw. Appl.* **2020**, *25*, 95–101. [CrossRef]

28. DJI Adds Airplane and Helicopter Detectors to New Consumer Drones, Newsroom News. *DJI*. 22 May 2019. Available online: https://www.dji.com/newsroom/news/dji-adds-airplane-and-helicopter-detectors-to-new-consumer-drones (accessed on 20 December 2021).
29. Environmental Protection Act 1994 (Qld). Available online: https://www.legislation.qld.gov.au/view/html/inforce/current/act-1994-062 (accessed on 20 December 2021).
30. Christian, A.; Cabell, R. Initial Investigation into the Psychoacoustic Properties of Small Unmanned Aerial System Noise. In Proceedings of the 23rd AIAA/CEAS Aeroacoustics Conference, Denver, CO, USA, 5–9 June 2017; pp. 1–2. [CrossRef]
31. Intaratep, N.; Alexander, W.; Devenport, W.; Grace, S.; Dropkin, A. Experimental Study of Quadcopter Acoustics and Performance at Static Thrust Conditions. In Proceedings of the 22nd AIAA/CEAS Aeroacoustics Conference, Lyon, France, 30 May–1 June 2016; pp. 1–14. [CrossRef]
32. Christiansen, F.; Rojano-Doñate, L.; Madsen, P.T.; Bejder, L. Noise Levels of Multi-Rotor Unmanned Aerial Vehicles with Implications for Potential Underwater Impacts on Marine Mammals. *Front. Mar. Sci.* **2016**, *3*, 277. [CrossRef]
33. Pomeroy, P.; O'Connor, L.; Davies, P. Assessing use of and reaction to unmanned aerial systems in gray and harbor seals during breeding and molt in the UK. *J. Unmanned Veh. Syst.* **2015**, *3*, 102–113. [CrossRef]
34. Stewart, P. Drone danger: Remedies for damage by civilian remotely piloted aircraft to person or property on the ground in Australia. *Torts Law J.* **2016**, *23*, 290–319. Available online: https://hdl.handle.net/10453/75976 (accessed on 20 December 2021).
35. Choi-Fitzpatrick, A.; Chavarria, D.; Cychosz, E.; Dingens, J.-P.; Duffey, M.; Koebel, K.; Siriphanh, S.; Yurika, T.-M.; Watanabe, H.; Juskauskas, T.; et al. *Up in the Air: A Global Estimate of Non-Violent Drone Use 2009–2015*; University of San Diego: San Diego, CA, USA, 2016. Available online: https://digital.sandiego.edu/gdl2016report/1 (accessed on 20 December 2021).
36. German Unmanned Aviation Association. *Analysis of the German Drone Market*; German Unmanned Aviation Association: Berlin, Germany, 2019. Available online: https://www.verband-unbemannte-luftfahrt.de/en/vul-markststudie_deutsch_final/ (accessed on 20 December 2021).
37. Unmanned Airspace. Special Report—Brazil DECEA to Integrate Drone Management within Its ATM System; Unmanned Airspace: UAS Traffic Management News 2018. Available online: https://www.unmannedairspace.info/uncategorized/special-report-brazils-decea-integrate-drone-management-within-atm-system/ (accessed on 20 December 2021).
38. Library of Parliament. *Civilian Drone Use in Canada*; Parliament of Canada: Ottawa, ON, Canada, 2017.
39. Canada Gazette. 2018. Available online: https://www.gazette.gc.ca/accueil-home-eng.html (accessed on 20 December 2021).
40. Wangshu, L. Number of UAV Pilots Takes off. China Daily. Available online: https://www.chinadaily.com.cn/a/201804/19/WS5ad7d557a3105cdcf651921f.html (accessed on 20 December 2021).
41. Single European Sky Air Traffic Management Research Joint Undertaking. *European Drones Outlook Study*; SESAR: Brussels, Belgium, 2016; p. 17.
42. Activity Report. *French Civil Aviation Safety Directorate*; Activity Report: Paris, France, 2018. Available online: https://www.ecologie.gouv.fr/sites/default/files/RapportDSAC-2018-EN-Web.pdf (accessed on 20 December 2021).
43. Mishra, M. India May Not Keep December 1 Date with Drones. *The Economic Times India*. 3 November 2018. Available online: https://economictimes.indiatimes.com/industry/transportation/airlines-/-aviation/india-may-not-keep-december-1-date-with-drones/articleshow/66483999.cms?from=mdr (accessed on 25 January 2022).
44. Federal Aviation Administration. *Unmanned Aircraft Systems*; Federal Aviation Authority: Washington, WA, USA, 2019. Available online: https://www.faa.gov/data_research/aviation/aerospace_forecasts/media/unmanned_aircraft_systems.pdf (accessed on 20 December 2021).
45. Civil Aviation Safety Regulations 1998 (Cwlth). Available online: https://www.legislation.gov.au/Details/F2021C00238 (accessed on 20 December 2021).
46. Civil Aviation Safety Authority, Advisory Circulars (Various). Available online: https://www.casa.gov.au/search-centre/advisory-circulars (accessed on 20 December 2021).
47. Civil Aviation Safety Regulations Part 101 (Unmanned Aircraft and Rockets) Manual of Standards 2019. Available online: https://www.legislation.gov.au/Details/F2020C00980 (accessed on 20 December 2021).
48. Canadian Aviation Regulations. Available online: https://tc.canada.ca/en/corporate-services/acts-regulations/list-regulations/canadian-aviation-regulations-sor-96-433 (accessed on 20 December 2021).
49. Transport Canada, Aeronautical Information Manual (TC AIM) RPA-Remotely Piloted Aircraft. 7 October 2021. Available online: https://tc.canada.ca/sites/default/files/2021-09/AIM-2021-2_ACCESS_E.pdf (accessed on 20 December 2021).
50. European Union Aviation Safety Authority. Available online: https://www.easa.europa.eu/ (accessed on 20 December 2021).
51. Commission Implementing Regulation (EU). 2019. Available online: https://eurlex.europa.eu/eli/regimpl/2019/2072/oj (accessed on 20 December 2021).
52. Commission Delegated Regulations (EU). 2019. Available online: https://eur-lex.europa.eu/legal-content/EN/TXT/?uri=uriserv:OJ.L_.2019.125.01.0004.01.ENG (accessed on 24 October 2021).
53. Air Navigation Order 2016 (UK). Available online: https://www.legislation.gov.uk/uksi/2016/765/contents/made (accessed on 20 December 2021).
54. CAP 722. Available online: https://publicapps.caa.co.uk/modalapplication.aspx?appid=11&mode=detail&id=415 (accessed on 20 December 2021).

55. Federal Aviation Administration Reauthorisation Act of 2018. Available online: https://www.faa.gov/about/reauthorization (accessed on 20 December 2021).
56. Office of the Australian Information Commissioner. *Australian Community Attitudes to Privacy Survey 2017*; Australian Government: Canberra, Australia, 2017. Available online: https://www.oaic.gov.au/updates/videos/australian-community-attitudes-to-privacy-survey-2017 (accessed on 20 December 2021).
57. The Personal Information Protection and Electronic Documents Act. *PIPEDA Fair Information Principles*; Office of the Privacy Commissioner of Canada: Ottawa, QC, Canada, 2019. Available online: https://www.priv.gc.ca/en/privacy-topics/privacy-laws-in-canada/the-personal-information-protection-and-electronic-documents-act-pipeda/p_principle/ (accessed on 20 December 2021).
58. Gatwickairport. Available online: https://www.gatwickairport.com/business-community/aircraftnoiseairspace/airspace/drone-safety/ (accessed on 20 December 2021).

remote sensing MDPI

Article

Automatic Filtering of Lidar Building Point Cloud in Case of Trees Associated to Building Roof

Fayez Tarsha Kurdi [1,*], Zahra Gharineiat [1], Glenn Campbell [1], Mohammad Awrangjeb [2] and Emon Kumar Dey [2]

[1] School of Surveying and Built Environment, Faculty of Health, Engineering and Sciences, University of Southern Queensland, Springfield Campus, Springfield, QLD 4300, Australia; Zahra.Gharineiat@usq.edu.au (Z.G.); Glenn.Campbell@usq.edu.au (G.C.)

[2] Institute of Integrated and Intelligent Systems, Griffith University, Nathan, QLD 4111, Australia; m.awrangjeb@griffith.edu.au (M.A.); emonkumar.dey@griffithuni.edu.au (E.K.D.)

* Correspondence: Fayez.TarshaKurdi@usq.edu.au

Abstract: This paper suggests a new algorithm for automatic building point cloud filtering based on the Z coordinate histogram. This operation aims to select the roof class points from the building point cloud, and the suggested algorithm considers the general case where high trees are associated with the building roof. The Z coordinate histogram is analyzed in order to divide the building point cloud into three zones: the surrounding terrain and low vegetation, the facades, and the tree crowns and/or the roof points. This operation allows the elimination of the first two classes which represent an obstacle toward distinguishing between the roof and the tree points. The analysis of the normal vectors, in addition to the change of curvature factor of the roof class leads to recognizing the high tree crown points. The suggested approach was tested on five datasets with different point densities and urban typology. Regarding the results' accuracy quantification, the average values of the correctness, the completeness, and the quality indices are used. Their values are, respectively, equal to 97.9%, 97.6%, and 95.6%. These results confirm the high efficacy of the suggested approach.

Keywords: LiDAR; classification; modelling; filtering; segmentation

Citation: Tarsha Kurdi, F.; Gharineiat, Z.; Campbell, G.; Awrangjeb, M.; Dey, E.K. Automatic Filtering of Lidar Building Point Cloud in Case of Trees Associated to Building Roof. *Remote Sens.* **2022**, *14*, 430. https://doi.org/10.3390/rs14020430

Academic Editor: Andrea Ciampalini

Received: 25 November 2021
Accepted: 13 January 2022
Published: 17 January 2022

1. Introduction

Light Detection and Ranging (LiDAR) data provide considerable advantages over other photogrammetric and remote sensing data sources. The data can be acquired at high speed and density during the day or night, and have automated processing potential and the possibility of simultaneous georeferenced detection of other supplementary data: laser intensity, RGB, and infrared images [1]. In addition, the laser scanning technique differs from other 3D data acquisition sensors such as optical cameras. The heavy demand for LiDAR data necessitates an increased need for automatic processing tools. Of these tools, the two main operations are the automatic classification and the automatic modelling of LiDAR data. In urban areas, a point cloud consists of several classes such as terrain, roads, buildings, vegetation, and other manmade objects. As each class has its own modelling demands, it is necessary to separate the city classes before starting the modelling step. In urban zones, the building class is of particular importance. Once this class is extracted, two families of modelling approaches are available: model-driven and data-driven [2]. In this context, the input data of any selected modelling algorithm is a building point cloud consisting of a set of points mainly covering the roof surfaces, as the scanning is usually achieved using an airborne vehicle. The building facades will be partially covered with low point density depending on the building location in comparison to the scanning trajectory and the orientation of the facade's planes. In any case, the roof points will be distributed into scanning lines formed according to the quality of the employed scanning system [3]. Moreover, the building point cloud may contain points covering the surrounded terrain in addition to the low and high vegetation that is associated with the building. Other inconsequential points may be present in the building point cloud such as those that are due to the objects located near the building and the decoration of the facade. However, the modelling approaches used for constructing the building model use the hypothesis that the building point cloud consists of only the roof surfaces (except for the vertical surfaces) [4], which is why the building cloud points need to be classified into two classes: the roof points and the nonroof (undesirable) points. In fact, the undesirable points sometimes represent an obstacle in the modelling algorithms, especially when their percentage exceeds certain limits. Sometimes they are considered as the main reason for automatic modelling failure or the source for some deformations in the obtained building model [5].

Automatic filtering of LiDAR building point cloud has become a very hot research topic. Shao et al. [6] develop a top-down strategy that starts by seed point sets' detection with semantic information. Then, the region-growing algorithm is applied to detect the building roof point cloud. Moreover, for filtering the LiDAR data and extracting the building class, Wen et al. [7] use Graph Attention Convolution Neural Network (GACNN). This technique is directly applied to the city point cloud. In another way, Varlik and Uray [8] prefer to project the LiDAR point cloud firstly onto a 2D plane before employing a U-net architecture deep-learning network. Alternatively, after detecting the building class, Hui et al. [9] apply multiscale progressive growth to optimize the obtained result.

Tarsha Kurdi et al. [10] suggested an algorithm for automatic extraction of the roof point cloud from the building point cloud by analysis of the Z histogram of the building point cloud. Unfortunately, this algorithm did not consider an important case, which is the case of high trees associated with the building roof. The importance of this case arises when the building mask is automatically extracted from the city point cloud by automatic classification. In this case, a considerable number of buildings will be extracted with their attached trees [11]. This paper suggests an extension of the filtering approach suggested by Tarsha Kurdi et al. [10] for considering the case of high trees that occlude the building roof. The suggested approach starts with the application of the algorithm suggested by Tarsha Kurdi et al. [10] that detects the combined roof and tree crown point cloud. To extract the building roof, three criteria are calculated and analyzed: the normal vector, the standard deviation, and the change of curvature factor. At this stage, the employed threshold values are automatically determined for each building point cloud independently (smart thresholds). This procedure increases the filtering success percentage because it considers the particularity of the point cloud of each building. Finally, the misclassified points are correctly reclassified by using an image-processing operation.

This paper consists of eight sections. The first section introduces the problem of the presence of high trees associated with the building roof which may be considered as a real obstacle in the automatic building-modelling algorithm. Second, similar and related approaches in the literature are discussed. Third, the approach suggested by Tarsha Kurdi et al. [10] is summarized. Fourth, the limitation of the previous approach is highlighted in the case of high trees associated with building roofs. Fifth, the suggested approach is detailed step by step. Sections 6 and 7 discuss the accuracy and the limitations of the suggested algorithm. Finally, the conclusion draws a panoramic budget and exposes future work.

2. Related Work

In LiDAR data, the aim of the building-extraction algorithm is the extraction of building class from the city point cloud [12]. Unfortunately, the majority of the classification approaches suggested in the literature cannot perfectly achieve the classification task because some misclassified points will be present in the classification result, e.g., the recognition of trees associated with building roofs still represents a challenge, as it is possible to detect a misclassified vegetation segment within the building class when the distribution of the vegetation segment points is similar to that of the building roof. That is why the detected building point cloud needs to be filtered before starting the modelling step.

In the case of building extraction by separating the off-terrain from the terrain classes, Demir [13] studied the case of low terrain slopes and detected the off-terrain objects via an elevation threshold. Thereafter, the nonbuilding objects were cancelled because they have no planar features. In the same context, modern approaches can use the machine-learning paradigm as suggested by Maltezos et al. [11]. This algorithm employed a deep-learning paradigm based on a convolutional neural network model for building detection from the city point cloud. The successful extraction percentage was about 80%. Moreover, the classification approach developed by Liu et al. [12] combined the Point Cloud Library (PCL) region growth segmentation and the histogram. For this purpose, they employed a PCL region growth algorithm to segment the point cloud and then calculated the normal vector for each cluster. The histograms of normal vector components were calculated to separate the building point cloud from the nonbuilding. Despite the classification result in this approach being better than the former ones, the obtained results still need improvement and filtering before starting the modelling step.

It is now understood why most of the modelling algorithms include a point cloud filtering step. Tarsha Kurdi et al. [14] calculated the building Digital Surface Model (DSM) by eliminating the undesirable points associated with vertical elements such as the building facade. In fact, DSM is a 2D matrix where the pixel value is equal to the interpolated Z coordinate of LiDAR points located in this pixel. This procedure cannot always eliminate all these points, which is why Park et al. [15] eliminate the unnecessary objects by realizing the cube operator to segment the building point cloud into roof surface patches, including superstructures, removing the unnecessary objects, detecting the boundaries of buildings and defining the model key points for building modelling. Zhang et al. [16] and Li et al. [17] used a region-growing algorithm based on the Triangulated Irregular Network (TIN) or grid data structures to detect the building roof point clouds.

In order to identify the roof points, some authors used additional data such as the ground plan [18] or the aerial images [19–21], while many others (Jung and Sohn [22] and Awrangjeb et al. [23]) considered only the extracted roof point cloud instead of the whole building point cloud to construct 3D building roof models.

Hu et al. [19] filter the building point cloud through projecting back the building points on imagery using the Normalized Difference Vegetation Index (NDVI). Then, the Euclidean clustering is applied to remove vegetation clusters with small areas. In another approach, the building point cloud is divided into two classes by using machine-learning techniques [24]. These classes are the roof and the façade classes. For achieving this goal, the TensorFlow neural network for deep learning is employed. Unfortunately, the result is not so promising where the classification percentages of the semantic classes are less than 85%. In the context of façade point class recognition, another technique is suggested by Martin-Jimenez et al. [25]. It suggests the calculation of the normal vector of the building point cloud. For this purpose, the covariance matrix is calculated for each point and its eight neighboring points. Then, the calculated normal vector is assigned to the tested point. This method does not consider the presence of decorative elements in the building facades in addition to the surrounding terrain point and small vegetation.

At this stage, it is important to note that the building point cloud may be divided into two main classes, which are roof and nonroof classes. The roof class involves a point set that is employed by the building construction algorithms, whereas the nonroof class is a point set of undesirable and useless points for the building modelling algorithms. The undesirable points represent the surrounding terrain, façade, vegetation, and noise. For the purposes of this paper, we will use the term undesirable points to describe the nonroof class, and the procedure of roof points extraction from the building point cloud will be referred to as filtering.

3. Filtering of Building Point Cloud Which Does Not Contain High Trees

This section summarizes the filtering approach suggested by Tarsha Kurdi et al. [10] because the suggested algorithm in this paper represents an extension of the previous one.

Building point clouds can be extracted through the manual or automatic classification of the city LiDAR point cloud. The approach suggested by Tarsha Kurdi et al. [10] detects the roof points from the building point cloud based on the analysis of the Z coordinate histogram. This algorithm supposes that the input building point cloud does not contain high trees (the same height or higher than the roof). The basic hypothesis adopted is that the point density of the roof surfaces is considerably greater than the building's facades. This algorithm starts by calculating the histogram of the Z coordinate of the building point cloud (see Figures 1b,e and 2b,e). This histogram consists of a set of consecutive 1m bins, each bar value representing the number of points within an altitude interval. The ideal histogram step value is dependent on the altimetry accuracy of the point cloud, the roof-texture thickness and the inclination of roof surfaces. For more details about the step value assignment in addition to all technical characteristics and analysis please see Tarsha Kurdi et al. [10].

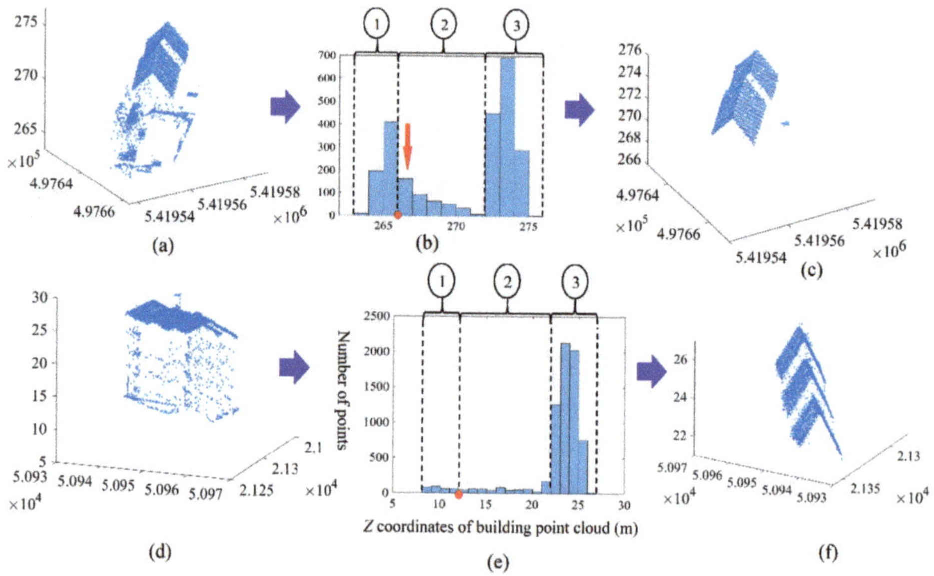

Figure 1. (**a**,**d**) Building point clouds in Vaihingen and Hermanni datasets; (**b**,**e**) histograms of Z coordinate of building point clouds consecutively of the last two buildings; (**c**,**f**) final results of building point clouds' filtering; The red dot in (**b**,**e**) is the PSTF; "1,2,3" in circles in (**b**,**e**) are building section numbers.

From Figures 1 and 2, the histogram of the Z coordinate of the building point cloud consists of three sections. Section 1 represents the terrain surrounding the building, low vegetation, noise, and the lower parts of the façade. Section 2 represents the points spread on the building façades, and Section 3 represents the roof point cloud. In order to determine the limits between these three sections, eight rules are adopted as follows:

Rule 1: The leftmost bar of the histogram belongs to the terrain section.

Rule 2: The rightmost bar of the histogram belongs to the building roof section.

Rule 3: Location of PSTF (point separating terrain and façade) (the red dot in Figure 1b). This point must be located within the four leftmost bars in the histogram. If the number of terrain points is not negligible (case of a flat terrain), the PSTF will be located after the bin with a significant population within the first four bars. If there is not a substantial number of terrain points within the building point cloud (case of significant terrain slope), the PSTF will be situated beside the fourth bar. In this case, the number of terrain bins will increase but their populations will drop down. Moreover, the first four bins will be considered as terrain bins whereas the other terrain bins will appear in the façade section. This misclassification is tolerated because it will not affect the final result, which is the detection of the roof point cloud.

Rule 4: Each bin with a considerable population (given the threshold) after the PSTF belongs to the roof surface section.

Rule 5: The roof points are presented in the histogram by one- or multi-bins situated after the PSTF. All bins to the right of the previous roof surface bin belong to the roof.

Rule 6: The bar of the façade section that is immediately located before the roof surface section may represent roof and façade points together. Therefore, the points belonging to the upper half of this bar must be added to the roof section.

Rule 7: All undesirable point bars situated after the PSTF have small values (given the threshold) in comparison to the values of roof surface bars.

Rule 8: In the histogram and after the PSTF, there are some bars with smaller values than the roof surface bars and greater values than the undesirable point bars (red arrows in Figure 1b). As these bars do not belong to the undesirable point or to the roof class, they are described as fuzzy bars. This is why it is necessary to check the presence of small roof planes amongst the points represented in the fuzzy bars.

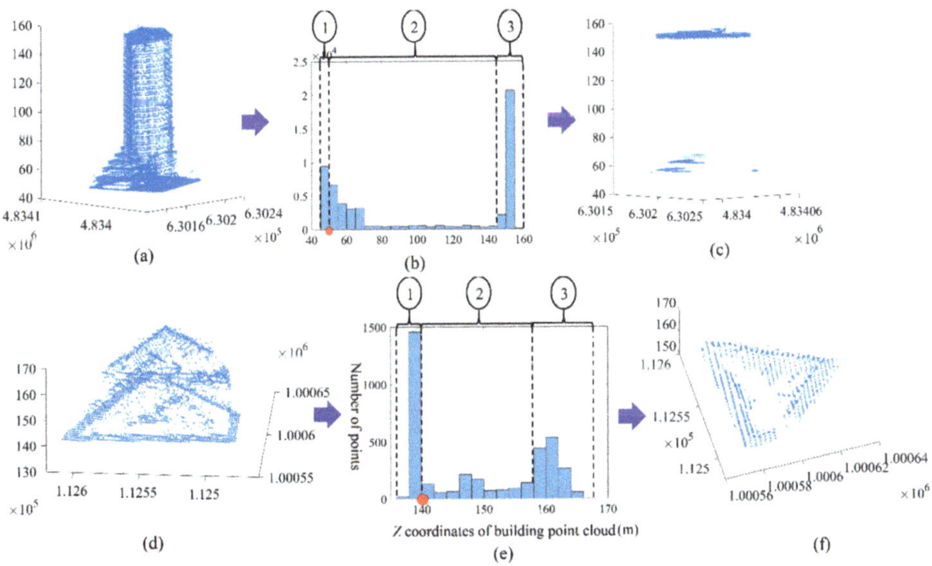

Figure 2. (**a**,**d**) Building point clouds, respectively, in Toronto and Strasbourg datasets; (**b**,**e**) histograms of Z coordinate of building point clouds consecutively of the last two buildings; (**c**,**f**) final results of building point clouds' filtering; The red dots in (**b**,**e**) are the PSTF; "1,2,3" in circles in (**b**,**e**) are building section numbers.

In fact, the idea of fuzzy bars solves the problem of the sensitivity of threshold value selection. Indeed, the point sets represented by fuzzy bars will be checked for the presence of roof planes within them. This approach is well-known in remote sensing as applying fuzzy logic [26]. For more details about this algorithm and the employed threshold values, see Tarsha Kurdi et al. [10].

This rule list could be written as a pseudocode which clarifies all employed equations, as follows:

$h_{i:1\ to\ n} = histogram\left(Z_{point\ cloud}\right)$
$h_{max} = \max(h_i); PSTF = 0$
For $i = 1$ ***To*** 3
  If $h_i - h_{i+1} > 0.5 \times h_{max}$ ***Or*** $h_i - h_{i+2} > 0.6 \times h_{max}$ ***Then*** $PSTF = i$ ***EndIf***
Next i
If $PSTF = 0$ ***Then*** $PSTF = 4$ ***EndIf***
$h_{mrb} = \max(h_i); i =$ ***From*** $PSTF$ ***To*** $length(h_i)$
$h_{rb} = zeros(n);\ S = 0; S_1 = 0$ (S_1 is the first roof bar, S is the last roof bar)
For $i = PSTF$ ***To*** n
  If $h_i \geq \frac{1}{3} \times h_{mrb}$ ***Then*** $h_{rb}(i) = h_i; s = i$
    If $S_1 = 0$ ***Then*** $S_1 = i$ ***EndIf***
  ElseIf $h_i \leq 0.1 \times h_{mrb}$ ***Then*** $h_{fbi} = h_i$ ***Else*** $h_{fuz} = h_i$ ***EndIf***
Next i
If $S < n$ ***Then For*** $i = S$ ***To*** n $h_{rbi} = h_i$ ***Next*** i ***EndIf***
If $h_{S1-1} \leq \frac{h_{mrb}}{10}$ ***Then*** $Z_{S1-0.5\ To\ S1-1} \in h_{rb}$ ***EndIf***

Where n is the number of bars in the histogram; PSTF: point separating terrain and façade; h_{rb}: roof bars; h_{fb}: façade bars; h_{fuz}: fuzzy bars.

4. Results of Filtering Algorithm Application

The experiments illustrated in this section not only prove the necessity of improving the previous filtering algorithm but also underline the differences between the extended and nonextended filtering algorithms. The algorithm described in the last section was tested with two sorts of building point cloud samples. The first ones represent building clouds that do not contain high trees that occlude the roof, whereas the second samples represent building clouds that contain high trees that occlude the roof. Figures 1 and 2 present the results of the last algorithm application on building clouds without associated trees and Figure 3 shows the results with attached trees adjacent to the roofs.

In Figures 1 and 2 where the building clouds do not contain trees that occlude the roof, the algorithm successfully eliminated all undesirable points and saved only the roof points (see Figures 1c,e and 2c,e). By contrast, in Figure 3 where the building clouds contain trees that occlude the roof, the algorithm eliminated all undesirable points except for the tree crowns, and saved the tree crowns in addition to the roof points (see Figure 3c,e,f).

Although the tree crowns are not excluded in the filtering results, the façade points which represent an obstacle in the building cloud filtering procedure are eliminated. This shows, in the case of the presence of trees that occlude the roof, a path towards cancelling the tree points based on the elimination of the vertical element points. At this stage, before improving the filtering algorithm, two remarks have to be considered:

- The analysis of fuzzy bars in the building point cloud histogram has to be postponed to the end of the tree point elimination. This choice is supported by the fact that the roof planes, perhaps presented by fuzzy bars, are low in comparison to the building roof level and have smaller areas. If these planes were to be extracted before eliminating the tree points, they may be eliminated during the tree point recognition step.
- The application of the first seven rules on the building point cloud creates a new point cloud that represents the building roof (without the planes of fuzzy bars) in addition to the tree crown. This cloud will be named the noisy roof cloud.

At this stage, it is useful to state that the results of extended filtering approach of the same samples mentioned in Figure 3 will be later illustrated in Figure 10, where the tree crowns are successfully distinguished thanks to the extended algorithm. The next section explains the suggested extension of the filtering algorithm.

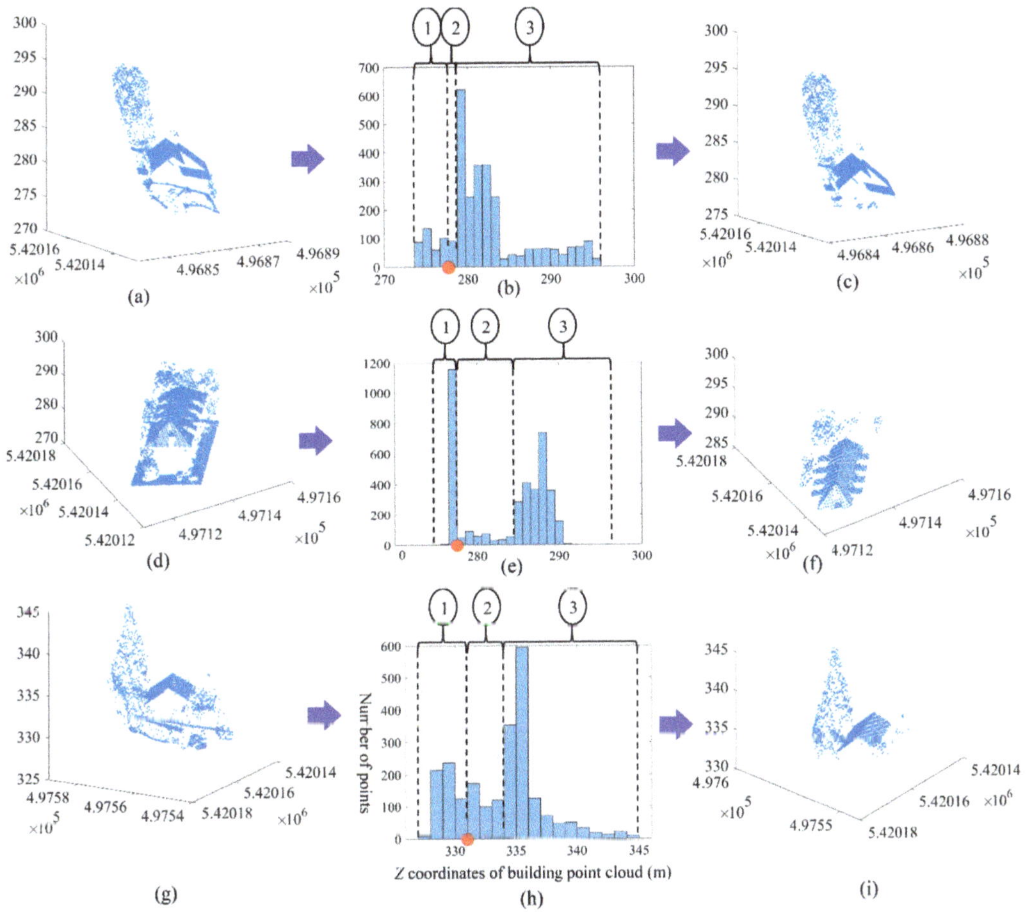

Figure 3. Results of filtering algorithm application in case of buildings which contain high trees that occlude the roof. (**a,d,g**) The original building point clouds; (**b,e,h**) histograms of building point clouds; (**c,f,i**) noisy roof point clouds; (**a–c**) building number 4; (**d–f**) building number 5; (**g–i**) building number 6; The red dots in (**b,e,h**) are the PSTF; "1,2,3" in circles in (**b,e,h**) are building section numbers.

5. Extended Filtering Algorithm

This paper suggests an extension of the filtering approach suggested by Tarsha Kurdi et al. [10] to consider the case of building point cloud containing high trees that occlude the roof. Figure 4 shows the workflow of the extended algorithm.

The extended algorithm starts by applying the first seven rules of histogram interpretation to the building point cloud (see Section 3). This operation allows the elimination of the majority of noisy points under the roof level except for the roof planes linked to the fuzzy bars. The eliminated noisy points represent the façade points, the surrounding terrain points, the low vegetation points, the high tree stem, and low branch points, and all other noisy points such as objects located near the building. The obtained point cloud represents the roof in addition to the associated tree crowns, which is why it is named the noisy roof point cloud. The next section exposes the calculation of the Neighborhood Matrix (N_m) of the noisy roof point cloud.

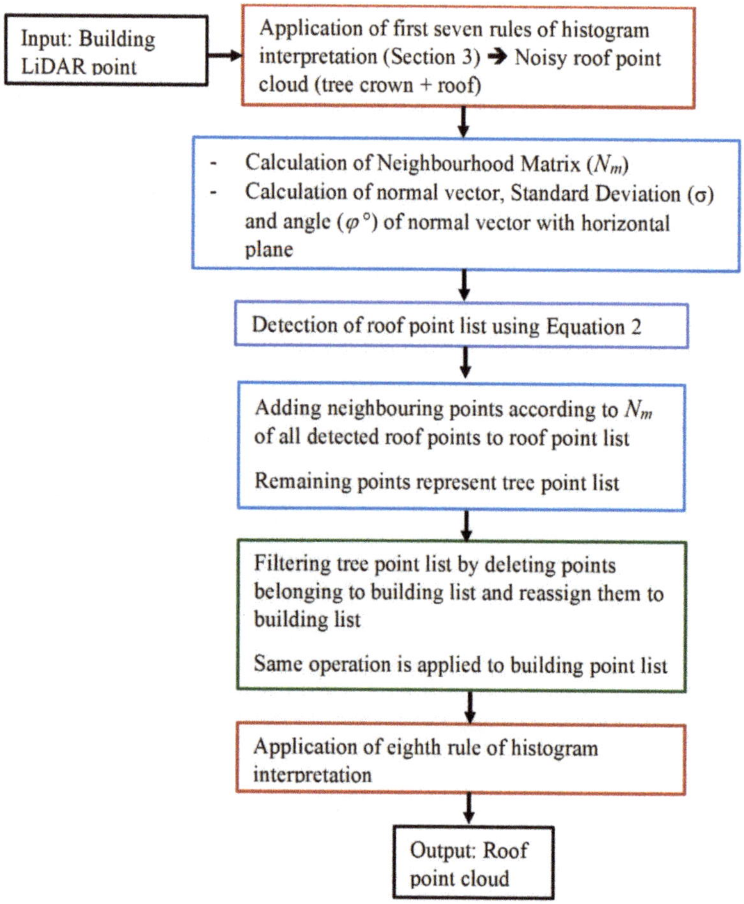

Figure 4. Workflow of the extended filtering algorithm where σ is the standard deviation of the fitted plane, $\varphi°$ is the angle of the normal vector with the horizontal plane, C_{cf} is the change of curvature factor, and $T_{h\sigma}$, $T_{h\varphi}$ and T_{hccf} are their respective thresholds.

5.1. Selection of Neighbouring Points

To calculate the normal vectors and roof features of the noisy point cloud, the approach suggested by Dey et al. [27] was adopted. This selection used the dynamic method for selecting the neighborhood of each point. Consequently, only the necessary minimum number of neighboring points were considered for each roof point. This leads to minimizing the error during the estimation of the normal vectors and the roof features values.

The approach suggested by Dey et al. [27] starts by selecting the initial minimum number of neighboring points by considering the case of a regular distribution of points where each point has eight neighboring points. To reliably calculate a normal vector to a plane, the point selection needs an evenly distributed sample of points from the plane. To ensure the neighborhood is not limited to a single scanning line, the considered point and its neighborhood are fitted to a 3D line where the standard deviation value represents an indicator factor reflecting the number of scanning lines that are included in the neighboring points. If the considered point and its neighborhood belong to more than one scanning line, the estimated values of the normal vector and the roof features will correctly describe the point's geometric position regarding the roof facets. Alternatively, if the considered point and its neighborhood belong to only one scanning line, the estimated values of the normal vector and the roof features may be inaccurate. In this context, an iteration loop is employed to increase the number of neighboring points in each iteration until the selected neighboring points are located on more than one scanning line. For more details about this algorithm, please see Kumar Dey et al. [27].

The neighborhood is selected using only X and Y coordinates instead of X, Y and Z. This choice is adopted because the 3D neighborhood selection leads to the detection of all points situated within a sphere around the considered point, whereas in 2D, the neighborhood selection allows the detection of all points situated within a vertical cylinder around the considered point. In the case of a point located on a roof plane, the two selection options provide almost the same result, but in the case of a point located on a tree, where the LiDAR points are distributed on the branches and leaves, the neighborhoods differ. Consequently, this choice helps to increase the difference between the roof and the tree points when the normal vector and roof features are calculated, as shown in the next section.

The operation described in this section allows us to define a new matrix named the Neighborhood Matrix (N_m). This matrix is determined from the neighboring points for each cloud point. This matrix consists of n rows (n is the number of points of noisy roof cloud) and N columns. The number N represents the maximum possible number of neighboring points (please see Dey et al. [27]). Furthermore, In the point cloud list, the order of each point is the point number, e.g., if the point number i has 45 neighboring points, then the first 45 entries of the row number i will contain the neighboring point numbers, and the rest of cells of this row will contain zeros. This matrix will be employed for calculating the normal vector in addition to the roof feature values. Moreover, it will be used later (see Section 5.4) to recognize the roof points.

Once the N_m is calculated, the normal vectors of the noisy roof cloud can be calculated.

5.2. Calculation of Normal Vectors and Change of Curvature Factor

Once the neighboring points are determined for each LiDAR point through the matrix N_m, the best plane passing through each point can be fitted and the normal vector then can be calculated. To achieve this objective, the plane equation will be fitted using the eigenvectors of the covariance matrix which was employed by Sanchez et al. [28]. In this paper, three criteria are employed for distinguishing the roof points from the tree ones. These criteria are the standard deviation (σ) of the fitted plane (in meters), the angle (φ°) of the normal vector with the horizontal plane and the change of curvature factor (C_{cf}) (unitless) which is calculated based on the eigenvalues according to Equation (1) [29].

$$C_{cf} = \frac{\lambda_3}{\lambda_1 + \lambda_2 + \lambda_3} \tag{1}$$

where λ_1, λ_2, λ_3 are the eigenvalues in descending order.

Once the target criteria are calculated for all cloud points, the next step is the filtering of the noisy roof points from the tree ones.

5.3. Filtering of Noisy Roof Point Cloud

In order to recognize the roof points in the noisy roof cloud, the three criteria calculated in the last section will be implemented. In this context, all points conforming to Equation (2) are selected.

$$\sigma_i \le T_{h\sigma} \; and \; \varphi_i \ge T_{h\varphi} \; and \; C_{cfi} \le T_{hccf} \tag{2}$$

where σ_i, φ_i and C_{cfi} are, respectively, the standard deviation, the angle (φ°) and the change of curvature factor of the point number i. Furthermore, $T_{h\sigma}$, $T_{h\varphi}$ and T_{hccf} are the assigned thresholds that are used for detecting the roof points. In this context, the employed threshold values change from one building to other, and a stochastic method is adopted for calculating the smart threshold values. This method relies on the histogram analysis as shown in the next section.

5.4. Calculation of Smart Threshold Values

For noisy roof cloud, Figure 5a,b and Figure 6 represent examples of the histograms of σ, $\varphi°$ and C_{cf}, respectively. The step values in these histograms are, respectively, equal to 0.2 m, 10° and 0.01 (unitless). These values are selected empirically depending on the range of each parameter and the sensitivity of the parameter variation to the filtering results, e.g., in Figures 5 and 6, the variation of $\varphi°$ values has less influence on the filtering result than the variation of C_{cf} parameter, which is why the number of bins of $\varphi°$ histogram is fewer than the number of bins of C_{cf} histogram. Though the high-quality results shown in Section 6 demonstrate that the selected values of the three histogram steps are successful, more investigations are needed to prove their optimal values, which can be applied to different point densities and urban typologies.

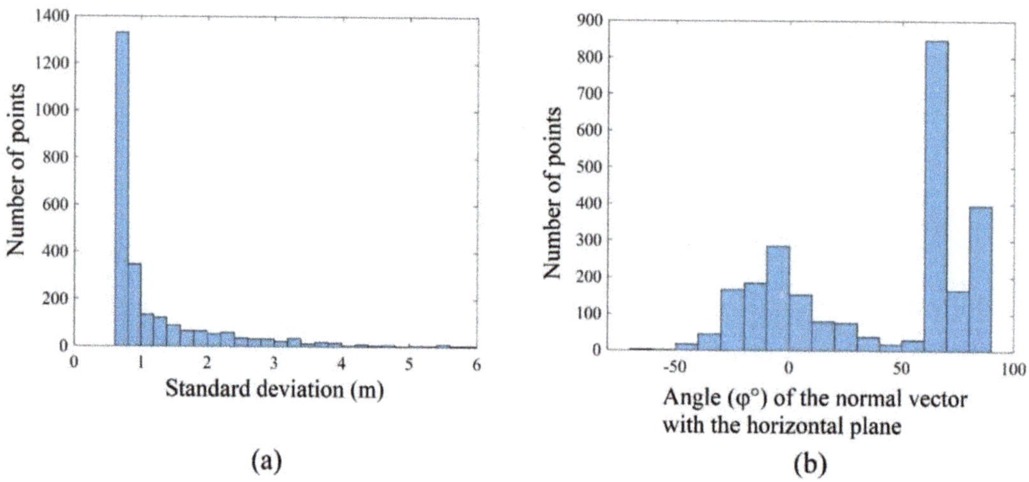

Figure 5. (**a**,**b**) Histograms of standard deviation (σ) and the angle ($\varphi°$) of normal vector with horizontal plane of noisy point cloud.

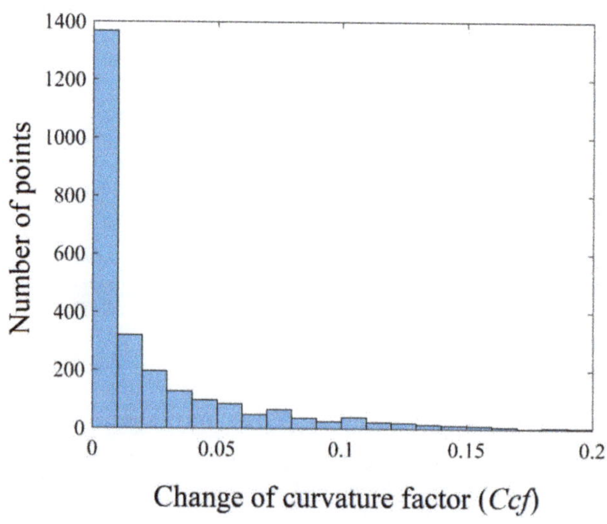

Figure 6. Histogram of change of curvature factor (C_{cf}) of noisy point cloud.

In Figure 5a, the longest bar, which is the leftmost bar, represents most of the points that belong to the roof. Indeed, the majority of roof point neighborhoods (except for the roof plane boundary points) have similar standard deviations. Moreover, seen in Figure 5a, the value of roof standard deviation is at a minimum in comparison to the plane boundaries, tree, and noisy points. In fact, the typical roof points' neighborhood standard deviation value is dependent on factors such as cloud altimetry accuracy and roof-texture thickness. At this stage, it is important to note that some of the points represented by the roof section of the histogram do not belong to the roof. This is because among the tree points it is possible to find some points that fit a plane with their neighborhood with minimal standard deviation. Moreover, some of the roof points such as the roof plane boundary points, the points of roof details such as chimneys and windows, and roof noise points, may be present in the other histogram bins. Despite the majority of roof points being represented by the roof section bins in the histogram, the percentage in these bins differs from one building to another. Indeed, the variation in texture thickness, the point density, and the roof complexity play an essential role in determining this percentage. Notwithstanding this, in all buildings, the majority of roof points will be represented by the roof section of the standard deviation histogram. Nevertheless, the use of only the standard deviation criterion is not enough to detect the roof points. Finally, the same analysis can be applied to the histogram of C_{cf} presented in Figure 6.

In Figure 5b, knowing that the histogram is oriented from the left to the right, all bars starting from the longest one to the right end of the histogram represent the majority of roof points. All remarks in the last paragraph can also be applied to this histogram.

At this stage, there are three remarks that must be made. First, each one of the three histograms illustrated in Figures 5 and 6 can be divided into two sections: a roof point section that represents the majority of roof points and a noise section that represents the majority of noise points. Second, in the roof point sections of the three histograms, the percentage of roof points is different from one histogram to another. Third, the number of points presented in the roof sections in the three histograms is not the same.

In order to smartly (automatically) determine the three threshold values employed in Equation (2), four rules are empirically adopted. Figure 7 illustrates the used algorithm for automatic calculation of the employed thresholds. The applied rule list could be summarized as follows:

Rule 1: In the case of the building represented by Figure 5, the most frequent value of the angle ($\varphi°$) according to Figure 5b is 70°. This value is very big with respect to the other values in the interval of angle value $[-60°, 90°]$. This is why a tolerance value has to be added in order to consider the small uncertainties of fitted plane orientations. These uncertainties normally occur due to the use of a minimum number of neighboring points for fitting the best plane. Consequently, the threshold $T_{h\varphi}$ is shifted to two bins to the left of the most frequent angle bin.

Rule 2: Rule 1 is applied for the $T_{h\sigma}$ threshold, except the threshold $T_{h\sigma}$ is shifted to only one bar to the right of the most frequent standard deviation bar.

Rule 3: The threshold T_{hccf} is considered to be the most frequent value in the histogram (Figure 6) without shifting.

Rule 4: The threshold (among the last three) that allows the selection of the maximum number of points is chosen, and the other two threshold values, are shifted so as to select a similar number of points using each threshold individually.

The application of this strategy allows the value of thresholds to change from one building to other (Table 1).

From Table 1, it can be noted that the threshold $T_{h\sigma}$ value is constant in eight buildings ($T_{h\sigma}$ = 0.8 m). In fact, the nine buildings belong to the same point cloud (Vaihingen city) and have the same urban typology and similar texture. On the other hand, the values of the thresholds $T_{h\varphi}$ and T_{hccf} are related to the architecture complexity level of the roof, e.g., Building 2 has the most architecturally complex roof. This is why the $T_{h\varphi}$ value is the smallest (30°) and T_{hccf} value is the largest (0.04).

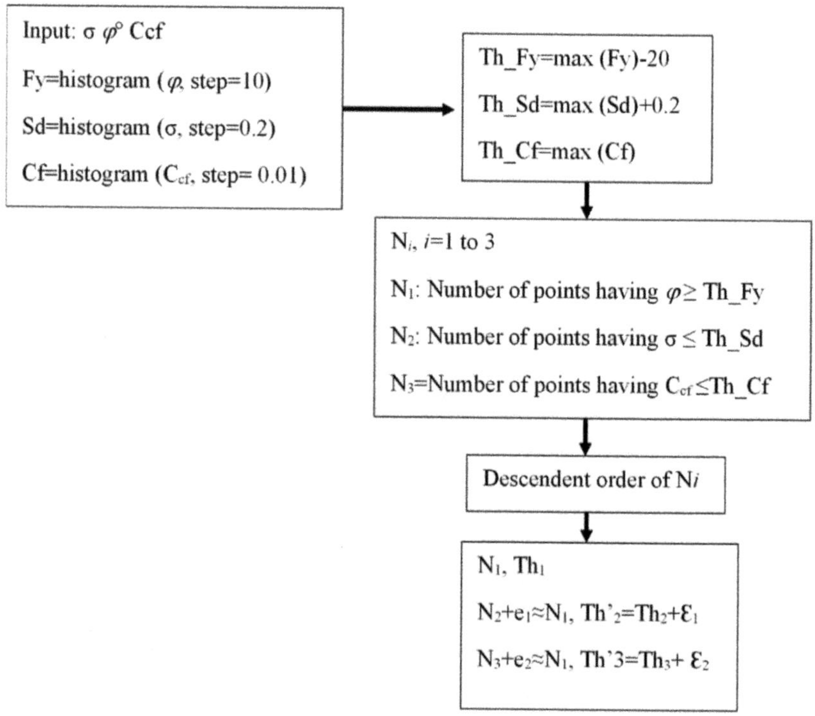

Figure 7. Algorithm flowchart for smart calculation of employed thresholds; Th_i (Th_Fy, Th_Sd, Th_Cf) are the used thresholds; ε_i are last shifts applied on thresholds; e_i: are difference of number of points detected by thresholds before applying the shifts ε_i.

Table 1. Values of calculated thresholds for the three buildings represented in Figure 3 in addition to five other buildings from Vaihingen point cloud.

	$\mathbf{T_{h\varphi}}$ **(Degree)**	$\mathbf{T_{h\sigma}}$ **(m)**	$\mathbf{T_{hccf}}$ **(Unitless)**
Building 1	40	0.8	0.01
Building 2	30	0.8	0.04
Building 3	50	0.8	0.02
Building 4	30	0.9	0.03
Building 5	30	0.8	0.02
Building 6	40	0.8	0.02
Building 7	30	0.8	0.01
Building 8	45	0.8	0.02
Building 9	30	0.8	0.02

Once the thresholds values are assigned, Equation (2) is applied for detecting the roof points. Figure 8a shows the results of the application of this procedure. It may be observed that all roof points are detected, except for the boundary points and the roof detail points. To complete the recognition of roof points, the N_m matrix is employed. For each detected point in Figure 8a, all its neighboring points have to be added to the roof class according to the matrix N_m. Figure 8b illustrates the result of adding the neighboring points. At this stage, all remaining points in the noisy point cloud will represent the tree class (red color in

Figure 8c). Inside the black circles in Figure 8c, some roof points that were misclassified as vegetation points can be seen. The process for improving this result is detailed in the next section.

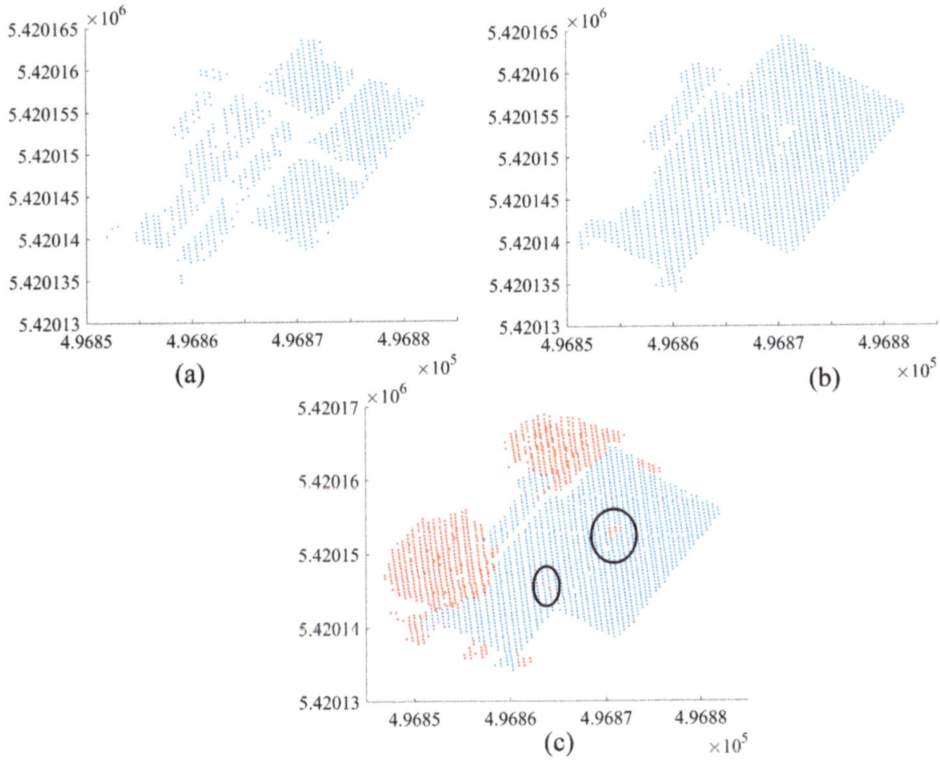

Figure 8. Separation roof points from high tree points. (**a**) Employment threshold result; (**b**) adding neighboring points to the result; (**c**) building points are in blue and tree points are in red; black circles refer to misclassified points.

5.5. Improvement of Filtering Results

The filtering algorithm presented previously separates the noisy roof cloud into two classes, which are the roof class and the noise or vegetation class. In the obtained result (see Figure 8c), there are some building points classified as noise and vice versa. In this section, an additional operation will be added to the filtering algorithm for improving the result. In this context, two normalized Digital Surface Models (nDSM) are calculated [30]. The first one is of the roof class (Figure 9b), whereas the second one is of the noise class (Figure 9a). The two nDSM models have the same dimensions which correspond to the dimensions of the whole noisy point cloud nDSM. At this stage, a new matrix is defined that is named Noisy Cloud 2D Matrix (NC2DM). To calculate this matrix, the noisy point cloud is superimposed over an empty matrix with the same dimensions as the last nDSM. Each cell in this matrix will contain a set of points that are located within the cell borders [30]. In fact, NC2DM has special importance because it serves as a link between the nDSM and the noisy point cloud, e.g., one nonzero pixel in nDSM corresponds to a list of points (belonging to the noisy point cloud) that are located in this pixel. Consequently, if any pixel of the roof nDSM is reassigned to the noise class, the list of points located in this pixel will be extracted from the roof cloud and reassigned to the noise cloud. In this context, the matrix NC2DM offers the possibility of directly recognizing the set of points that is moved from one class to another.

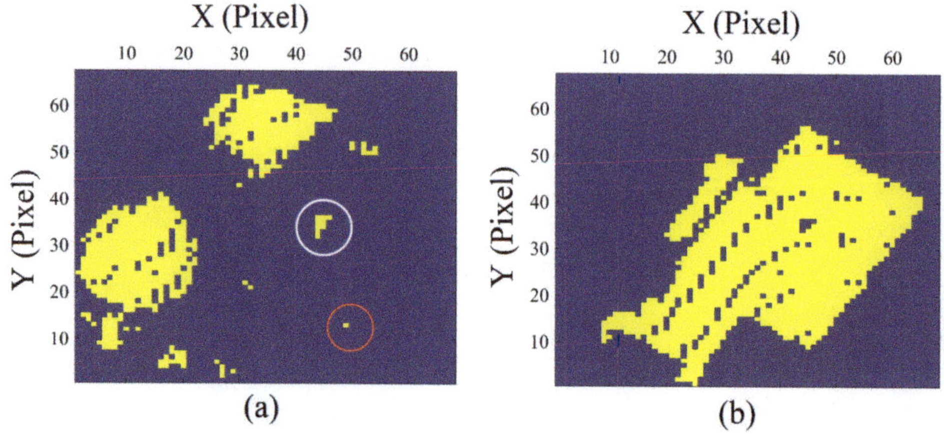

Figure 9. Visualization of normalized Digital Surface Models (nDSM) before improvement; (**a**) vegetation class; (**b**) roof class.

Starting with the noise nDSM, a region-growing algorithm is applied for segmenting the noise nDSM. For each segment, if it is entirely located within the roof perimeter then it is considered as part of the roof, e.g., the segment situated within the white circle in Figure 9a is located inside the roof perimeter, so it is reassigned to the roof class. Moreover, a segment with an area smaller than 5 pixels and tangent to the roof is considered as part of the roof, e.g., the segment within the red circle in Figure 9a is tangent to the roof perimeter and its area is equal to 1 pixel, so it is reassigned to the roof class. The last two types of pixels are cancelled from the noise class and reassigned to the roof class. Inversely, the same operation is applied to the roof class.

Once the improved filtering algorithm is completed, rule 8 (Section 3) is applied to add the result of fuzzy bars analysis. Figure 10 shows the final results of filtering the three building point clouds illustrated in Figure 3. In this figure, all trees (in green color) are recognized, and the building roof (in red color) is detected accurately.

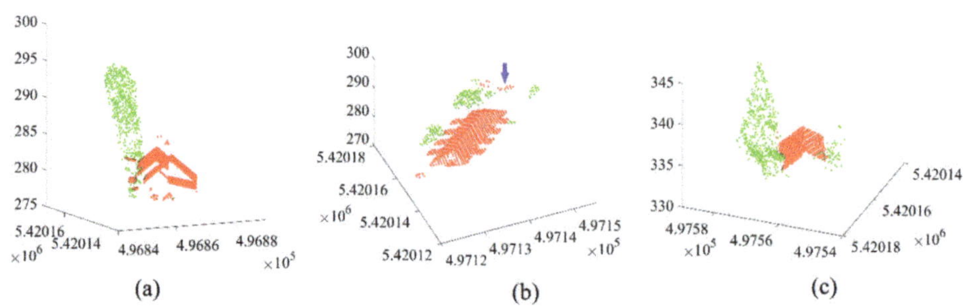

Figure 10. Final filtering results of three building point clouds illustrated in Figure 3; (**a**) Building 1; (**b**) Building 2; (**c**) Building 3. Green color represents vegetation and red color represents roof.

At this stage, it is important to note that the extended filtering algorithm can be applied in the case of existing high trees associated with the building as well as in the case of nonexistent high trees in the building point cloud. The difference between the two cases is the processing time. Of course, if there are no high trees occluding the roof, the application of the original filtering algorithm will be faster because the processing time consumed by the extended algorithm is greater than the original algorithm. In the next section, the accuracy of the suggested filtering algorithm will be estimated using the five point clouds exposed in Section 6.1.

6. Results

Before presenting the results of the suggested approach, it is unavoidable to detail the used dataset samples. The next section introduces the employed point clouds.

6.1. Datasets

To evaluate the efficiency of the suggested approach, five datasets were used (Hermanni, Strasbourg, Toronto, Vaihingen and Aitkenvale) (See Figure 11 and Table 2). These five test sites contain 12, 56, 72, 68 and 28 buildings, respectively. The selected sites represent different urban typologies, and their point clouds have different point densities. While some of them contain high trees associated with the building, others are vegetation-free. This choice is adopted for estimating the reliability of the filtering approach in these two cases. Table 2 shows the characteristics of the selected point cloud samples such as their acquisition dates and height, the employed sensor and the point density.

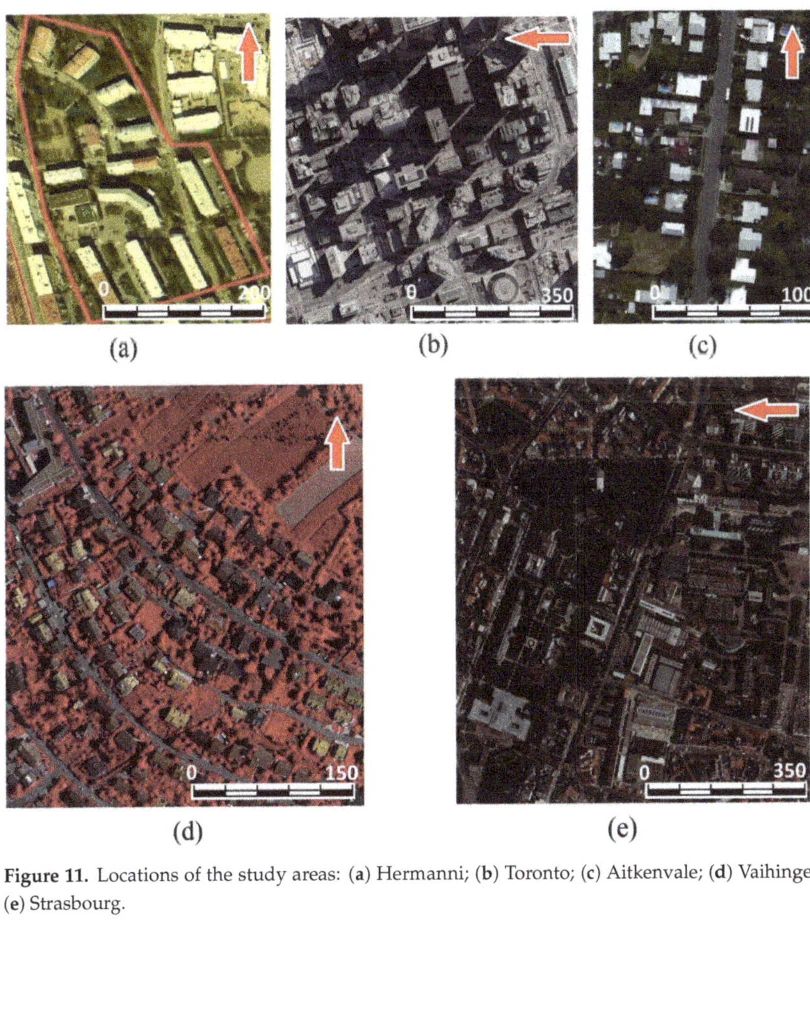

(a) (b) (c)

(d) (e)

Figure 11. Locations of the study areas: (**a**) Hermanni; (**b**) Toronto; (**c**) Aitkenvale; (**d**) Vaihingen; (**e**) Strasbourg.

Table 2. Tested datasets.

	Hermanni	Strasbourg	Toronto	Vaihingen	Aitkenvale
Acquisition	June 2002	September 2004	February 2009	August 2008	–
Sensor	TopoEye	TopScan (Optech ALTM 1225)	Optech ALTM-ORION M	Leica Geosystems (Leica ALS50)	–
Point density (point/m^2)	7–9	1.3	6	4–6.7	29–40
Flight height (m)	200	1440	650	500	>100

The Hermanni dataset represents a housing area in Helsinki city, where multistorey buildings are enclosed by trees. This data belongs to the building extraction project of EuroSDR [31].

The Strasbourg point cloud in France represents Victory boulevard. This sample has special importance because it has a low point density and it represents several urban typologies. Toronto point cloud [32] in Canada, contains both low and high-storey buildings with significant architectural variety.

The fourth dataset is selected in Germany, in Vaihingen city [32]. The scanned area covers small detached houses within a zone rich in vegetation. The trees associated with buildings represent a high variation of quality and volume. The last dataset is of Aitkenvale in Queensland, Australia [33]. It contains 28 buildings with a high point density (29–40 point/m^2). It covers an area of 214 m $\times$ 159 m and it contains residential buildings and tree coverage that partially covers buildings. In terms of topography, it is a flat area.

6.2. Accuracy Estimation

To estimate the accuracy of the extended filtering algorithm, the roof point clouds of the chosen building samples were manually extracted (point per point), since manual extraction is supposed to be more accurate than automatic extraction [1], and then considered as the reference.

Figure 12 compares the number of points between the building point clouds, the reference roof clouds, and the filtered roof clouds of the Vaihingen dataset. The first 51 buildings have trees associated with the roofs and the last 17 buildings are high-vegetation-free. It can be seen that the filtering algorithm performs the task of recognizing high trees associated with the roofs well. Consequently, the results in the case of high-tree-occluded buildings are comparable to the case of high-tree-free buildings. Indeed, the number of points of the filtered roof clouds is similar to that of the reference models.

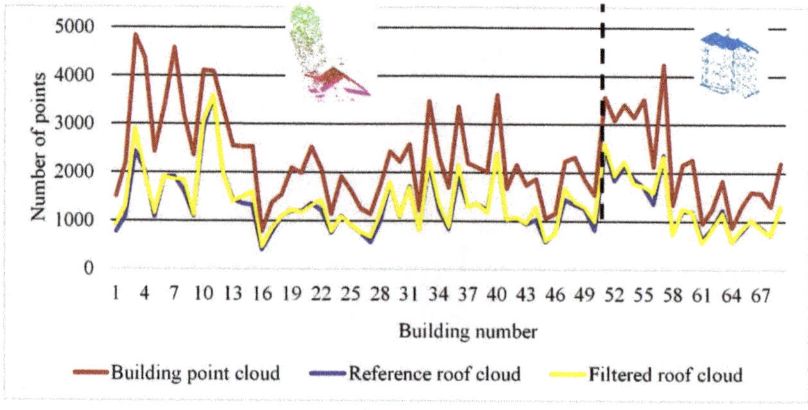

Figure 12. Comparison between number of points of building point clouds, reference roof clouds and filtered roof clouds of Vaihingen dataset; the first 51 buildings have trees associated with roofs and the last 17 buildings have no trees associated with roofs.

In the context of error estimation, the confusion matrix is used for calculating the correctness (Corr), the completeness (Comp), and the quality indices (Q) (Equations (3)–(5) [1]):

$$\text{Comp} = \frac{TP}{TP + FN} \quad (3)$$

$$\text{Corr} = \frac{TP}{TP + FP} \quad (4)$$

$$\text{Q} = \text{Corr} \times \text{Comp} \quad (5)$$

where *TP* (true positive) means the common points of both reference and detected roof, *FN* (false negative) means the points of the reference not found in the detected roof and *FP* (false positive) means the points of the detected roof not found in the reference.

Table 3 shows the correctness, completeness and quality indices for the five datasets used. It can be noted that the undesirable point proportion in the building point cloud is considerable (average = 26.7%). The high accuracy of the suggested filtering algorithm can be observed from the values of the correctness and completeness indices that were almost always greater than 95%.

Table 3. Average values of three datasets' filtering accuracy elements.

Dataset	Number of Buildings	Number of Points	Undesirable Points (%)	Corr (%)	Comp (%)	Q (%)
			Average Values			
Hermanni	12	7380	13.37	99.69	99.76	99.45
Strasbourg	56	1433	18.05	98.44	95.63	94.24
Toronto	72	14,700	23.25	98.56	95.57	94.23
Vaihingen	68	2542	41.65	94.85	98.39	93.37
Aitkenvale	28	9822	37.1	98.1	98.75	96.93
Average		7175	26.7	97.9	97.6	95.6

Corr: Correctness; Comp: Completeness; Q: Quality

The quality value for the Toronto data (94.23%) is smaller in comparison to the Hermanni data (99.45%). In fact, as mentioned in Section 6.1, the Toronto data represents tall buildings with a wide variety of rooftop and façade structures and textures, e.g., in the building illustrated in Figure 13a–c, a set of high-density points are detected inside the building (see the red arrow in Figure 13a). These points are perhaps due to the quality of the façade texture on that level. On one hand, as these points fit a plane and have a high density, they are detected by the filtering algorithm. On the other hand, these points were confusing for the operator who achieved the filtering manually because it is unusual to find this quantity of points inside a building. To summarize, the cascading flat roof property and texture variety in the data for Toronto play a major role in this result, and it is possible that the quality value may be underestimated in this case.

From another viewpoint, the quality evaluation of Vaihingen, Toronto, Strasbourg, and Aitkenvale filtering results are all smaller also than that of Hermanni dataset. Indeed, the developed algorithm suffers from three limitations that decrease the accuracy of the final filtering results. These limitations are:

- In the case of high buildings with a high point density, the small roof planes that are not located in the main roof level may not be detected.
- In the case of buildings that consist of several masses with different heights, the portions of façade points with the same altitudes of the low roof planes will not be cancelled.
- Some of high tree points will not be eliminated.

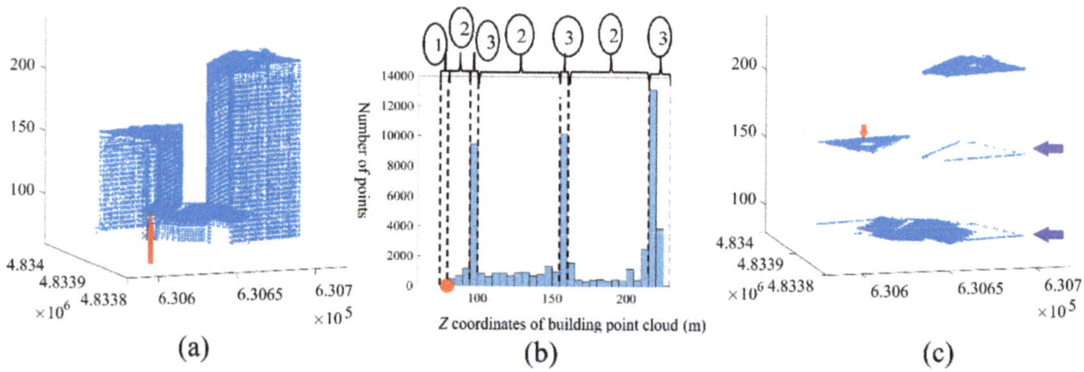

Figure 13. (**a**) Building point cloud in Toronto dataset; (**b**) histogram of Z coordinate of building point cloud; (**c**) final result of building point cloud filtering; the red dot in (**b**) is PSTF; the blue arrow in (**c**) façade points that are saved; the red arrow in (**c**) the missed small and low plane; "1,2,3" in circles in (**b**) are building section numbers.

On one hand, these limitations may adequately explain why the FN values are slightly greater than zero. On the other hand, the higher FN values may be due to the accuracy of the manual extraction of the reference roof clouds. Indeed, in the manual extraction, the presence of confusing points attached to the transparent surfaces of buildings sometimes makes the operator unsure of the proper classification. Regarding the importance of the last three points, they will be detailed in Section 7.

However, despite the difference in the typology, the point density and the undesirable point percentage in these five point-clouds, the excellent results show the capacity of the suggested filtering algorithm to process the majority of building and point cloud qualities.

7. Discussion

In this section, four ideas are discussed which are the accuracy discussion, accuracy comparison between the suggested approach and similar ones, comparison with deep-learning-based methods and the ablation study applied on a building point cloud containing high trees.

7.1. Filtering Accuracy Discussion

Concerning the algorithm limitations, if the roof area is large and the point density is high, the threshold of the number of the façade points according to Rule 6 (Section 3) will have a considerable relative size (10% of the greatest roof bar), e.g., if a flat roof area = 500 m^2 and the mean point density = 6 points/m^2, the threshold of the number of the façade points according to Rule 6 =300 points. Consequently, the detection probability of a small and low roof plane (for example: area = 5 m^2 and the number of points = 30 points) will be low because the number of the plane points in addition to the façade points belonging to the same bar may be smaller than the selected threshold. In the same context, when the building is high, this probability will decrease more because the point density of the roof surface will be greater than the mean point density. Figure 13c shows that one low and small roof plane is missed by the filtering algorithm (the red arrow points to a missed roof plane).

The second case of probable filtering error is when the building consists of several masses of differing heights. The portions of façade points that have the same altitudes of the lower roof planes will not be cancelled (blue arrows in Figure 13c). Indeed, the saved façade points in this case may belong to the same bins of the lower roof planes.

In the case of trees associated with roofs, despite the suggested algorithm's ability to recognize the majority of tree points, some tree points with the same characteristics of roof points (σ, φ and C_{cf}) may not be eliminated (see blue arrow in Figure 10b). In fact, the percentage of these points in all tested buildings, in comparison to the number of the tree points, was negligible (less than 1%), which is why their influence on the modelling step was neglected.

Another limitation of the suggested algorithm is that it was tested only on the Lidar point clouds. Therefore, more tests are required in the future for other point cloud types such as the photogrammetric point clouds from oblique areal imaging, oblique scanners or from Unmanned Aerial Vehicle (UAV) imagery.

In conclusion, despite the last limitations not being common, the algorithm could be improved by considering them. Although the suggested filtering algorithm has the previous limitations, it always conserves the main roof planes and eliminates the majority of undesirable points (Q = 95.6%). The testing results on a great number of buildings, illustrated in Figure 12, shows the proficiency of the suggested approach, where the successful filtering rate is equal to 98% despite the variation in typology and point density of selected samples.

7.2. Accuracy Comparison

In the literature, the building point cloud filtering procedure is applied in two contexts. First, in the context of automatic classification of a point cloud when the classification approach aims to detect the building point clouds, the filtering operation is applied on the building mask to improve its quality [9,11]. Second, in the building modelling algorithm, the filtering operation is applied to the building point cloud before starting the construction of the building model. Hence, the filtering algorithm aims to eliminate the undesirable points before starting the modelling step [34–36]. In the two last cases, the filtering algorithm is merged with the classification or modelling approaches. This is why it is rare to find independent accuracy estimations of a filtering algorithm in the literature. Despite this, the influence of the employed filtering operation will appear in the quality of the final product (building mask or building model). This is why the accuracy of the suggested approach is compared with the accuracy of the approaches [7,9,11,34–36] in the context of comparing the efficacity of the suggested approach with other similar approaches suggested in the literature

To clarify how this comparison can be achieved, let us take this example. First, the result of an urban point cloud classification is a building mask named m_1. According to the classification accuracy, a certain amount of nonbuilding points and noisy points may be present in this mask. Second, a region-growing algorithm is applied to this mask to detect each building independently and is then filtered by the suggested algorithm in this paper. This procedure allows for generating a filtered building mask named m_2. Third, a 2D outlines modelling algorithm is applied to m_1, which produces a new model named m_3. Finally, the three calculated masks can be compared with a reference model to estimate their accuracy through the confusion matrix. If the deformations generated by the modelling algorithm in m_3 are neglected, a comparison between the last three models' accuracies can be achieved, to judge which model is more faithful to the reference model.

According to Griffiths and Boehm [37], the high trees associated with the building roof are considered the main source of incorrect results among the several LiDAR data procedures such as automatic building extraction using convolutional neural networks [7,11] and building outlines' modelling [34–36]. In this context, it is useful to compare the accuracy of the suggested approach with the accuracy of six previous studies (Table 4).

At this stage, it is more appropriate to compare algorithms as they perform on common datasets. This is why all selected approaches in Table 4 employ the Vaihingen dataset (Table 2) provided by the ISPRS (International Society for Photogrammetry and Remote Sensing).

From Table 4, it can be noticed that correctness and the quality factors in the suggested approach (97.9% and 95.6%) are considerably greater than the other six approaches, where the maximum values are equal to 96.8% and 93.2% in Zhao et al. [36]. These two measures consider the FP value (the points of the detected roof not found in the reference) which represent, in this case, the high trees occluding to the roofs. The completeness factor does not consider the FP value, which is why the values of completeness of the previous approaches (99.4% in Widyaningrum et al. [35]) are sometimes greater than the same measure in the suggested approach (97.6%). From this comparison, the importance of elimination of the high trees associated with the roofs can be deduced.

Table 4. Accuracy comparison of suggested approach with previous studies.

	Corr (%)	Comp (%)	Q (%)
Maltezos et al. [11]	85.3	93.8	80.8
Widyaningrum et al. [35]	90.1	99.4	89.6
Huang et al. [34]	96.8	96.2	93.2
Zhao et al. [36]	91.0	95.0	86.8
Hui et al. [9]	91.61	93.61	85.74
Wen et al. [7]	95.1	91.2	86.7
Suggested approach	97.9	97.6	95.6

It is important to note that the accuracy of the classification approaches which use the convolutional neural networks [7,11] is lower than the accuracy of the suggested approach (Table 4) when applied to the same Vaihingen dataset. This fact emphasizes the necessity of recognition and elimination of the high trees associated with building roofs which sometimes play the main role of accuracy dropping in the classification algorithms.

Finally, it is important to remember that the approach suggested by Tarsha Kurdi et al. [10] did not consider the case of tree presence that is associated with the building roof, which is why it gives the erroneous result shown in Figure 3, when there are trees attached to the building roof.

7.3. Comparison with Deep-Learning-Based Methods

At this stage, a specific comparison with deep-learning-based methods can clarify the importance of the suggested approach, which is a rule-based approach. One deep-learning-based approach consists of three main steps: training, validation, and testing [11]. The training operation permits calculating the architecture deep-learning network parameters that allow for classifying the labelled input point cloud. For this purpose, the employed labelled point cloud is divided into training and validation data. The same network with the calculated parameters will be applied to the validation point cloud. Once the learning parameters are validated, they can be used for testing other datasets. Otherwise, the rule-based filtering method is based on the physical behavior of the point cloud [38], and consists of a list of operations connected through a suggested workflow network. On one hand, the application of the deep-learning-based approach to LiDAR data envisages three issues, which are the input feature selection, the training- and validation-data labelling, and overfitting. On the other hand, its advantages regarding rule-based approaches are the possibility of simultaneous application in different types of land, no need for parameter adjustment, and fast performance for high-volume data [38]. In fact, the suggested rule-based approach does not suffer from the issues presented in deep-learning-based approaches. Furthermore, the parameter adjustment problem is minimized by using the idea of fuzzy bars among the Z-coordinate histogram analysis (see Section 3) and the employment of smart thresholds (see Section 5.4). Finally, in order to be fair, the deep-learning-based approaches for LiDAR data filtering nowadays represent a hot research topic that can profit from all available rule-based approaches to recognize the efficient input features.

7.4. Ablation Study

To show the contributions of the improved modules, an ablation study is applied to the building point cloud shown in Figure 10a. As shown in Table 5, the total building point cloud contains 2936 points and consists of three classes: roof, tree and undesirable points. The filtering algorithm before the extension eliminates only the undesirable points (433 points). Thereafter, for removing the high tree crown points, the employment of three features together ($\varphi°$, σ and *ccf*) with the smart thresholds enable recognizing the roof points accurately. Consequently, the results illustrated in Figure 5 underline the necessity of the extension achieved on the original filtering algorithm.

Table 5. Number of points detected in each step of suggested algorithm.

Total Build	Z_His	$\varphi°$	σ	*ccf*	$\varphi°+\sigma+ccf$	N_m	Imp	Fuz	Roof	Tree	Und
2936	2503	1456	1329	1368	1065	597	14	55	1727	776	433

Total Build: Building point cloud; Z_his: Z histogram analysis; N_m: neighborhood matrix; Imp: improvement of filtering algorithm; Fuz: fuzzy bars analysis; Und: undesirable points; +: and.

8. Conclusions and Perspective

This paper suggested a new approach to automatic building point cloud filtering for detecting the roof point cloud. This algorithm is important as the majority of building-modelling algorithms try to construct the building models by focusing on the building roof point cloud. Indeed, eliminating the undesirable points from the building point cloud helps to improve the quality of the constructed building model and increases the probability of success of the modelling algorithm, especially when the quantity of the undesirable points is considerable. In this context, the suggested approach considers the general case when there are trees occluding the building roof. Although the suggested algorithm suffers from some limitations that may slightly reduce the accuracy of some results, the expounded filtering algorithm conserves the building's main roof cloud and eliminates most of the undesirable points. Nevertheless, more investigations are envisaged in the future to make allowance for the limitation cases. Although this algorithm was tested using five datasets with different typologies and point densities, it still needs to be tested and extended in future research to include other point cloud types, such as photogrammetric point clouds from oblique areal imaging or from Unmanned Aerial Vehicle (UAV) imagery.

Author Contributions: Conceptualization: F.T.K.; Methodology: F.T.K.; Software: F.T.K.; Validation: F.T.K., M.A., Z.G., G.C., E.K.D.; Formal analysis: F.T.K.; Investigation: F.T.K., E.K.D., M.A., Z.G., G.C; Resources: F.T.K., M.A., Z.G., G.C., E.K.D.; Data curation: F.T.K., E.K.D., M.A., Z.G., G.C.; Writing—original draft preparation: F.T.K.; Writing—review and editing: F.T.K., M.A., E.K.D., Z.G., G.C.; Visualization: F.T.K., M.A., Z.G., G.C., E.K.D.; Supervision: F.T.K., M.A., Z.G., G.C. All authors have read and agreed to the published version of the manuscript.

Funding: This research received no external funding.

Institutional Review Board Statement: Not applicable.

Informed Consent Statement: Not applicable.

Data Availability Statement: ISPRS datasets was provided by the German Society for Photogrammetry, Remote Sensing and Geoinformation (DGPF) [32]. Hermanni dataset was provided by EuroSDR [31]. Dataset of Aitkenvale in Queensland, Australia is provided by Awrangjeb et al. [33]. Strasbourg dataset is provided by Tarsha Kurdi et al. [2].

Acknowledgments: We would like to thank the German Society for Photogrammetry, Remote Sensing and Geoinformation (DGPF) for providing Vaihingen and Toronto datasets [32].

Conflicts of Interest: The authors declare no conflict of interest.

References

1. Shan, J.; Yan, J.; Jiang, W. *Global Solutions to Building Segmentation and Reconstruction. Topographic Laser Ranging and Scanning: Principles and Processing*, 2nd ed.; Shan, J., Toth, C.K., Eds.; CRC Press: Boca Raton, FL, USA, 2018; pp. 459–484.
2. Tarsha Kurdi, F.; Landes, T.; Grussenmeyer, P.; Koehl, M. Model-driven and data-driven approaches using Lidar data: Analysis and comparison. In Proceedings of the ISPRS Workshop, Photogrammetric Image Analysis (PIA07), International Archives of Photogrammetry, Remote Sensing and Spatial Information Systems. Munich, Germany, 19–21 September 2007; 2007; Volume XXXVI, pp. 87–92, Part3 W49A.
3. Wehr, A.; Lohr, U. Airborne laser scanning—An introduction and overview. *ISPRS J. Photogramm. Remote Sens.* **1999**, *54*, 68–82. [CrossRef]
4. He, Y. Automated 3D Building Modelling from Airborne Lidar Data. Ph.D. Thesis, University of Melbourne, Melbourne, Australia, 2015.
5. Tarsha Kurdi, F.; Awrangjeb, M. Comparison of Lidar building point cloud with reference model for deep comprehension of cloud structure. *Can. J. Remote Sens.* **2020**, *46*, 603–621. [CrossRef]
6. Shao, J.; Zhang, W.; Shen, A.; Mellado, N.; Cai, S.; Luo, L.; Wang, N.; Yan, G.; Zhou, G. Seed point set-based building roof extraction from airborne LiDAR point clouds using a top-down strategy. *Autom. Constr.* **2021**, *126*, 103660. [CrossRef]
7. Wen, C.; Li, X.; Yao, X.; Peng, L.; Chi, T. Airborne LiDAR point cloud classification with global-local graph attention convolution neural network. ISPRS J. Photogramm. *Remote Sens.* **2021**, *173*, 181–194. [CrossRef]
8. Varlik, A.; Uray, F. Filtering airborne LIDAR data by using fully convolutional networks. *Surv. Rev.* **2021**, 1–11. [CrossRef]
9. Hui, Z.; Li, Z.; Cheng, P.; Ziggah, Y.Y.; Fan, J. Building extraction from airborne LiDAR data based on multi-constraints graph segmentation. *Remote Sens.* **2021**, *13*, 3766. [CrossRef]
10. Tarsha Kurdi, F.; Awrangjeb, M.; Munir, N. Automatic filtering and 2D modelling of Lidar building point cloud. *Trans. GIS J.* **2021**, *25*, 164–188. [CrossRef]
11. Maltezos, E.; Doulamis, A.; Doulamis, N.; Ioannidis, C. Building extraction from LiDAR data applying deep convolutional neural networks. *IEEE Geosci. Remote Sens. Lett.* **2019**, *16*, 155–159. [CrossRef]
12. Liu, M.; Shao, Y.; Li, R.; Wang, Y.; Sun, X.; Wang, J.; You, Y. Method for extraction of airborne LiDAR point cloud buildings based on segmentation. *PLoS ONE* **2020**, *15*, e0232778. [CrossRef]
13. Demir, N. Automated detection of 3D roof planes from Lidar data. *J. Indian Soc. Remote Sens.* **2018**, *46*, 1265–1272. [CrossRef]
14. Tarsha Kurdi, F.; Awrangjeb, M.; Munir, N. Automatic 2D modelling of inner roof planes boundaries starting from Lidar data. In Proceedings of the 14th 3D GeoInfo 2019, Singapore, 26–27 September 2019; pp. 107–114. [CrossRef]
15. Park, S.Y.; Lee, D.G.; Yoo, E.J.; Lee, D.C. Segmentation of Lidar data using multilevel cube code. *J. Sens.* **2019**, *2019*, 4098413. [CrossRef]
16. Zhang, K.; Yan, J.; Chen, S.C. A framework for automated construction of building models from airborne Lidar measurements. In *Topographic Laser Ranging and Scanning: Principles and Processing*, 2nd ed.; Shan, J., Toth, C.K., Eds.; CRC Press: Boca Raton, FL, USA, 2018; pp. 563–586.
17. Li, M.; Rottensteiner, F.; Heipke, C. Modelling of buildings from aerial Lidar point clouds using TINs and label maps. *ISPRS J. Photogramm. Remote Sens.* **2019**, *154*, 127–138. [CrossRef]
18. Vosselman, G.; Dijkman, S. 3D building model reconstruction from point clouds and ground plans. *Int. Arch. Photogramm. Remote Sens. Spat. Inf. Sci.* **2001**, *34*, 37–44.
19. Hu, P.; Miao, Y.; Hou, M. Reconstruction of Complex Roof Semantic Structures from 3D Point Clouds Using Local Convexity and Consistency. *Remote Sens.* **2021**, *13*, 1946. [CrossRef]
20. Nguyen, T.H.; Daniel, S.; Guériot, D.; Sintès, C.; Le Caillec, J.-M. Super-Resolution-Based Snake Model—An Unsupervised Method for Large-Scale Building Extraction Using Airborne LiDAR Data and Optical Image. *Remote Sens.* **2020**, *12*, 1702. [CrossRef]
21. Zhang, W.; Wang, H.; Chen, Y.; Yan, K.; Chen, M. 3D building roof modelling by optimizing primitive's parameters using constraints from Lidar data and aerial imagery. *Remote Sens.* **2014**, *6*, 8107–8133. [CrossRef]
22. Jung, J.; Sohn, G. Progressive modelling of 3D building rooftops from airborne Lidar and imagery. In *Topographic Laser Ranging and Scanning: Principles and Processing*, 2nd ed.; Shan, J., Toth, C.K., Eds.; CRC Press: Boca Raton, FL, USA, 2018; pp. 523–562.
23. Awrangjeb, M.; Gilani, S.A.N.; Siddiqui, F.U. An effective data-driven method for 3D building roof reconstruction and robust change detection. *Remote Sens.* **2018**, *10*, 1512. [CrossRef]
24. Pirotti, F.; Zanchetta, C.; Previtali, M.; Della Torre, S. Detection of building roofs and façade from aerial laser scanning data using deep learning. *ISPRS-Int. Arch. Photogramm. Remote Sens. Spat. Inf. Sci.* **2019**, *4211*, 975–980. [CrossRef]
25. Martin-Jimenez, J.; Del Pozo, S.; Sanchez-Aparicio, M.; Laguela, S. Multi-scale roof characterization from Lidar data and aerial orthoimagery: Automatic computation of building photovoltaic capacity. *J. Autom. Constr.* **2020**, *109*, 102965. [CrossRef]
26. Sibiya, M.; Sumbwanyambe, M. An algorithm for severity estimation of plant leaf diseases by the use of colour threshold image segmentation and fuzzy logic inference: A proposed algorithm to update a "Leaf Doctor" application. *AgriEngineering* **2019**, *1*, 205–219. [CrossRef]
27. Dey, K.E.; Tarsha Kurdi, F.; Awrangjeb, M.; Stantic, B. Effective Selection of Variable Point Neighbourhood for Feature Point Extraction from Aerial Building Point Cloud Data. *Remote Sens.* **2021**, *13*, 1520. [CrossRef]
28. Sanchez, J.; Denis, F.; Coeurjolly, D.; Dupont, F.; Trassoudaine, L.; Checchin, P. Robust normal vector estimation in 3D point clouds through iterative principal component analysis. *ISPRS J. Photogramm. Remote Sens.* **2020**, *163*, 18–35. [CrossRef]

29. Thomas, H.; Goulette, F.; Deschaud, J.; Marcotegui, B.; LeGall, Y. Semantic Classification of 3D Point Clouds with Multiscale Spherical Neighborhoods. In Proceedings of the International Conference on 3D Vision (3DV), Verona, Italy, 5–8 September 2018; pp. 390–398. [CrossRef]
30. Tarsha Kurdi, F.; Landes, T.; Grussenmeyer, P. Joint combination of point cloud and DSM for 3D building reconstruction using airborne laser scanner data. In Proceedings of the 4th IEEE GRSS / WG III/2+5, VIII/1, VII/4 Joint Workshop on Remote Sensing & Data Fusion over Urban Areas and 6th International Symposium on Remote Sensing of Urban Areas, Télécom, Paris, France, 11–13 April 2007; p. 7.
31. Eurosdr. Available online: www.eurosdr.net (accessed on 23 November 2021).
32. Cramer, M. The DGPF test on digital aerial camera evaluation—Overview and test design. *Photogramm.-Fernerkund.-Geoinf.* **2010**, *2*, 73–82. [CrossRef]
33. Awrangjeb, M.; Zhang, C.; Fraser, C.S. Automatic extraction of building roofs using LiDAR data and multispectral imagery. *ISPRS J. Photogramm. Remote Sens.* **2013**, *83*, 1–18. [CrossRef]
34. Huang, R.; Yang, B.; Liang, F.; Dai, W.; Li, J.; Tian, M.; Xu, W. A top-down strategy for buildings extraction from complex urban scenes using airborne LiDAR point clouds. Infrared Phys. *Technol.* **2018**, *92*, 203–218. [CrossRef]
35. Widyaningrum, E.; Gorte, B.; Lindenbergh, R. Automatic Building Outline Extraction from ALS Point Clouds by Ordered Points Aided Hough Transform. *Remote Sens.* **2019**, *11*, 1727. [CrossRef]
36. Zhao, Z.; Duan, Y.; Zhang, Y.; Cao, R. Extracting buildings from and regularizing boundaries in airborne LiDAR data using connected operators. *Int. J. Remote Sens.* **2016**, *37*, 889–912. [CrossRef]
37. Griffiths, D.; Boehm, J. Improving public data for building segmentation from Convolutional Neural Networks (CNNs) for fused airborne lidar and image data using active contours. *ISPRS J. Photogramm. Remote Sens.* **2019**, *154*, 70–83. [CrossRef]
38. Ayazi, S.M.; Saadat Seresht, M. Comparison of traditional and machine learning base methods for ground point cloud labeling. *Int. Arch. Photogramm. Remote Sens. Spatial Inf. Sci.* **2019**, *XLII-4/W18*, 141–145. [CrossRef]

remote sensing

MDPI

Review

Spatial Validation of Spectral Unmixing Results: A Systematic Review

Rosa Maria Cavalli

Research Institute for Geo-Hydrological Protection (IRPI), National Research Council (CNR), 06128 Perugia, Italy; rosa.maria.cavalli@irpi.cnr.it; Tel.: +39-075-501-422

Abstract: The pixels of remote images often contain more than one distinct material (mixed pixels), and so their spectra are characterized by a mixture of spectral signals. Since 1971, a shared effort has enabled the development of techniques for retrieving information from mixed pixels. The most analyzed, implemented, and employed procedure is spectral unmixing. Among the extensive literature on the spectral unmixing, nineteen reviews were identified, and each highlighted the many shortcomings of spatial validation. Although an overview of the approaches used to spatially validate could be very helpful in overcoming its shortcomings, a review of them was never provided. Therefore, this systematic review provides an updated overview of the approaches used, analyzing the papers that were published in 2022, 2021, and 2020, and a dated overview, analyzing the papers that were published not only in 2011 and 2010, but also in 1996 and 1995. The key criterion is that the results of the spectral unmixing were spatially validated. The Web of Science and Scopus databases were searched, using all the names that were assigned to spectral unmixing as keywords. A total of 454 eligible papers were included in this systematic review. Their analysis revealed that six key issues in spatial validation were considered and differently addressed: the number of validated endmembers; sample sizes and sampling designs of the reference data; sources of the reference data; the creation of reference fractional abundance maps; the validation of the reference data with other reference data; the minimization and evaluation of the errors in co-localization and spatial resampling. Since addressing these key issues enabled the authors to overcome some of the shortcomings of spatial validation, it is recommended that all these key issues be addressed together. However, few authors addressed all the key issues together, and many authors did not specify the spatial validation approach used or did not adequately explain the methods employed.

Keywords: mixed pixels; spectral unmixing; spatial validation; accuracy

Citation: Cavalli, R.M. Spatial Validation of Spectral Unmixing Results: A Systematic Review. *Remote Sens.* **2023**, *15*, 2822. https://doi.org/10.3390/rs15112822

Academic Editors: Jorge Delgado García, Fayez Tarsha Kurdi and Tarig Ali

Received: 30 March 2023
Revised: 22 May 2023
Accepted: 24 May 2023
Published: 29 May 2023

1. Introduction

1.1. Background

A pixel that contains more than one "land-cover type" is defined as a mixed pixel, and its spectrum is formed by combining the spectral signatures of these "land-cover types" [1]. The presence of mixed pixels in the image constrains the techniques that can be carried out to analyze, characterize, and classify the remote sensing images [2,3]. To retrieve mixed-pixel information from remote sensing images, a shared research effort allowed developing several methods (e.g., spectral unmixing, probabilistic, geometric-optical, stochastic geometric, and fuzzy models [1]). However, the literature shows that, for over 40 years, spectral unmixing has been the most commonly used method for discrimination, detection, and classification of superficial materials [4–6].

The spectral unmixing was defined as the "procedure by which the measured spectrum of a mixed pixel is decomposed into a collection of constituent spectra, or endmembers, and a set of corresponding fractions, or abundances, that indicate the proportion of each endmember present in the pixel" [6]. It is important to point out that many names were given to the spectral unmixing procedure: hyperspectral unmixing [7,8], linear mixing [9], nonlinear spectral mixing models [10,11], semi-empirical mixing model [12], spectral mixing models [13–15], spectral mixture analysis [16–22], spectral mixture modeling [23,24], and spectral unmixing [19,25,26]. In this paper, the term spectral unmixing was chosen.

The first studies that introduced the spectral unmixing procedure were carried out about 40 years ago (Table 1). In order to study Moon minerals, Adams & McCord [27] observed nonlinear behavior of the spectra of Apollo 11 and 12 samples that were measured in the laboratory. In order to analyze the spectra of Mars, Singer & McCord [28] assumed that the spectrum of the mixed pixel was a bilinear combination of the spectra of its two constituent materials, and it was weighted by their abundances in the mixed pixel; their model required two constraints: the sum of the weighing factors must be one, and their values must not be negative. Hapke [29] proposed a nonlinear mixing model that was called "isotropic multiple scattering approximation" by Heylen et al. [8]. Johnson et al. [12] and Smith et al. [13] combined "spectral mixing model" with the modified Kubelka–Munk model and principal component analysis, respectively. In order to analyze the spectra of Mars, Adams & Smith [23] improved the "bilinear model", which was proposed by Singer & McCord [28], considering more than two constituent materials of the mixed pixel and adding the residual error.

Table 1. Studies that introduced spectral unmixing procedure.

Paper	Publication Year	Study Area	Spectral Range	Name Given to Spectral Unmixing Procedure	Citations in Google Scholar
Adams & McCord [27]	1971	Lunar	0.35–2.5 µm	-	136
Singer & McCord [28]	1979	Mars	0.35–2.5 µm	-	347
Hapke [29]	1981	Planets		-	2200
Johnson et al. [12]	1983	Minerals	0.35–2.5 µm	Semi-empirical mixing model	288
Smith et al. [13]	1985	Minerals	0.60–2.20 µm	Spectral mixing model	454
Adams et al. [23]	1986	Mars	0.35–2.5 µm	Spectral mixture modeling	1634
Adams et al. [16]	1989	-	1.2–2.4 µm	Spectral mixture analysis	131

Adams et al. [16] decomposed the "spectral mixture analysis" in two consecutive steps: the first step decomposes the spectrum of each mixed pixel into a collection of constituent spectra (called endmembers), and the second step determines the proportion of every endmember present in the pixel. The literature highlighted two main models for performing the first step: linear and nonlinear mixture models. To estimate the proportion of every endmember (called fractional abundances), many solutions were proposed (e.g., Gram–Schmidt Orthogonalization [30], Least Square Methods [31], Minimum Variance Methods [6], Singular Value Decomposition [32], Variable Endmember Methods [6]).

1.2. Reviews on the Spectral Unmixing Procedure

In order to more effectively understand the importance of spectral unmixing, a quantification of the works that have studied, implemented, and applied this procedure since 1971 were provided. For this purpose, all names that were given to the spectral unmixing procedure were exploited as terms in the search strategy. A total of 5768 and 5852 papers were identified using Web of Science and Scopus search engines, respectively (accessed on 19 May 2023). Among these papers, 19 reviews offered the status of spectral unmixing (Table 2).

An interesting overview of the "linear models" developed up to 1996 was offered by Ichoku & Karneili [1], who compared this method with four other unmixing models: probabilistic, geometric-optical, stochastic geometric, and fuzzy models. The authors summarized that evaluated spatial accuracies were not representative of the real accuracies at the level of individual pixels because the spatial validation was performed for a few test pixels.

Table 2. Reviews on the spectral unmixing procedure.

Paper	Publication Year	Publication Title	Number of References Cited in the Review	Citations in Google Scholar [1]
Ichoku & Karneili [1]	1996	A review of mixture modelling techniques for subpixel land cover estimation	57	281
Heinz & Chein-I-Chang [33]	2001	Fully Constrained Least Squares Linear Spectral Mixture Analysis Method for Material Quantification in Hyperspectral Imagery	39	1955
Keshava & Mustard [6]	2002	Spectral unmixing	40	2761
Keshava [34]	2003	A Survey of Spectral Unmixing Algorithms	3	641
Martinez et al. [35]	2006	Endmember extraction algorithms from hyperspectral images	16	67
Veganzones & Grana [36]	2008	Endmember Extraction Methods: A Short Review	23	82
Bioucas-Dias & Plaza [7]	2010	Hyperspectral unmixing: Geometrical, statistical, and sparse regression-based approaches	97	77
Parente & Plaza [37]	2010	Survey of geometric and statistical unmixing algorithms for hyperspectral images	53	124
Bioucas-Dias & Plaza [38]	2011	An overview on hyperspectral unmixing: geometrical, statistical, and sparse regression based approaches	51	78
Somer et al. [39]	2011	Endmember variability in Spectral Mixture Analysis: A review	179	660
Bioucas-Dias et al. [40]	2012	Hyperspectral Unmixing Overview: Geometrical, Statistical, and Sparse Regression-Based Approaches	96	2597
Quintano et al. [41]	2012	Spectral unmixing: a review	163	141
Ismail & Bchir [42]	2014	Survey on Number of Endmembers Estimation Techniques for Hyperspectral Data Unmixing	22	1
Heylen et al. [8]	2014	A Review of Nonlinear Hyperspectral Unmixing Methods	201	452
Shi & Wang [43]	2014	Incorporating spatial information in spectral unmixing: A review	106	197
Drumetz et al. [44]	2016	Variability of the endmembers in spectral unmixing: recent advances	26	34
Wang et al. [45]	2016	A survey of methods incorporating spatial information in image classification and spectral unmixing	280	75
Wei & Wang [5]	2020	An Overview on Linear Unmixing of Hyperspectral Data	74	17
Borsoi et al. [4]	2021	Spectral Variability in Hyperspectral Data Unmixing	317	63

[1] Accessed on 31 January 2023.

Heinz & Chein-I-Chang [33] focused on the second constraint of linear spectral mixture analysis (i.e., the fractional abundances of each mixed pixel must be positive), which is very difficult to implement in practice. Reviewing the literature, the authors pointed out that because most research did not know in detail the spectra present in the image scene, their results did not necessarily reflect the true abundance fractions of the materials [33].

Keshava [42] exploited the hierarchical taxonomies to facilitate comparison of the wide variety of methods used for spectral unmixing and revealed their similarities and differences. Furthermore, the author restated that most of the methods developed to solve problems were due to lack of detailed knowledge of ground truth. In their extensive description of spectral unmixing methodology, Keshava and Mustard [6] focused on the processing chain of linear unmixing methods applied to hyperspectral data. The authors highlighted that the shortcomings in spatial validation were due to the lack of detailed ground-truth knowledge; for this reason, the main focus of the research was on determining endmembers, rather than recovering fractional abundance maps [6].

Bioucas-Dias et al. [36] aimed to update the previous review, which was proposed by Keshava and Mustard [6] 10 years earlier. Therefore, the authors extensively described the methods that were proposed from 2002 to 2012 to improve the mathematical validity of the spectral unmixing. Bioucas-Dias & Plaza [7,38], Parente & Palza [37], Veganzones & Grana [40], and Martinez et al. [41] provided brief, but comprehensive reviews of methods for statistical and geometric extraction of endmembers. Somers et al. [39] provided a comprehensive and extensive review of the methods to address the temporal and spatial variability of the endmembers in the spectral unmixing.

An introduction to nonlinear unmixing methods and an overview of the most commonly used approaches were provided by Heylen et al. [8]. These authors also pointed out the lack of detailed ground truths for accurate validation of the spectral unmixing procedures [8]. After performing a general review of spectral unmixing, Quintano et al. [41] provided an interesting summary of its applications. Moreover, the authors pointed out the difficulty in spatially validating the results of spectral unmixing results and identified two main reasons: "(1) it is difficult to collect ground truth as scale directly corresponding to remotely sensed data resolution; (2) traditional classification accuracy analysis measurement tools may not be suitable for mixed pixel analysis" [41].

Wei & Wang [5] presented an overview of four aspects of the spectral unmixing (i.e., geometric method, nonnegative matrix factorization (NMF), Bayesian method, and sparse unmixing), whereas an overview of the methods that estimated the number of endmembers was provided by Ismail & Bchir [39]. Shi & Wang [43] provided a comprehensive review of the methods that combined spatial and spectral information for the spectral unmixing; the authors called them "spatial spectral unmixing" [43]. To extract endmembers, select endmember combinations, and estimate endmember fraction abundances, these methods exploited the correlation between neighboring pixels [43]. Wang et al. [45] provided an overview of the methods that incorporated the spatial information not only in spectral unmixing, but also in the all image classifiers. The authors underlined that most of the spatial accuracy was based on "the idea of area-weighted accuracy" because it was derived from some validation samples.

The most recent review was offered by Borsoi et al. [4], who provided a comprehensive review of the methods to solve the spectral variability problem in hyperspectral data. The spectral variability is mainly due to atmospheric, illumination, and environmental conditions [46,47]. Starting from the availability or non-availability of spectral libraries, the authors organized the "Spectral Unmixing algorithms" "according to a practitioner's point of view, based on the necessary amount of supervision and the computational cost" and highlighted that the algorithms with less supervision (i.e., Fuzzy Unmixing, MESMA—Multiple Endmember Spectral Mixture Analysis—and variants, Bayesian models) are the methods with high computational cost [4]. Moreover, the authors pointed out the difficulty of assessing the accuracy of these methods due to the lack of detailed ground truths [4]. A review of four of these methods, which address the spectral variability problem, was also provided by Drumetz et al. [44].

It is important to mention that the spatial accuracy of spectral unmixing results can be evaluated using images and/or in situ data and/or maps, and the spectral accuracy of spectral unmixing results can be evaluated using spectral signatures that were acquired in situ and/or in the laboratory and/or obtained from images [4,6,8,33,45]. However, an independent validation dataset is required (i.e., the spectral library and/or the reference maps) [48].

1.3. Objectvives

In conclusion, since 1971 many methods have been introduced to improve the mathematical validity of the spectral unmixing procedure, but the validation of the results still needs much improvement, especially the spatial validation. In particular, the lack of detailed ground-truth knowledge is the main reason of the many shortcomings in the spatial validation of the spectral unmixing results. However, no author provided an overview focusing on the spatial validation of the spectral unmixing results.

Therefore, this systematic review aims to provide readers with (a) an overview of how the previous authors approached spatial validation of spectral unmixing results and (b) recommendations for overcoming the many shortcomings of spatial validation and minimizing its errors. The systematic review was carried out in accordance with the Preferred Reporting Items for Systematic reviews and Meta-Analysis (PRISMA) statement [49,50]. The methodological approach employed in this systematic literature review is explained in Section 2, whereas the results, discussion, and conclusions are presented in Sections 3 and 4.

2. Materials and Methods

2.1. Identification Criteria

This systematic literature review aims to provide readers with an overview of the approaches applied for spatial validation of spectral unmixing results and does not claim to be exhaustive since too many works have studied, implemented, and applied this technique since 1971. Therefore, the papers published in 2022, 2021, and 2020 were chosen to analyze the current status, whereas those published not only in 2011 and 2010, but also in 1996 and 1995 were selected to assess the progress over time. The year 1995 was chosen as the initial time for the systematic review, because in this year, spectral unmixing and other "mixture modeling techniques" were well implemented and, thus, commonly employed [1,6,51–54]. The Web of Science (WoS) and Scopus search engines were used to identify the papers that spatially validated the spectral unmixing results and were published in 2022, 2021, 2020, 2011, 2010, 1996, and 1995.

Initially, the papers that named the spectral unmixing in the titles, abstracts, and keywords were identified. For this purpose, all the names assigned to spectral unmixing (i.e., hyperspectral unmixing, linear mixing, nonlinear spectral mixing models, semi-empirical mixing model, spectral mixing models, spectral mixture analysis, spectral mixture modeling, spectral unmixing) were employed as unique query strings (first yellow box in Figure 1).

The total records identified from these databases was 2999. The subject areas of the search engines were checked to refine the identification of the papers. Therefore, "4.169 Remote Sensing", "4.174 Digital Signal Processing", "4.17 Computer Vision & Graphics", "5.250 Imaging &Tomography", "5.20 Astronomy & Astrophysics", "5.191 Space Sciences", "8.8 Geochemistry, Geophysics & Geology", "8.93 Archaelogy", "8.19 Oceanography, Meteorology & Atmospheric", "8.140 Water Resources", "8.124 Environmental Sciences", "3.40 Forestry", and "3.45 Soil Science" were "Citation Topics" selected in the WoS database, whereas "Earth and Planetary Sciences", "Physics and Astronomy", and "Environmental Science" were the subject areas selected in the Scopus database. After refining the subject areas, the identified papers became 2034 (second yellow box in Figure 1): 1396 were the papers published in 2022, 2021, and 2020; 538 were the papers published in 2011 and 2010; 100 were the papers published in 1996 and 1995.

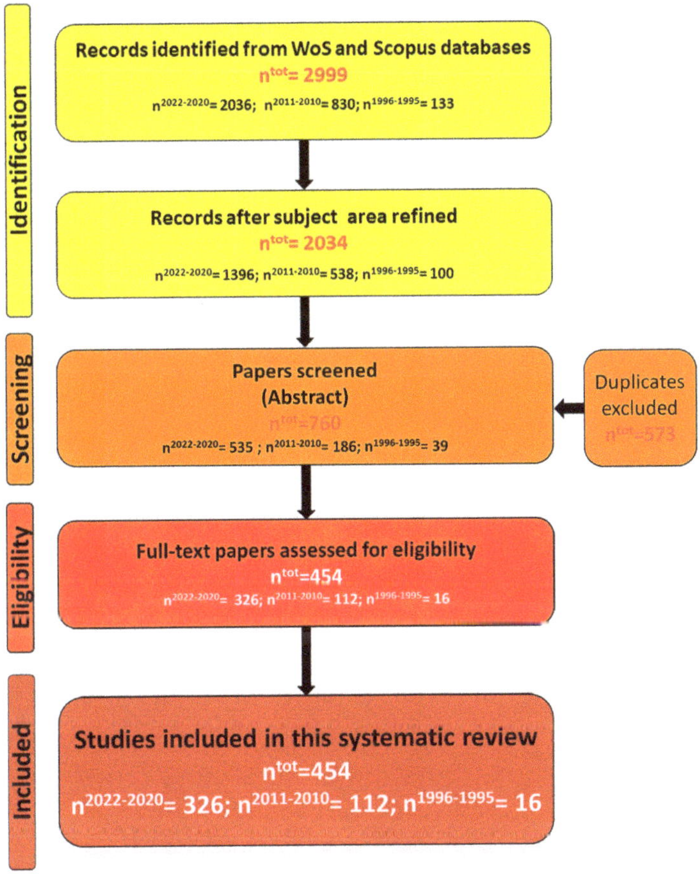

Figure 1. PRISMA flow chart showing the different steps of the dataset creation, where n^{tot} was the total number of papers; $n^{2022-2020}$ was the number of papers that were published in 2022, 2021, and 2020; $n^{2011-2010}$ was the number of papers that were published in 2011 and 2010; $n^{1996-1995}$ was the number of papers that were published in 1996 and 1995.

2.2. Screening and Eligible Criteria

Reading the abstracts of the identified papers was conducted to select only those that applied spectral unmixing to remote images. Excluding the duplicates, 760 papers were selected with the first screening (orange box in Figure 1): 535 were the papers published in 2022, 2021, and 2020; 186 were the papers published in 2011 and 2010; 100 were the papers published in 1996 and 1995.

Reading the full text of the screened papers was conducted to identify only those that spatially validated the spectral unmixing results (bright red box in Figure 1). The last analysis identified the eligible papers: 326 were the papers published in 2022, 2021, and 2020; 112 were the papers published in 2011 and 2010; 16 were the papers published in 1996 and 1995.

In conclusion, 454 eligible papers were included in this systematic review. In Appendix A, the Tables A1–A7 summarize the characteristics of the eligible papers that were published in 2022, 2021, 2020, 2011, 2010, 1996, and 1995, respectively.

3. Results

3.1. Spatial Validation of Spectral Unmixing Results

The screening carried out showed that the number of studies that spatially validated the results of spectral unmixing has significantly increased over the selected years (bright red box in Figure 1): about 100 research papers per year were published in the past 3 years; about 50 research papers per year were published in 2011 and 2010; about 10 research papers per year were published in 1996 and 1995. The screening carried out showed also that the number of studies that applied spectral unmixing has significantly increased over the selected years (orange box in Figure 1): about 180 research papers per year were published in the past 3 years; about 90 research papers per year were published in 2011 and 2010; about 20 research papers per year were published in 1996 and 1995. In order to assess the importance of spatial validation in the spectral unmixing procedure, the papers that applied spectral unmixing to remote imaging were analyzed (orange box in Figure 1). Figure 2 shows the percentage of these papers that were not validated (the percentage in grey wedges), spectrally validated (the percentage in yellow wedges), spatially validated (the percentage in blue wedges), and spatially and spectrally validated (the percentage in green wedges) the spectral unmixing results. Therefore, spatial validation was carried out alone (blue wedges in Figure 2) or together with spectral validation (green wedges in Figure 2).

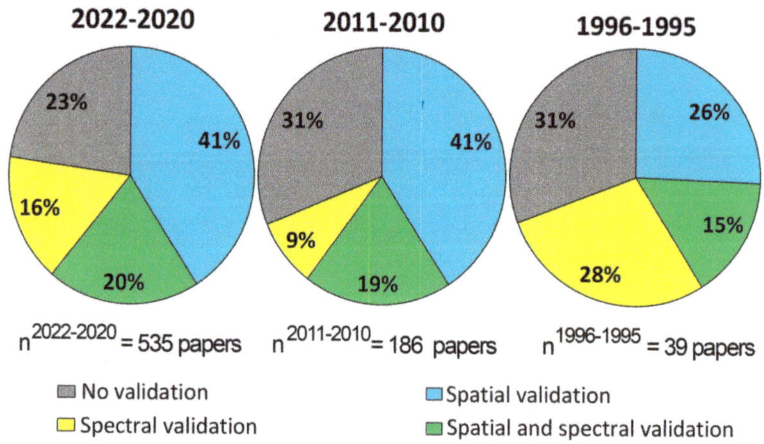

Figure 2. Distribution of the papers that applied the spectral unmixing to remote images (orange box in the Figure 1) according to different ways in which their results were validated, where $n^{2022-2020}$ was the number of papers that were published in 2022, 2021, and 2020; $n^{2011-2010}$ was the number of papers that were published in 2011 and 2010; $n^{1996-1995}$ was the number of papers that were published in 1996 and 1995.

Considering all papers that performed spatial validation (blue and green wedges in Figure 2), the percentage of these research published in 2022, 2021, and 2020 (61% of a total of 326 papers) was comparable to that of the papers that were published in 2011 and 2010 (60% of a total of 112 papers), whereas these percentages were greater than those of the papers that were published in 1996 and 1995 (41% of a total of 16 papers). Moreover, the percentage of the research published in 2022, 2021, and 2020 that did not validate the results (23%) was smaller than those of the papers that were published in the other 2 groups of years (31%). In conclusion, these values highlighted not only the increasing application of spectral unmixing over these years, but also the high priority given to the spatial validation.

3.2. Remote Images

The eligible papers published in 2022, 2021, 2020, 2011, 2010, 1996, and 1995 are summarized in Tables 3–9, according to the remote images to which spectral unmixing was applied. Authors who applied only spatial validation were cited in the fourth columns of Tables 3–9, whereas those who applied both spatial and spectral validation were cited in the fifth columns.

Table 3. Eligible papers published in 2022.

Remote Image Analyzed	Time Series	Study Area Scale	Spatial Validation Carried Out	Spatial and Spectral Validation Carried Out
AMMIS * (0.5 m) [55]	No	Local	[56,57]	
Apex * (2.5 m) [58]	No	Local	[59]	
ASTER (15–30–90 m) [60]	No	Regional [1]	[61]	
ASTER (15–30 m)	Yes [2]	Local		[62]
AVHRR (1–5 km) [63]	Yes [1]	Regional [1]	[64]	[65,66]
AVIRIS * (10/20 m) [67]	No	Local	[57,68–87]	[88–98]
AVIRIS-NG * (5 m) [99]	No	Local	[100]	
CASI * (2.5 m) [101]	No	Local	[59,78]	
DESIS * (30 m) [102]	Yes [1]	Regional [1]		[103]
DESIS * (30 m)	No	Local		[104]
EnMap * (30 m) [105]	No	Local	[69]	
GaoFen-6 (2–8–16 m) [106]	No	Regional [1]		[107]
GaoFen-2 (3.2 m)	Yes [1]	Regional [1]	[108]	
GaoFen-1 (2–8–16 m)	No	Local	[109]	
HYDICE * (10 m) [110]	No	Local	[59,68,76,77,79,81,82,85,86,90,111]	[89,96,97]
Hyperion * (30 m) [112]	Yes [1]	Local	[75]	
Hyperion * (30 m)	No	Local	[113]	[114–116]
HySpex * (0.6–1.2 m) [104]	No	Local	[104,117]	
Landsat (15–30 m) [118]	Yes [1]	Continental [1]	[119]	
Landsat (15–30 m)	Yes [1]	Regional [1]	[108,120–133]	[134,135]
Landsat (15–30 m)	No	Regional [1]		[107,136,137]
Landsat (15–30 m)	Yes [1]	Local	[138,139]	[62]
Landsat (15–30 m)	No	Local [2]	[140,141]	
Landsat (15–30 m)	No	Local	[109,142]	
M3 hyperspectral image * [143]	No	Moon		[143]
MIVIS * (8 m) [144]	No	Local		[145]
MERIS (300 m) [146]	Yes [1]	Local		[147]
MODIS (0.5–1 km) [148]	Yes [1]	Continental [1]		[149]
MODIS (0.5 km)	Yes [1]	Regional [1]	[108,150–152]	[137]
MODIS (0.5 km)	No	Local	[153]	
NEON * (1 m) [154]	No	Local	[154]	
PRISMA * (30 m) [155]	No	Local		[114,156–158]
ROSIS * (4 m) [159]	No	Local	[56,57,78,81,85]	
Samson * (3.2 m) [59]	No	Local	[59,72]	[89,97]
Sentinel-2 (10–20–60 m) [160]	Yes [1]	Regional [1]	[108,133,161–163]	
Sentinel-2 (10–20–60 m)	No	Regional [1]	[136]	[107,164,165]
Sentinel-2 (10–20–60 m)	Yes [1]	Local	[166,167]	[168]
Sentinel-2 (10–20–60 m)	No	Local [2]		[104]
Sentinel-2 (10–20–60 m)	No	Local		[169]
Specim IQ * [170]	Yes [1]	Laboratory	[170]	
SPOT (10–20 m) [171]	No	Local [2]	[140]	
WorldView-2 (0.46–1.8 m) [172]	No	Local	[166]	
WorldView-3 (0.31–1.24–3.7 m)	No	Local	[166]	

* Hyperspectral sensor; [1] Multiple images acquired from same sensor; [2] Multiple images acquired from different sensors.

Table 4. Eligible papers published in 2021.

Remote Image Analyzed	Time Series	Study Area Scale	Spatial Validation Carried Out	Spatial and Spectral Validation Carried Out
ASTER (15–30–90 m)	No	Regional [1]		[173]
AVIRIS *	No	Local	[174–201]	[202–225]
AVIRIS-NG * (5 m)	No	Local		[226]
CASI *	No	Local	[174,227]	
Simulated EnMAP *	Yes [1]	Regional [1]	[228]	
GaoFen-5 * (30 m)	No	Local		[229]
HYDICE * (10 m)	No	Local	[192,230–232]	[204,212,214,216,218]
HyMap * (4.5 m)	Yes	Local	[233]	
Hyperion * (30 m)	No	Local		[212,234,235]
Hyperion * (30 m)	Yes [1]	Local	[236,237]	
HySpex	No	Local		[238]
Landsat (30 m)	Yes [1]	Regional [1]	[239–244]	
Landsat (30 m)	Yes [1]	Local [2]	[245–253]	
Landsat (30 m)	No	Local	[227,254–259]	
Landsat (30 m)	No	Regional [1]	[260]	
MODIS (0.5–1 km)	No	Local	[254,261]	
MODIS (0.5–1 km)	Yes [1]	Regional [1]	[262–264]	
PRISMA * (30 m)	No	Local		[265]
ROSIS * (4 m)	No	Local	[191,200,266]	[217,267]
Samson * (3.2 m)	No	Local	[188,232,268]	[207,210,211,214,224,225,267]
Sentinel-2 (10–20–60 m)	No	Local	[255,258]	[226,269]
Sentinel-2 (10–20–60 m)	Yes [1]	Local	[243,253,270]	[229,271,272]
Sentinel-2 (10–20–60 m)	No	Regional [1]		[273]
Sentinel-2 (10–20–60 m)	Yes [1]	Regional [1]	[244]	
UAV multispectral image [274]	No	Local	[274]	
WorldView-2 (0.46–1.8 m)	Yes [1]	Local		[275]
WorldView-3 (0.31–1.24–3.7 m)	No	Local [2]	[276]	
ZY-1-02D * (30 m) [228]	No	Local		[228]

* Hyperspectral sensor; [1] Multiple images acquired from same sensor; [2] Multiple images acquired from different sensors.

Table 5. Eligible papers published in 2020.

Remote Image Analyzed	Time Series	Study Area Scale	Spatial Validation Carried Out	Spatial and Spectral Validation Carried Out
AISA Eagle II airborne hyperspectral scanner * [277]	No	Local	[277]	
ASTER (15–30–90 m)	No	Regional [1]	[278]	
ASTER (15–30–90 m)	Yes [1]	Local [2]		[279,280]
AVIRIS *	No	Local	[281–298]	[299–327]
AVIRIS NG *	No	Local	[291]	
AWiFS [328]	Yes [1]	Local [2]	[328]	
CASI *	No	Local	[329]	
Simulated EnMAP * (30 m)	No	Regional [1]	[330]	
GaoFen-1 WFV	Yes [1]	Local	[331]	
GaoFen-1 WFV	Yes [1]	Local [2]	[332]	[333]
GaoFen-2	No	Local [2]	[332]	
HYDICE * (10 m)	No	Local	[292,293,298,334,335]	[299,307,309,310,316,318,321,322,324]
HyMAP *	No	Local [2]		[280]
HyMAP *	No	Local	[336]	
HySpex * (0.7 m)	No	Local	[337]	
Hyperion * (30 m)	No	Local	[336]	[338]
Landsat (30 m)	Yes [1]	Local [2]	[332]	[280,339]
Landsat (30 m)	Yes [1]	Local	[252,340–347]	
Landsat (30 m)	Yes [1]	Continental [1]	[348]	
Landsat (30 m)	Yes [1]	Regional [1]	[349–355]	[356]
Landsat (30 m)	No	Regional [1]	[357]	
MODIS (0.5–1 km)	Yes [1]	Local	[340,358–361]	[333]
MODIS (0.5–1 km)	Yes [1]	Regional [1]	[362,363]	
MODIS (0.5–1 km)	Yes [1]	Local [2]	[364,365]	[279]
PlanetScope (3 m) [366]	Yes [1]	Local [2]	[366]	
PROBA-V (100 m) [367]	Yes [1]	Regional [1]	[353,368–371]	
ROSIS * (4 m)	No	Local	[285,372]	[373]

Table 5. *Cont.*

Remote Image Analyzed	Time Series	Study Area Scale	Spatial Validation Carried Out	Spatial and Spectral Validation Carried Out
Samson * (3.2 m)	No	Local	[284,374,375]	[301,303,305,315,320,323,324]
Sentinel-2 (10–20–60 m)	No	Local [2]	[332,376]	[280,339]
Sentinel-2 (10–20–60 m)	Yes [1]	Local	[328,340,377–382]	[333,383]
Suomi NPP-VIIRS [354]	Yes [1]	Regional [1]	[353]	
UAV hyperspectral data * [384]	Yes [1]	Local		[384]
WorldView-2	Yes [1]	Local	[342]	
WorldView-2	Yes [1]	Local [2]		[385]
WorldView-3	Yes [1]	Local [2]		[385]

* Hyperspectral sensor; [1] Multiple images acquired from same sensor; [2] Multiple images acquired from different sensors.

Table 6. Eligible papers published in 2011.

Remote Image Analyzed	Time Series	Study Area Scale	Spatial Validation Carried Out	Spatial and Spectral Validation Carried Out
AHS * [386]	No	Local		[386]
ASTER	No	Local		[387–389]
ASTER	Yes [1]	Local	[390,391]	
AVIRIS *	No	Local	[307,392–403]	[387,404–417]
CASI *	No	Local		[418]
MERIS (300 m)	No	Local	[419]	
MODIS (0.5–1 km)	Yes [1]	Local	[420–423]	
HYDICE *	No	Local	[392,424]	[414,415,425]
HyMAP *	No	Local	[392,426]	[427]
Hyperion * (30 m)	No	Local		[387,428]
HJ-1 * (30 m) [429]	No	Local	[429,430]	
Landsat (30 m)	Yes [1]	Local	[431–433]	[387]
Landsat (30 m)	No	Local	[434,435]	
Landsat (30 m)	Yes [1]	Local [2]	[436–438]	
Landsat (30 m)	No	Local [2]	[423,439]	
QuickBird (0.6–2.4 m) [440]	No	Local	[441,442]	
SPOT (10–20 m)	No	Local [2]	[439,441]	

* Hyperspectral sensor; [1] Multiple images acquired from same sensor; [2] Multiple images acquired from different sensors.

Table 7. Eligible papers published in 2010.

Remote Image Analyzed	Time Series	Study Area Scale	Spatial Validation Carried Out	Spatial and Spectral Validation Carried Out
Airborne hyper-spectral image * (about 1.5 m) [443]	No	Regional [1]		[443]
AHS * (2.4 m)	No	Local	[444]	
ASTER (15–30–90 m)	Yes [1]	Local	[445,446]	
ASTER (15–30–90 m)	Yes [1]	Regional [1]	[447]	
ATM (2 m) [101]	No	Local [2]		[101]
AVHRR (1 km)	Yes [1]	Regional [1]	[448]	
AVIRIS * (20 m)	No	Local	[449–457]	[458–463]
CASI * (2 m)	No	Local		[101]
CASI *	No	Laboratory		[464,465]
CHRIS * (17 m) [466]	No	Local	[467]	
DAIS * (6 m) [464]	No	Local	[465]	
DESIS *	No	Local		[468,469]
HYDICE *	No	Local	[455,470,471]	[458,463]
HyMAP *	No	Local	[471]	
Hyperion * (30 m)	No	Local	[472–474]	
HJ-1 * (30 m)	No	Local	[475,476]	

Table 7. *Cont.*

Remote Image Analyzed	Time Series	Study Area Scale	Spatial Validation Carried Out	Spatial and Spectral Validation Carried Out
Landsat (30 m)	Yes [1]	Regional [1]	[477–483]	
Landsat (30 m)	No	Regional [1]	[484–489]	[490]
Landsat (30 m)	No	Local [2]	[491,492]	
Landsat (30 m)	No	Local	[493]	
MIVIS * (3 m)	No	Regional [1]	[494]	
MODIS (0.5–1 km)	Yes [1]	Regional [1]	[495]	
MODIS (0.5–1 km)	Yes [1]	Continental [1]	[496]	
QuickBird (2.4 m)	No	Local [2]	[491]	
QuickBird (2.4 m)	No	Local	[497,498]	
SPOT (10–20 m)	Yes [1]	Regional [1]	[480]	
SPOT (2.5–10–20 m)	No	Local [2]	[486,491,492]	
SPOT (2.5–10–20 m)	No	Local	[499]	[500]

* Hyperspectral sensor; [1] Multiple images acquired from same sensor; [2] Multiple images acquired from different sensors.

Table 8. Eligible papers published in 1996.

Remote Image Analyzed	Time Series	Study Area Scale	Spatial Validation Carried Out	Spatial and Spectral Validation Carried Out
AVIRIS *	No	Local	[501,502]	[503]
GERIS * [504]	No	Local	[504]	
Landsat (30 m)	No	Local	[14,505]	[506]
SPOT (2.5–10–20 m)	No	Local	[507]	

* Hyperspectral sensor.

Table 9. Eligible papers published in 1995.

Remote Image Analyzed	Time Series	Study Area Scale	Spatial Validation Carried Out	Spatial and Spectral Validation Carried Out
AVHRR (1–5 km)	Yes [1]	Regional [1]		[508]
AVIRIS * (20 m)	No	Local	[509]	[510,511]
Landsat (30 m)	No	Local	[512]	[513]
MIVIS * (4 m)	No	Local	[514]	
MMR * [515]	Yes [1]	Local	[515]	

* Hyperspectral sensor; [1] Multiple images acquired from same sensor.

The first columns of Tables 3–9 and the second columns of Tables A1–A7 show the sensor name and the spatial resolution of the images. Considering all eligible papers, 27 hyperspectral sensors and 16 multispectral sensors were employed. Hyperspectral sensors were highlighted in the first columns of Tables 3–9 with an asterisk. The literature often combined spectral unmixing with hyperspectral data because the number of bands must be greater than the number of endmembers [4,5,42,44]. However, the percentage of papers that employed hyperspectral data (57% of a total of 458 papers) is slightly higher than the percentage of papers that employed multispectral data (43% of a total of 458 papers). The second columns of Tables 3–9 show the papers that performed the time series studies, whereas the third columns of these tables show the papers that performed the local, regional, or continental studies.

The analysis of these data showed that most studies that analyzed hyperspectral images were performed at the local scale and did not carry out the multitemporal studies, whereas most studies that analyzed multispectral images were performed at the regional or continental scale and carried out the multitemporal studies (more than one image was analyzed). Therefore, the spectral unmixing is widely applied to multispectral images, despite their smaller number of bands than hyperspectral images, because these data are characterized by greater spatial and temporal availability than those of the hyperspectral data.

Moreover, the spectral unmixing was also applied to some hyperspectral and multispectral images that were characterized with high spatial resolutions (e.g., AMMIS image with spatial resolution equal to 0.5 m [56] and WorldView-3 image with spatial resolution of 0.31 m [166]). These papers confirm that, no matter how high the spatial resolution might be, no image pixel results were completely homogeneous in spectral characteristics [9,516,517].

3.3. Accuracy Metrics

Accuracy, which is defined as "the degree of correctness of the map", is usually assessed by comparing the "ground truth" with the map retrieved from remote images [518,519]. Because no map can fully and completely map the territory [520], ground truth is more correctly called reference data [521]. To assess the differences between the reference data and results of the spectral unmixing, the eligible papers exploited different metrics. Figure 3 shows the pie chart of the distribution of the metrics that were adopted by eligible papers.

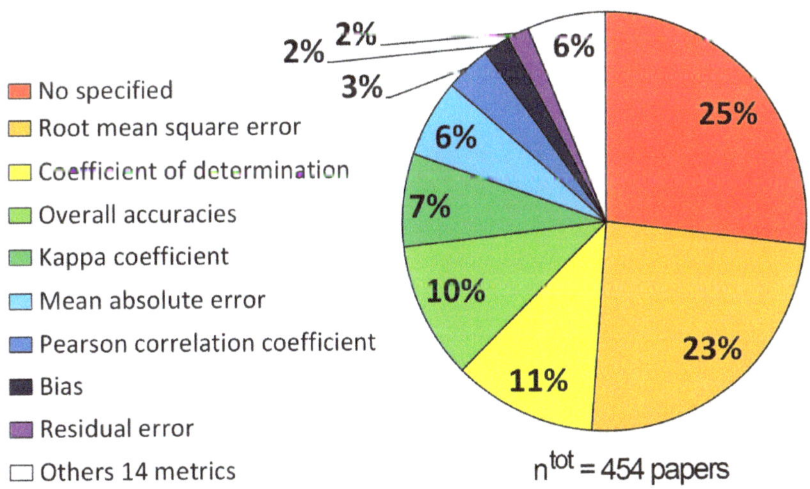

Figure 3. Distribution of the eligible papers according to the metrics employed to evaluate the spatial accuracy.

The other 14 metrics were average accuracy [522], correct labeling percentage for the unchanged pixels [141], correlation coefficient [150], Kling–Gupta efficiency [523], mean abundance error [117], mean error [169], mean relative error [169], normalized average of spectral similarity measures [524], producer's accuracy [153], Receiver Operating characteristic Curves (ROC) method [525], relative mean bias [165], separability spectral index [526], signal-to-reconstruction error [56], and systematic error [109].

In conclusion, the authors of 454 eligible papers employed 22 different metrics, and most authors employed more than 1 metric. Overall, 25% of the eligible papers did not specify the accuracy metrics used. It is very important to note that some standard accuracy assessments, such as the kappa coefficient, "assume implicitly that each of the testing samples is pure"; therefore, some of these metrics were inappropriate for evaluating the accuracy of the fractional abundance maps [41,518].

3.4. *Key Issues in the Spatial Validation*

Since the literature highlighted many sources of error in accuracy assessment of retrieved maps [518,519,521], the authors identified and carried out several "key issues" to address and minimize these errors. Figure 4 and Tables A1–A7 summarize the key issues that were identified.

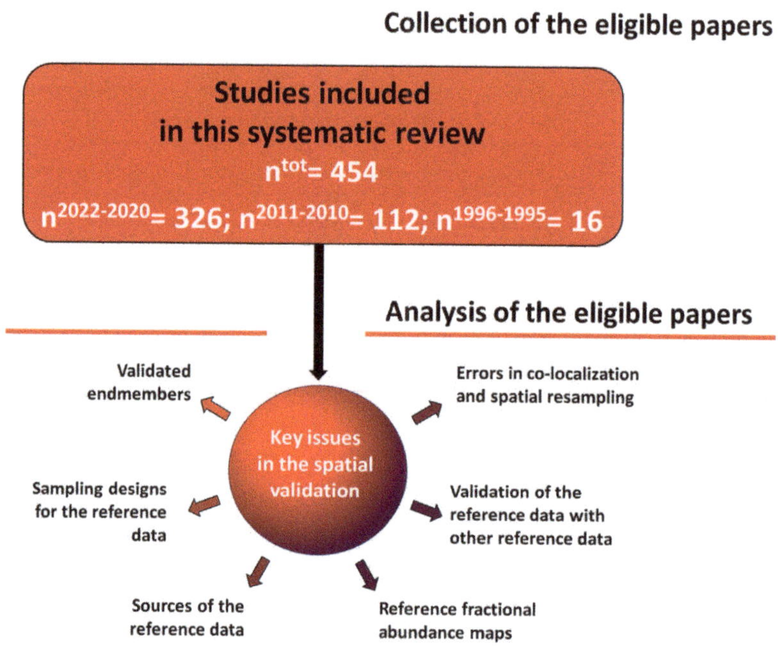

Figure 4. Key issues in the spatial validation that were addressed by the eligible papers.

3.4.1. Validated Endmembers

Before analyzing the endmembers that were validated, it is necessary to remember that the number of endmembers that were determined with the images must be less than the number of sensor bands; therefore, the number of endmembers that were determined with the multispectral data is less than the number of endmembers that were determined with the hyperspectral data [6,23,527]. Therefore, the authors who elaborated the multispectral images employed smaller levels of model complexity than authors who elaborated the hyperspectral images [528,529]. For example, the VIS model was used to map only three endmembers (Vegetation, Impervious surfaces, and Soil) in many urban areas that were retrieved from multispectral data (e.g., [109,152,477,493]).

The third columns of Tables A1–A7 list the endmembers that were determined using spectral unmixing; the fourth columns of these showed the number of these endmembers that were validated. It is interesting to note that some authors validated smaller number of endmembers than the number of the endmembers that were determined (i.e., 40 eligible papers). Dividing the works that analyzed hyperspectral images from those that analyzed multispectral data, Figure 5 shows the percentage of studies that validated the total or partial number of endmembers. It is important to highlight that, since 4 eligible papers analyzed both hyperspectral and multispectral data [104,227,231,281], the sum of papers that analyzed hyperspectral data and papers that analyzed multispectral data (i.e., 458) is greater than the number of eligible papers (i.e., 454).

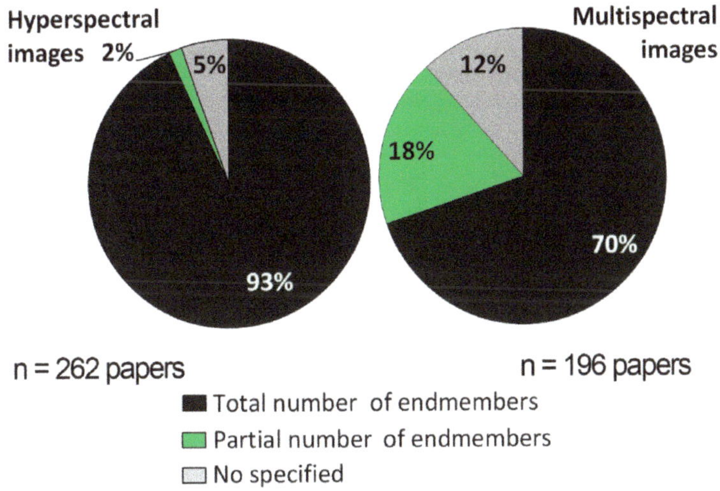

Figure 5. Distribution of the eligible papers that fully or partially validated endmembers determined with hyperspectral images (**right**) or multispectral images (**left**), where n was the number of papers considered in each pie chart.

Therefore, only 2% of the studies that elaborated hyperspectral images partially validated the determined endmembers, whereas 18% of the studies that elaborated multispectral images partially validated the determined endmembers. As mentioned above, hyperspectral images were used to carry out non-repeated surveys over time and at local-scale studies (252 papers of a total of 262), whereas most multispectral images were used to carry out regional- or continental-scale studies that were or were not repeated over time (180 papers of a total of 196). Therefore, some of these authors, who analyzed more than one image, chose to spatially validate only the materials or groups of materials on which they focused their study. For example, Hu et al. [149] spatially validated only blue ice fractional abundance maps that were retrieved from MODIS images covering the period 2000–2021 in order to present a FABIAN (Fractional Austral-summer Blue Ice over Antarctica) product. It should be noted that 5 and 12% of the papers that analyzed hyperspectral or multispectral data, respectively, did not specify which endmembers were validated.

3.4.2. Sampling Designs for the Reference Data

The literature demonstrated that a possible source of error in spatial validation is due to the choice of the sampling design for the reference data [518,519,521,530]. The sampling design mainly includes the definition of the sample size and the sampling design of the reference data [518]. Authors of eligible papers chose three kinds of sample sizes: the whole study area; the representative area; small sample sizes (pixels, plots, and polygons samples). The eighth columns of Tables A1–A7 show the different sample sizes that were adopted by every eligible paper, and Table 10 shows the number of papers that adopted the different sample sizes.

Table 10. Sample sizes of the reference data that were employed by the eligible papers.

Sample Sizes of the Reference Data	Papers Published in 2022, 2021, and 2020	Papers Published in 2011 and 2010	Papers Published in 1996 and 1995
Whole study area	172	55	10
Small sample sizes	78	38	1
Representative area	21	7	0
Not specified	59	12	5

Most authors of the eligible papers chose to validate the whole study areas, followed, in descending order, by the choice to employ the different number of small sample sizes and then the representative areas. It is also important to note the high percentages of the papers that did not specify the sample size of the reference data: 18, 11, and 31%, respectively.

The literature also pointed out that the sampling designs for spatially validating maps at local scale cannot be the same as the designs for spatially validating maps at regional or continental scale [518,530]. As mentioned above, most of the studies that analyzed the hyperspectral data were performed at local scale (252 papers of a total of 262), whereas the studies that analyzed the multispectral images performed at regional or continental scale (180 papers of a total of 196). Therefore, the eligible papers that analyzed hyperspectral images were analyzed separately from those that analyzed multispectral images (Figure 6 on the right and left, respectively), not only to analyze the different sampling designs adopted from the hyperspectral and multispectral data, but also to highlight the different sampling designs chosen for local or regional/continental scale studies. Figure 6 shows the percentage of the eligible papers that employed the different sample sizes and the percentage of the eligible papers that employed a different number of small sample sizes.

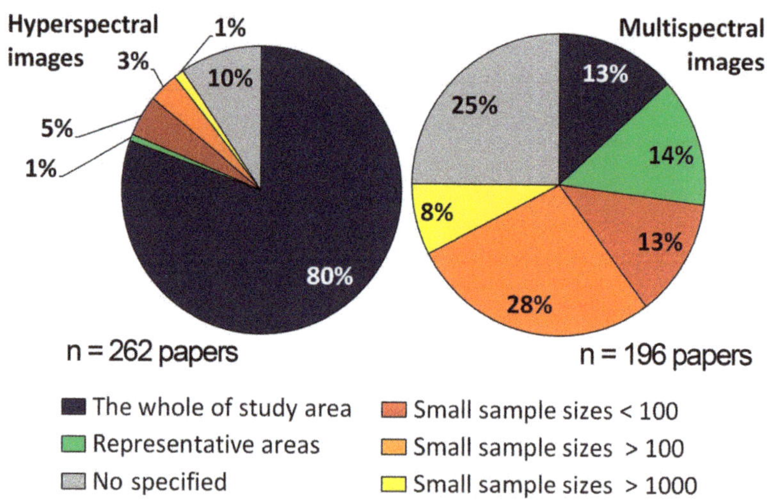

Figure 6. Distribution of the eligible papers according to the sample sizes and the number of the small sample sizes that were chosen to analyze hyperspectral (**right**) or multispectral (**left**) images, where n was the number of papers considered in each pie chart.

Most papers that processed hyperspectral images validated the whole study area (212 papers), whereas most papers that processed multispectral images employed small sample sizes (94 papers).

The authors of eligible papers that employed small sample sizes adopted three different sampling designs of reference data: partial, random, and uniform. The ninth columns of Tables A1–A7 show the sampling designs of every eligible paper. Most authors who published in 2022, 2021, and 2020 and published in 2011 and 2010 chose the random distribution of reference data (78% for a total of 326 papers and 76% for a total of 110 papers, respectively), whereas the authors who published in 1996 and 1995 did not specify the sampling designs employed. Stehman and Foody [519] highlighted that "the most commonly used designs" that were chosen to assess the land cover products were "simple random, stratified random, systematic, and cluster" designs. Therefore, these results confirmed that random designs were the most commonly used approaches.

3.4.3. Sources of the Reference Data

Eligible papers employed four different sources of reference data to spatially validate spectral unmixing results: images, in situ data, maps, and previous reference maps. Table 11 shows the number of the eligible papers that employed these reference data sources, whereas the fifth columns of Tables A1–A7 detail the sources of the reference data.

Table 11. Reference data sources employed by the eligible papers.

Sources of Reference Data	Papers Published in 2022, 2021, and 2020	Papers Published in 2011 and 2010	Papers Published in 1996 and 1995
Maps	13	2	8
In situ data	55	35	2
Images	106	31	6
Previous reference maps	156	44	0

The number of authors who chose to utilize geological, land use, or land cover maps as reference maps is the smallest (5% of the total eligible papers), followed, in ascending order, by the number who chose to create the reference maps using in situ data (20% of the total eligible papers), and then by the number of authors who chose to create the reference maps using other images (31% of the total eligible papers). Firstly, the number of authors who chose to use the previous reference maps is the largest (44% of the total eligible papers).

As regards the authors who chose to create the reference maps using other images, most of them employed images at higher spatial resolutions than those of the remote images analyzed (95% of a total of 143 papers). To create the reference maps from the images, 47% of the eligible papers did not specify the method used to map the endmembers, 29% employed the photo-interpretation, 21% classified the images, 2% used the vegetation indexes, and 2% used the mixed approach by classifying and/or photo-interpreting and/or applying vegetation indexes (e.g., [114,145,531]). As regards the classification methods, there are four works that applied the same classification procedure to analyze the remote images and to create the reference maps [65,66,149,261]. Among these, the authors of 3 papers compared the fractional abundance maps that were retrieved from the multispectral images at moderate spatial resolutions (10, 30, and 60 m) with the fractional abundance maps that were retrieved from the multispectral data at coarse spatial resolutions (0.5 and 1 km) [65,66,149].

Moreover, the reference data sources that were chosen to validate the results of the hyperspectral images were analyzed separately from those that were chosen to validate the results of the multispectral images. Figure 7 shows the percentage of the papers that adopted the different sources of the reference data to validate the results of hyperspectral (right) and multispectral data (left).

As regards the papers that analyzed the multispectral data, most of the authors chose to create the reference maps from the other images, whereas most of the authors that analyzed the hyperspectral data chose to employ the previous reference maps. It is important to emphasize that 97% of these reference maps are available online together with hyperspectral images and/or reference spectral libraries (e.g., [532–535] Figure 8). Therefore, these images were well known: Cuprite (NV, USA, e.g., [70,458]), Indian Pines (IN, USA, e.g., [78,458]), Jasper Ridge (CA, USA, e.g., [68,97]), Salinas Valley (CA, USA, e.g., [75,78]) datasets that were acquired with AVIRIS sensors; Pavia (Italy, e.g., [81,85]) datasets that were acquired with the ROSIS sensor; Samson (FL, USA, e.g., [59,89]) dataset that was acquired with the Samson sensor; University of Houston (TX, USA, e.g., [59,78]) dataset that was acquired with the CASI-1500 sensor; Urban (TX, USA, e.g., [59,68]) and Washington DC Mall (Washington, DC, USA, e.g., [81,90]) datasets that were acquired with the HYDICE sensor.

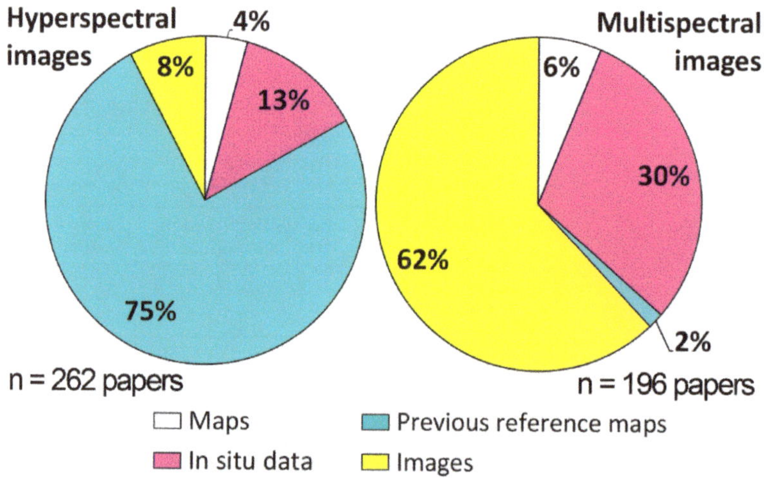

Figure 7. Distribution of the eligible papers according to the reference data sources that were chosen to analyze hyperspectral (**right**) or multispectral (**left**) images, where n was the total number of papers considered in each pie chart.

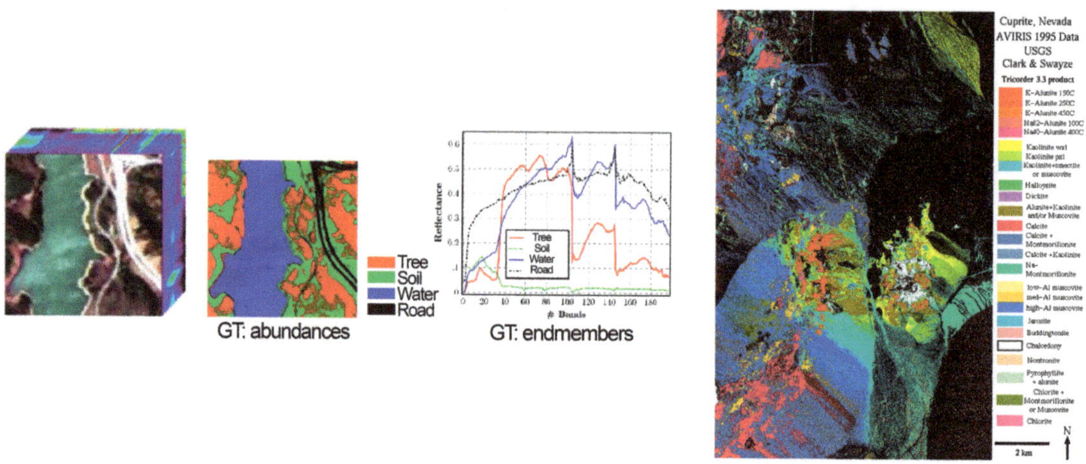

(**a**) (**b**)

Figure 8. *Cont.*

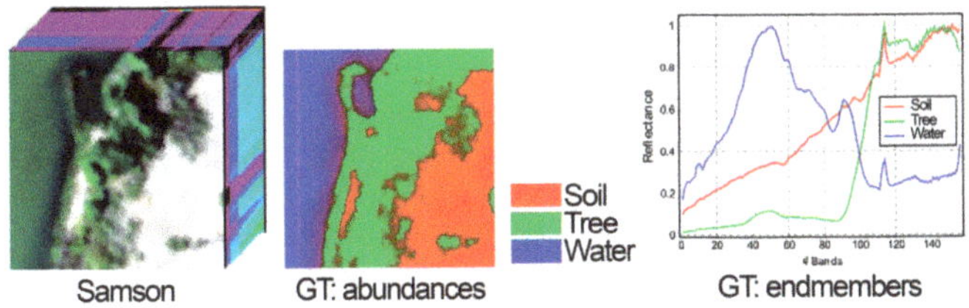

(c)

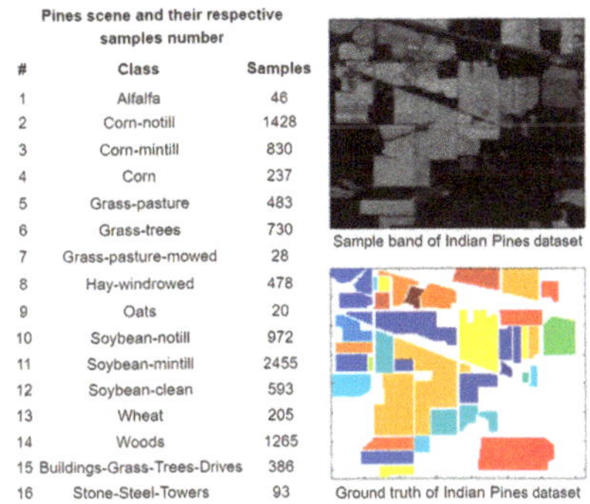

(d)

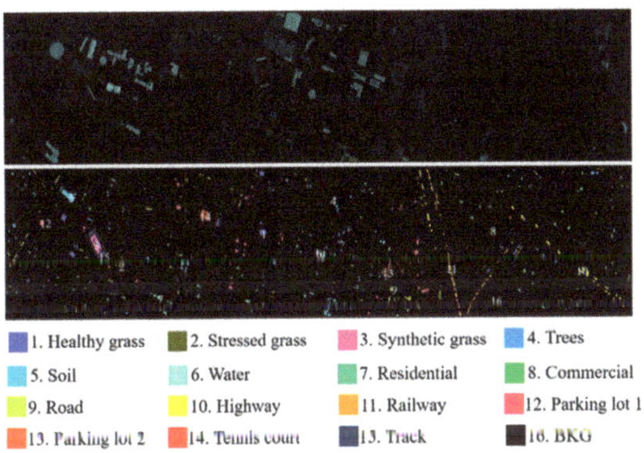

(e)

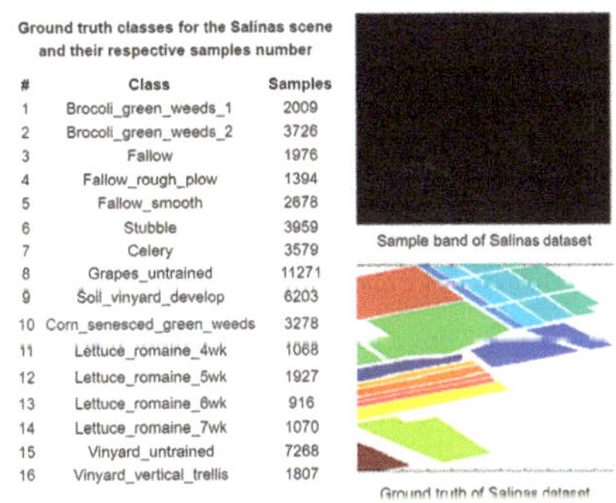

(f)

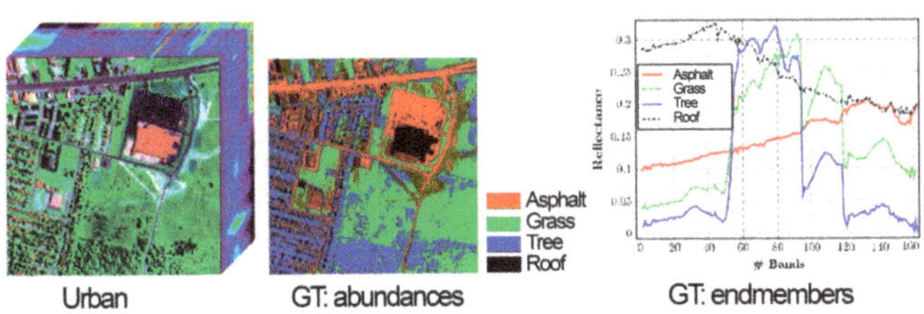

(g)

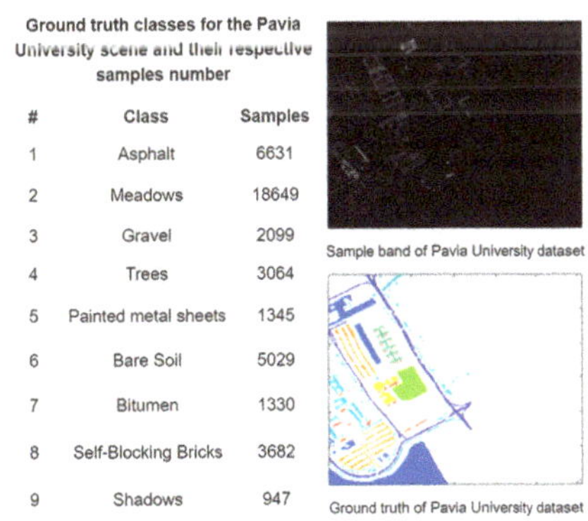

(h)

Figure 8. *Cont.*

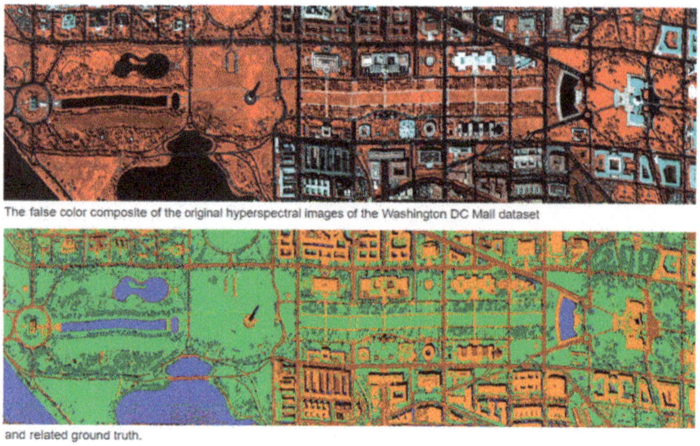

(i)

Ground truth classes for the Pavia center scene and their respective samples number

#	Class	Samples
1	Water	824
2	Trees	820
3	Asphalt	816
4	Self-Blocking Bricks	808
5	Bitumen	808
6	Tiles	1260
7	Shadows	476
8	Meadows	824
9	Bare Soil	820

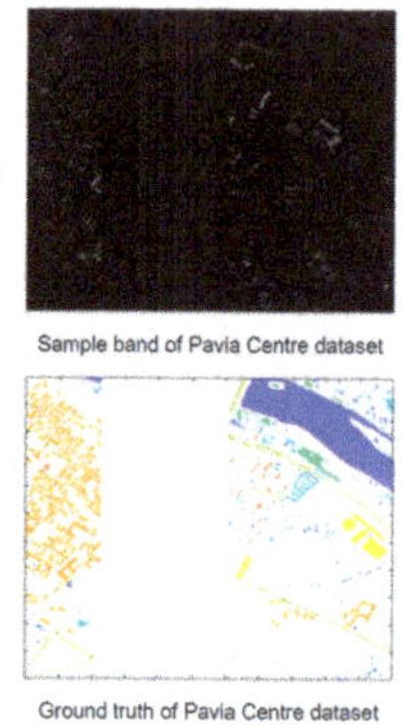

(j)

Figure 8. Reference data available online together with hyperspectral images: (**a**) Jasper Ridge reference map and spectral library [535]; (**b**) Cuprite reference map [536]; (**c**) Samson reference map and spectral library [535]; (**d**) Indian Pines reference map [535]; (**e**) University of Houston reference map [535]; (**f**) Salinas Valley reference map [535]; (**g**) Urban reference map [535]; (**h**) Pavia University reference map [535]; (**i**) Washington DC reference map [535]; (**j**) Pavia center reference map [535].

Moreover, 93% of these papers proposed a method and tested it not only on these "real" hyperspectral data, but also on created synthetic images. Borsoi et al. [4] highlighted that in order to overcome "the difficulty in collecting ground truth data", some authors generated synthetic images. However, the authors complained because "there is not a clearly agreed-upon protocol to generate realistic synthetic data" [4].

3.4.4. Reference Fractional Abundance Maps

"Misclassifications" of the reference data or "misallocations of the reference data" are another possible source of error in spatial validation, defined as "imperfect reference data" by [519] or "error magnitude" by [518]. The authors highlighted that these errors can be caused also by the use of "standard" reference maps to validate the spectral unmixing results (i.e., the fractional abundance maps) [41,518,519]. The difference between standard reference maps and reference fractional abundance maps is that each pixel of the standard reference map is assigned to a corresponding land cover class, whereas each pixel of the reference fractional abundance map is labeled with the fractional abundances of each endmember that is present in that pixel. Therefore, the values of the standard reference map are equal to 0 or 1, whereas the values of the reference fractional abundance map are greater than 2 and vary between 0 and 1 (100 values are able to fully validate the fractional abundance of endmembers [114]).

The reference fractional abundance maps were employed by 133 eligible papers that were published in 2022, 2021, and 2020; by 62 eligible papers that were published in 2011 and 2010; and by 13 eligible papers that were published in 1996 and 1995 (45% of the total eligible papers). Moreover, among these works, 87, 47, and 8 papers estimated the full range of abundances using 100 values (31% of the total eligible papers), whereas 41, 10, and 5 works partially estimated the fractional abundances using less than 100 values (12% of the total eligible papers). It is important to note that 7% of the total eligible papers did not specify if they used the standard reference maps or the reference fractional abundance maps.

The eligible papers were separately analyzed according to reference data sources that were adopted in order to find out how fractional abundances were estimated. In the four parts of Figure 9, the eligible papers that were clustered according to the reference data sources are shown, and each part of Figure 9 shows the percentage of the papers that did not specify the reference maps used and the number of the papers that fully or partially estimated the reference fractional abundance maps.

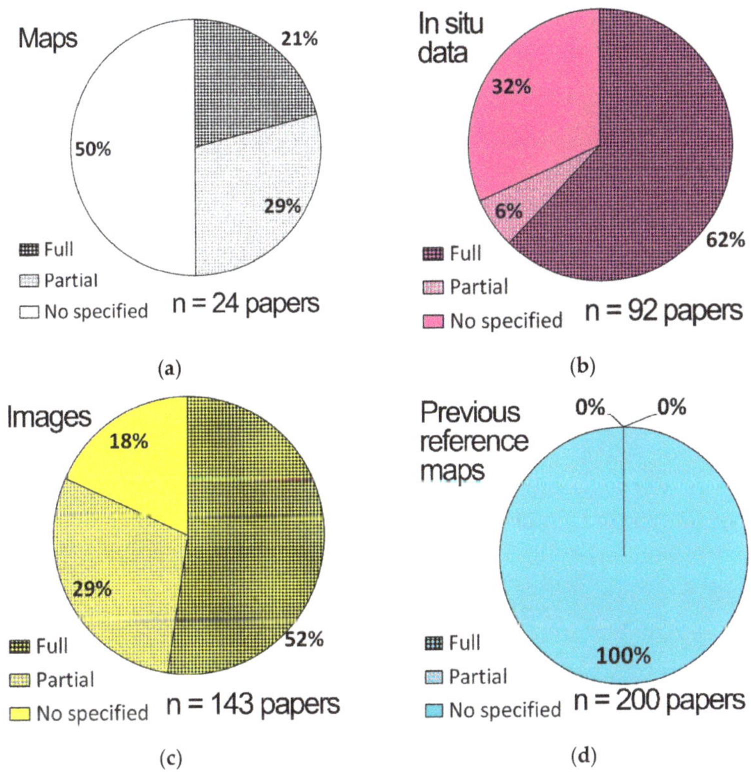

Figure 9. Distribution of the eligible papers that did not specify the reference maps used, fully and partially estimated fractional abundances according to the reference data sources, where n was the total number of papers that were clustered according to the reference data sources and included in the pie charts: (**a**) The papers that employed the maps; (**b**) The papers that employed in situ data; (**c**) The papers that employed the images; (**d**) The papers that employed the previous reference maps.

High-spatial-resolution images were the most widely employed to make the reference fractional abundance maps (81% of the total papers that employed the images), followed by in situ data (68% of the total papers that employed in situ data), and then the maps (50% of the total papers that employed maps). Moreover, in situ data were the most widely employed to estimate the full range of fractional abundances (62% of the total papers that employed in situ data), followed by high-spatial-resolution images (52% of the total papers that employed the images), and then the maps (21% of the total papers that employed the images). The previous reference maps were not employed to make the reference fractional abundance maps.

Many authors highlighted that it is not easy to create the reference fractional abundances maps (e.g., [4,6,518,519]). Cavalli [145] implemented a method that was proposed by [537] in order to create the reference fractional abundance maps. This method is able to create the reference fractional abundance maps by varying the spatial resolution of the high-resolution reference maps several times, and the range of fractional abundances can be fully estimated according to the spatial resolution of the reference maps [114].

3.4.5. Validation of the Reference Data with Other Reference Data

In order to further minimize the errors due to "misclassifications" or "misallocations of the reference data" [518,519], some authors validated the reference data using other reference data: 61 eligible papers published in 2022, 2021, and 2020; 21 eligible papers published in 2011 and 2010; 4 eligible papers published in 1996 and 1995. Therefore, 81% of the total eligible papers did not take into consideration that the reference map may not be "ground truth" and may be "imperfect" [519,520].

It is very important to point out that some authors took advantage of the online availability of reference data to validate reference data (e.g., [114,123,127,140,145,152,231,448,496]). Many efforts are being made to create the networks of accurate validation data [48,538–540]. For example, Zhao et al. [140] exploited in situ measurements of the Leaf Area Index (LAI) that were provided by the VALERI project [540], whereas Halbgewachs et al. [123], Lu et al. [423], Shimabukuro et al. [353], and Tarazona Coronel [127] utilized validation data that were provided by the Program for Monitoring Deforestation in the Brazilian Amazon (PRODES) [541].

3.4.6. Error in Co-Localization and Spatial Resampling

The key issues described above addressed only the errors in the thematic accuracy of the spectral unmixing results [518,519], whereas this key issue aimed to address the geometric errors due to the comparison of remote images with reference data [542]. The impact of co-localization and spatial resampling errors was minimized and/or evaluated by 6% of the eligible papers: 20 eligible papers published in 2022, 2021, and 2020; 8 eligible papers published in 2011 and 2010; 1 eligible paper published in 1996. In order to minimized the errors, Arai et al. [368], Cao et al. [164], Li et al. [107], Soenen et al. [500], and Zurita-Milla et al. [419] carefully chose the size of the reference maps; Bair et al. [254], Cavalli [114,145], Ding et al. [152], Fernandez-Garcia et al. [256], Hamada et al. [441], Hajnal et al. [169], Lu et al. [435], Ma & Chan [78], Rittger et al. [262], Sun et al. [263], Yang et al. [488], and Yin et al. [151] spatially resampled the reference fractional abundance maps; Estes et al. [447] compared different windows of pixels (i.e., 3×3, 7×7, 11×11, 15×15, and 21×21); Pacheco & McNairn [480] selected the size and the spatial resolution of the reference maps; Ben-dor et al. [507], Fernandez-Guisuraga et al. [342], Kompella et al. [328], Laamarani et al. [343], and Plaza & Plaza [465] carefully co-localized the reference fractional abundance maps on the reference maps; Wang et al. [366] expanded the windows of the field sample size; Zhu et al. [64] resampled at "four kinds of grids" (i.e., 1100×1100 m, 2200×2200 m, 4400×4400 m, and 8800×8800 m) the reference fractional abundance map and compared the results. Bair et al. [254], Binh et al. [341], Cavalli [114,145], Cheng et al. [543], and Ruescas et al. [448] evaluated the errors in co-localization and spatial-resampling due to the comparison of different data at different spatial resolutions. Moreover, Cavalli [145] proposed a method to minimize the errors: the comparison of the histograms of the reference fractional abundance values with the histograms of the retrieved fractional abundance values.

It is important to point out that 94% of the total papers did not address the geometric errors due to the comparison of remote images with reference data.

4. Conclusions

The term validation is defined as "the process of assessing, by independent means, the quality of the data products derived from the system outputs" by the Working Group on Calibration and Validation (WGCV) of the Committee on Earth Observing Satellites (CEOS) [48]. Since 1969, research has been involved to establish shared key issues to validate the land cover products that were retrieved from the remote images [518,519,539,544]. These products can be obtained by applying classifications called "hard", because they extract information only from "pure pixels," and classifications called "soft", because they also extract information from "mixed pixels" [519,544]. However, not only the literature related to the spatial validation, but also every review on the spectral unmixing procedure (i.e., a soft classification) highlighted that the key issues in the spatial validation of soft classification results have yet to be clearly established and shared (e.g., [4,6,518,519]).

Since no review was performed on this fundamental topic, this systematic review aims (a) to identify and analyze how the authors addressed the spatial validation of spectral unmixing results and (b) to provide readers with recommendations for overcoming the many shortcomings of spatial validation and minimizing its errors. The papers published in 2022, 2021, and 2020 were considered to analyze the current status of spatial validation, and the papers published not only in 2011 and 2010, but also in 1996 and 1995, were considered to analyze its progress over time. Since the literature on spectral unmixing is extensive, only papers published in these seven years were considered. A total of 454 eligible papers were included in this systematic review and showed that the authors addressed 6 key issues in the spatial validation. In this text, the order in which the key issues were presented is not an order of importance.

1. The first key issue concerned the number of the endmembers validated. Some authors chose to focus on only one or two endmembers, and only these were spatially validated. This key issue was designed to facilitate the conduct of regional- or continental-scale studies and/or multitemporal analysis. It is important to note that 8% of the eligible papers did not specify which endmembers were validated.
2. The second key issue concerned the sampling designs for the reference data. The authors who analyzed hyperspectral images preferred to validate the whole study area, whereas those who analyzed multispectral images preferred to validate small sample sizes that were randomly distributed. It is important to point out that 16% of the eligible papers did not specify the sampling designs for the reference data.
3. The third key issue concerned the reference data sources. The authors who analyzed hyperspectral images primarily used the previously referenced maps and secondarily created reference maps using in situ data, whereas the authors who analyzed multispectral images chose to create reference maps primarily using high-spatial-resolution images and secondarily using in situ data.
4. The fourth key issue was, perhaps, the one most closely related to the spectral unmixing procedure; it concerned the creation of the reference fractional abundance maps. Only 45% of the eligible papers created the reference fractional abundance maps to spatially validate the fractional abundance maps retrieved. These mainly employed high-resolution images and secondarily in situ data. Therefore, 55% of the eligible papers did not specify the employment of the reference fractional abundance maps.
5. The fifth key issue concerned the validation of the reference data with other reference data; it was addressed only by 19% of the eligible papers. Therefore, 81% of the eligible papers did not validate the reference data.
6. The sixth key issue concerned the error in co-localization and spatial resampling data, which was minimized and/or evaluated only by 6% of the eligible papers. Therefore, 94% of the eligible papers did not address the error in co-localization and spatial resampling data.

In conclusion, to spatially validate the spectral unmixing results and minimize and/or evaluate its errors, six key issues were considered not only from the eligible papers published in 2022, 2021 and 2020, but also from those published in 2010, 2011, 1996, and 1995. In addition, the results obtained from both hyperspectral and multispectral data were spatially validated considering all key issues, but these were addressed in different ways. All six key issues addressed together enabled rigorous spatial validation to be performed. Therefore, this systematic review provided readers with the most suitable tool to rigorously address spatial validation of the spectral unmixing results and minimize its errors.

The key difference between reference data suitable for hard and soft classifications is that the latter reference maps must have higher spatial resolution than the resolutions of the image pixels [6,114,518]. The optimal scale would be that 100 times larger than the image pixel resolution [114]. However, many hyperspectral data were validated using the previous reference maps at the same spatial resolution as the remote image, so these standard reference maps can only create reference fractional abundance maps with the help of other reference data. The employment of the standard reference maps instead of the reference fractional abundance maps was also evidenced by the employment of metrics to assess spatial accuracy that "assume implicitly that each of the testing samples is pure" [37,217].

However, only 4% of eligible papers addressed every key issue, and many authors did not specify which approach they employed to spatially validate the spectral unmixing results. Moreover, most of the authors who specified the approach employed did not adequately explain the methods used and the reasons for their choices. Six "good practice criteria to guide accuracy assessment methods and reporting" were identified by [519]. Therefore, these papers did not fully meet three good practice criteria: "reliable", "transparent", and "reproducible" [519].

Funding: This research received no external funding.

Data Availability Statement: Not applicable.

Conflicts of Interest: The authors declare no conflict of interest.

Appendix A

In accordance with the PRISMA statement [49,50], 454 eligible papers were identified, screened, and included in this systematic review: 326 eligible papers were published in 2022, 2021, and 2020; 112 eligible papers were published in 2011 and 2010; 16 eligible papers were published in 1996 and 1995. The eligible criterion was that the results of the spectral unmixing were spatially validated. Analyzing these papers, six key issues were identified that were differently addressed to spatially validate the spectral unmixing results. The different ways in which the key issues were addressed by the eligible papers published in 2022, 2021, 2020, 2011, 2010, 1996, and 1995 are summarized in Tables A1–A7, respectively.

Table A1. Main characteristics of the eligible papers that were published in 2022.

Paper	Remote Image	Determined Endmembers	Validated End-members	Sources of Reference Data	Method for Mapping the Endmembers	Validation of Reference Data with Other Reference Data	Sample Sizes and Number of Small Sample Sizes	Sampling Designs	Reference Data	Estimation of Fractional Abundances	Error in Co-Localization and Spatial Resampling
Abay et al. [62]	ASTER (15–30 m) Landsat OLI (30 m)	Goethite, hematite	All	Geological map	-	In situ observations	-	-	Reference map	-	-
Ambarwulan et al. [147]	MERIS (300 m)	Several total suspended matter concentrations	All	In situ data	-	-	71 samples	-	-	-	-
Benhalouche et al. [156]	PRISMA (30 m)	Hematite, magnetite, limonite, goethite, apatite	All	In situ data	-	-	-	-	-	-	-
Bera et al. [120]	Landsat TM, ETM+, OLI (30 m)	Vegetation, impervious surface, soil	All	Google Earth images	Photointerpretation	Soil map	01 polygons	Uniform	Reference fractional abundance maps	Partial	-
Brice et al. [121]	Landsat TM, OLI (30 m)	Water, wetland vegetation, trees, grassland	1	Planet images (4 m)	Photointerpretation	In situ observations	427 wetlands	-	Reference fractional abundance map	Partial	-
Cao et al. [164]	Sentinel-2 (10–20–60 m)	Vegetation, high albedo impervious surface, low albedo impervious surface, soil	All	GaoFen-2(0.8–3.8 m)	Photointerpretation	In situ observations	300 squares (100 × 100 m)	Stratified random	Reference fractional abundance maps	Partial	Polygon size
Cavalli [114]	Hyperion (30 m) PRISMA (30 m)	Lateritic tiles, lead plates, asphalt, limestone, trachyte rock, grass, trees, lagoon water	All All	Panchromatic IKONOS image (1 m) Synthetic Hyperion and PRISMA images (0.30 m)	Photointerpretation The same spectral unmixing procedure performed to real images	In situ observations and shape files provided by the city and lagoon portal of Venice (Italy)	The whole study area	The whole study area	Reference fractional abundance maps	Full	Spatial resampling the reference maps and evaluation of the errors Evaluation of the errors in co-localization and spatial-resampling
Cavalli [145]	MIVIS (8m)	Lateritic tiles, lead plates, vegetation, asphalt, limestone, trachyte rock	All All	Panchromatic IKONOS image (1 m) Synthetic MIVIS image (0.30 m)	Photointerpretation The same spectral unmixing procedure performed to real image	In situ observations and shape files provided by the city and lagoon portal of Venice (Italy)	The whole study area	The whole study area	Reference fractional abundance maps	Partial	Spatial resampling the reference maps and evaluation of the errors Evaluation of the errors in co-localization and spatial-resampling
Cerra et al. [104]	DESIS (30 m) HySpex (0.6–1.2 m) Sentinel-2 (10–20–60 m)	PV panels, 2 grass, 2 forest, 2 soil, 2impervious surfaces	1	Reference map	-	-	The whole study area	The whole study area	-	-	-
Cipta et al. [137]	Landsat OLI (30 m) MODIS (500 m)	Rice, non-rice	All	In situ data	-	-	10 samples	-	-	-	-
Compains Iso et al. [134]	Landsat TM, OLI (30 m)	Forest, shrubland, grassland, water, rock, bare soil	All	Orthophoto (≤0.5 m)	Photointerpretation	-	50 squares (30 × 30 m)	Random	Reference fractional abundance maps	Partial	-
Damarjati et al. [157]	PRISMA (30 m)	A. obtusifolia, sand, wetland vegetations	All	In situ data	-	-	-	-	Reference maps	-	-
Dhaini et al. [70]	AVIRIS (20 m)	Andradite, chalcedony, kaolinite, jarosite, montmorillonite, nontronite Road, trees, water, soil Asphalt, dirt, tree, roof	All	Reference map	-	-	The whole study area	The whole study area	Reference maps	-	-
Ding et al. [122]	Landsat TM, OLI (30 m)	Vegetation, impervious surface, soil	All	Google satellite images (1 m)	Photointerpretation	-	100 points	Random	Reference maps	-	-
Ding et al. [152]	MODIS (250–500 m)	Vegetation, non-vegetation	All	Landsat (30 m)	K-means unsupervised classified method	Google map	Landsat images	Representative areas	Reference fractional abundance maps	Partial	Spatial resampling the reference maps
Fang et al. [71]	AVIRIS (20 m)	Road, 2building, trees, grass, soil Road, trees, water, soil	All	Reference map	-	-	The whole study area	The whole study area	Reference maps	-	-
Fernández-Guisuraga et al. [161]	Sentinel-2 (10–20 m)	Soil, green vegetation, non-photosynthetic vegetation	1	Photos	-	-	60 situ plots (20 × 20 m)	Stratified random	Reference fractional abundance map	Full	-
Gu et al. [98]	AVIRIS (20 m)	Vegetation, soil, road, river soil, water, vegetation	All	Reference map	-	-	The whole study area	The whole study area	Reference maps	-	-
Guan et al. [86]	AVIRIS (20 m) HYDICE (10 m)	Trees, water, dirt, road Asphalt, grass, trees, roofs	All All	Reference map Reference map	- -	- -	The whole study area The whole study area	The whole study area The whole study area	Reference maps Reference maps	- -	- -
Hadi et al. [68]	AVIRIS (20 m) HYDICE (10 m)	Trees, water, dirt, road Asphalt, grass, trees, roofs	All All	Reference map Reference map	- -	- -	The whole study area The whole study area	The whole study area The whole study area	Reference maps Reference maps	- -	- -
Hajnal et al. [169]	Sentinel-2 (10–20–60 m)	Vegetation, impervious surface, soil	All	APEX image (2 m), High-resolution land cover map	Support vector classification	-	APEX image	Representative areas	Reference fractional abundance maps	Full	Spatial resampling the reference maps
Halbgewachs et al. [123]	Landsat TM, OLI (30 m), TIRS (60 m)	Forest, non-Forest (non-photosynthetic vegetation, soil, shade)	2	Annual classifications of the Program for Monitoring Deforestation in the Brazilian Amazon (PRODES)	-	Official truth-terrain data from deforested and non-deforested areas prepared by PRODES	494 samples	Stratified random	Reference maps	-	-

Table A1. *Cont.*

Paper	Remote Image	Determined Endmembers	Validated Endmembers	Sources of Reference Data	Method for Mapping the Endmembers	Validation of Reference Data with Other Reference Data	Sample Sizes and Number of Small Sample Sizes	Sampling Designs	Reference Data	Estimation of Fractional Abundances	Error in Co-Localization and Spatial Resampling
He et al. [56]	AMMIS (0.5 m)	Urban surface materials	All	Reference map	-	-	The whole study area	The whole study area	Reference maps	-	-
	ROSIS (4 m)	Urban surface materials	All	Reference map	-	-	The whole study area	The whole study area	Reference maps	-	-
Hong et al. [69]	AVIRIS (20 m)	Trees, water, dirt, road, roofs	All	Reference map	-	-	The whole study area	The whole study area	Reference maps	-	-
	EnMAP (30 m)	Asphalt, soil, water, vegetation	All	Reference map	-	-	The whole study area	The whole study area	Reference maps	-	-
Hu et al. [149]	MODIS (0.5–1 km)	Blue ice, coarse-grained snow, fresh snow, bare rock, deep water, slush, wet snow	1	Sentinel-2 images	The same spectral unmixing procedure performed to MODIS images	Five auxiliary datasets	Six test areas identified as blue ice areas in the Landsat-based LIMA product	Representative areas	Reference fractional abundance maps	Full	-
Hua et al. [72]	AVIRIS (10 m)	Dirt, road	All	Reference map	-	-	The whole study area	The whole study area	Reference maps	-	-
	Samson (3.2 m)	-	All	Reference map			The whole study area	The whole study area	Reference maps		
Jamshid Moghadam et al. [115]	Hyperion (30 m)	Kaolinite/smeetite, sepiolite, lizardite, chorite	All	Geological map	-	-	The whole study area	The whole study area	Reference maps	-	-
Jin et al. [143]	M3 hyperspectral image	Lunar surface materials	All	Lunar Soil Characterization Consortium dataset	-	-	-	-	Reference fractional abundance maps	Full	-
Jin et al. [73]	AVIRIS (10 m)	Road, soil, tree, water	All	Reference map	-	-	The whole study area	The whole study area	Reference maps	-	-
	Samson (3.2 m)	Water, tree, soil	All	Reference map			The whole study area	The whole study area	Reference maps		
Kremezi et al. [166]	Sentinel-2 (10–20–60 m) WorldView-2 (0.46–1.8 m) WorldView-3 (0.31–1.24–3.7 m)	PET-1.5 l bottles, LDPE bags, fishing nets	All	In situ data	-	-	3 squares (10 × 10 m)	-	Reference map	-	-
Kuester et al. [111]	HYDICE (10 m)	Urban surface materials	All	Reference map	-	-	The whole study area	The whole study area	Reference maps	-	-
Kumar et al. [113]	Hyperion (30 m)	Sal-forest, teak-plantation, scrub, grassland, water, cropland, mixed forest, urban, dry riverbed	All	Google Earth images	-	-	Same squares (30 × 30 m)	-	Reference fractional abundance maps	Partial	-
Lathrop et al. [124]	Landsat 8 OLI (15–30 m)	Mud, sandy mud, muddy sand, sand	All	In situ data	-	-	805 circles (250 m radius)	Uniform	Reference fractional abundance map	Partial	-
Legleiter et al. [103]	DESIS (30 m)	12 cyanobacteria genera, water	All	In situ data	-	-	-	-	-	-	-
Li et al. [75]	AVIRIS (10 m)	Vegetation, bare soil, vineyard, etc.	All	Field reference data	-	-	The whole study area	The whole study area	Reference maps	-	-
	Hyperion (30 m)	-	All	Hyperion (30 m) image	-	-	The whole study area	The whole study area	Reference map	-	-
Li et al. [74]	AVIRIS (10 m)	Tree, water, dirt, road	All	Reference map	-	-	The whole study area	The whole study area	Reference maps	-	-
Li et al. [107]	GaoFen-6 (2–8–16 m) Landsat 8 OLI (15–30 m) Sentinel-2 (10–20–60 m)	Green vegetation, bare rock, bare soil, non-photosynthetic vegetation	All	Photo acquired with drones	Classification	In situ measurements of fractional vegetation cover and bare rock	285 polygons	Random	Reference fractional abundance maps	Full	Polygon size
Li et al. [76]	AVIRIS (10 m)	Andradite, chalcedony, kaolinite, jarosite, montmorillonite, nontronite	All	Reference map	-	-	The whole study area	The whole study area	Reference maps	-	-
	HYDICE (10 m)	Asphalt, grass, trees, roofs	All	Reference map	-	-	The whole study area	The whole study area	Reference maps	-	-
Luo et al. [77]	AVIRIS (10 m)	Andradite, chalcedony, kaolinite, jarosite, montmorillonite, nontronite	All	Reference map	-	-	The whole study area	The whole study area	Reference maps	-	-
	HYDICE (10 m)	Asphalt, grass, trees, roofs	All	Reference map	-	-	The whole study area	The whole study area	Reference maps	-	-
Lyngdoh et al. [100]	AVIRIS (20 m) AVIRIS-NG (5 m)	Trees, water, dirt, road Red soil, black soil, crop residue, built-up areas, bituminous roads, water	All	Reference map	-	-	The whole study area	The whole study area	Reference maps	-	-
Ma & Chang [78]	AVIRIS (10 m)	-	All	Reference map	-	-	The whole study area	The whole study area	Reference maps	-	Spatial resampling the reference maps
	CASI (2.5 m)	Urban surface materials	All	Reference map	-	-	The whole study area	The whole study area	Reference maps	-	-
	ROSIS (4 m)	Urban surface materials	All	Reference map	-	-	The whole study area	The whole study area	Reference maps	-	-
Matabishi et al. [469]	DESIS (30 m)	Roof materials	All	VHR images	-	Field validation data	1053 ground reference points	-	Reference fractional abundance maps	Full	-
Meng et al. [163]	Sentinel-2 (10–20–60 m)	Vegetation, non-vegetation	1	Google Earth Pro image (1 m)	-	-	10535 squares (10 × 10 m)	Stratified random	Reference fractional abundance maps	Partial	-
Nill et al. [125]	Landsat TM, OLI (30 m)	Shrubs, coniferous trees, herbaceous plants, lichens, water, barren surfaces	All	RGB camera (0.4–8 cm) Orthophotos (10–15 cm)	-	Field validation data	216 validation pixels	Stratified random	Reference fractional abundance maps	Full	-
Ouyang et al. [126]	Landsat-8 OLI (30 m)	Impervious surface, evergreen vegetation, seasonally exposed soil	1	Land use and land cover maps (0.5 m)	-	-	264 circles (1 km radius)	Random	Reference fractional abundance map	Partial	-

Table A1. *Cont.*

Paper	Remote Image	Determined Endmembers	Validated End-members	Sources of Reference Data	Method for Mapping the Endmembers	Validation of Reference Data with Other Reference Data	Sample Sizes and Number of Small Sample Sizes	Sampling Designs	Reference Data	Estimation of Fractional Abundances	Error in Co-Localization and Spatial Resampling
Ozer & Leloglu [167]	Sentinel-2 (10–20–60 m)	Soil, vegetation, water	All	Aerial images (30 cm)	-	-	-	-	Reference fractional abundance map	Partial	-
P et al. [61]	ASTER (90 m)	Iron Oxide	1	In situ data	-	-	13 samples	-	-	-	-
Palsson et al. [59]	Apex	Asphalt, vegetation, water, roof	All	Reference map	-	-	The whole study area	The whole study area	Reference maps	-	-
	AVIRIS (10 m)	Road, soil, tree, water	All	Reference map	-	-	The whole study area	The whole study area	Reference maps	-	-
	CASI (2.5)	Urban surface materials	All	Reference map	-	-	The whole study area	The whole study area	Reference maps	-	-
	HYDICE (10 m)	Asphalt, grass, trees, roofs	All	Reference map	-	-	The whole study area	The whole study area	Reference maps	-	-
	Samson (3.2 m)	Soil, tree, water	All	Reference map	-	-	The whole study area	The whole study area	Reference maps	-	-
Pan & Jiang [65]	AVHRR (1–5 km)	Snow, bare land, grass, forest, shadow	All	Landsat7 TM+ image (30 m)	The same procedure performed to AVHRR image	-	Landsat image	Representative area	Reference fractional abundance maps	Full	-
Pan et al. [66]	AVHRR (1–5 km)	Snow, bare land, grass, forest, shadow	All	Landsat5 TM image (30 m)	The same procedure performed to AVHRR image	The land use/land cover	Landsat image	Representative area	Reference fractional abundance maps	Full	-
Paul et al. [470]	DESIS (30 m)	PV panel, vegetation, sand	All	VHR image	-	-	-	Random	Reference fractional abundance maps	Full	-
Pervin et al. [154]	NEON (1 m)	Tall woody plants, herbaceous and low stature vegetation, bare soil	All	NEON AOP image (0.1 m)	Supervised classification	Drone imagery (0.01 m)	3 sets of 10 pixels	Random	Reference fractional abundance maps	Partial	-
Qi et al. [89]	AVIRIS (10 m)	Road, soil, tree, water	All	Reference map	-	-	The whole study area	The whole study area	Reference maps	-	-
	HYDICE (10 m)	Asphalt, grass, trees, roofs	All	Reference map	-	-	The whole study area	The whole study area	Reference maps	-	-
	Samson (3.2 m)	Soil, tree, water	All	Reference map	-	-	The whole study area	The whole study area	Reference maps	-	-
Rajendran et al. [116]	Hyperion (30 m)	Chlorophyll-a	1	WorldView-3 image (0.31–1.24–3.7 m)		Field validation data	-	-	Reference fractional abundance maps	Full	-
Ronay et al. [170]	Specim IQ	Weed species	All	In situ data	-	-	The whole study area	The whole study area	Reference fractional abundance maps	Full	-
Santos et al. [131]	Landsat MSS, TM, OLI (30 m)	Natural vegetation, anthropized area, burned, water	All	In situ data	-	-	samples	Random	Reference maps	-	-
Shaik et al. [158]	PRISMA (30 m)	Broadleaved forest, Coniferous forest, Mixed forest, Natural grasslands, Sclerophyllous vegetation	All	Land use and land cover map	-	Field validation data	-	-	Reference maps	-	-
Shao et al. [109]	Landsat-8 OLI (15–30 m) GaoFen-1 (2–8–16 m)	Vegetation, soil impervious surfaces (high albedo; low albedo), water	1	GaoFen-1 image (2 m)	Object-based classification and photointerpretation of the results.	Ground-based measurements	300 pixels	Uniform	Reference fractional abundance map	Partial	-
Shi et al. [90]	AVIRIS (10 m)	Road, soil, tree, water	All	Reference map	-	-	The whole study area	The whole study area	Reference maps	-	-
	HYDICE (10 m)	Road, roof, soil, grass, trail, tree, water	All	Reference map	-	-	The whole study area	The whole study area	Reference maps	-	-
Shi et al. [79]	AVIRIS (10 m)	Road, soil, tree, water	All	Reference map	-	-	The whole study area	The whole study area	Reference maps	-	-
	HYDICE (10 m)	Road, roof, soil, grass, trail, tree, water	All	Reference map	-	-	The whole study area	The whole study area	Reference maps	-	-
Shimabukuro et al. [132]	Landsat TM, OLI (30 m)	Forest plantation	All	MapBiomas annual LULC map collection 6.0	-	-	20000 samples	Stratified random	Reference maps	Partial	-
Silvan-Cardenas et al. [139]	Landsat (30 m)	-	-	In situ data	-	-	samples	-	Reference maps	-	-
Sofan et al. [135]	Landsat-8 OLI (15–30 m)	Vegetation, smoldering, burnt area	All	PlanetScope images (3 m)	Photointerpretation	-	-	Random	-	-	-
Song et al. [153]	MODIS (0.5 km)	Water, urban, tree, grass	All	GlobalLand30 maps (GLC30) produced based on Landsat (30 m)	-	-	-	-	Reference fractional abundance maps	Full	-
	AVIRIS (10 m)	-	-	Reference map	-	-	The whole study area	The whole study area	Reference maps	-	-
	HYDICE (10 m)	Road, roof, soil, grass, trail, tree, water	-	Reference map	-	-	The whole study area	The whole study area	Reference maps	-	-
Sun et al. [80]	AVIRIS (10 m)	Andradite, chalcedony, kaolinite, jarosite, montmorillonite, nontronite	All	Reference map	-	-	The whole study area	The whole study area	Reference fractional abundance maps	Full	-
Sun et al. [165]	Sentinel-2 (10–20–60 m)	Rice residues, soil, green moss, white moss	1	Photos (1.5 m)	Photointerpretation	In situ observations	30 samples	Random	Reference fractional abundance maps	Partial	-

Table A1. *Cont.*

Paper	Remote Image	Determined Endmembers	Validated Endmembers	Sources of Reference Data	Method for Mapping the Endmembers	Validation of Reference Data with Other Reference Data	Sample Sizes and Number of Small Sample Sizes	Sampling Designs	Reference Data	Estimation of Fractional Abundances	Error in Co-Localization and Spatial Resampling
Sutton et al. [119]	Landsat TM, OLI (30 m)	Drylands, semi-arid zone, arid zone	All	In situ data	-	-	4207 samples	No-uniform	-	-	-
Tao et al. [91]	AVIRIS (10 m)	Andradite, chalcedony, kaolinite, jarosite, montmorillonite, nontronite	All	Reference map	-	-	The whole study area	The whole study area	Reference maps	-	-
Tarazona Coronel [127]	Landsat TM, OLI (30 m)	Vegetation	1	Landsat (15–30 m) and Sentinel-2 (10–20–60 m) images	Photointerpretation	Official truth-terrain data from deforested and non-deforested areas prepared by PRODES	300 samples	Stratified random	Reference fractional abundance maps	Partial	-
van Kuik et al. [133]	Landsat TM, OLI (30 m) Sentinel-2 (10–20–60 m)	Blowouts to sand, water, vegetation	1	Unoccupied Aerial Vehicle (UAV) orthomosaics (1 m)	Photointerpretation	-	-	-	Reference fractional abundance maps	Partial	-
Viana-Soto et al. [138]	Landsat TM, OLI (30 m)	Tree, shrub, background (herbaceous, soil, rock)	1	Orthophotos	Photointerpretation	Validation samples	-	Uniform	Reference fractional abundance maps	Full	-
Wang et al. [87]	AVIRIS (10 m)	-	-	Reference map	-	-	The whole study area	The whole study area	Reference maps	-	-
Wang et al. [142]	Landsat-8 OLI (30 m)	Impervious surfaces (high albedo, low albedo), forest, grassland, soil	1	QuickBird image (0.6 m)	Spectral angle mapping classification	In situ observations	13,080 points	Random	Reference fractional abundance maps	Partial	-
Wang et al. [150]	MODIS (0.5 km)	Vegetation, non-vegetation	All	Landsat image (30 m)	K-means-based unsupervised classification	-	Landsat image	Representative area	Reference fractional abundance maps	Partial	-
Wang et al. [92]	AVIRIS (10 m)	Andradite, chalcedony, kaolinite, jarosite, montmorillonite, nontronite	All	Reference map	-	-	The whole study area	The whole study area	Reference map	-	-
Wu & Wang [85]	AVIRIS (10 m)	Urban surface materials	All	Reference map	-	-	The whole study area	The whole study area	Reference maps	-	-
	HYDICE (10 m)	Road, roof, soil, grass, trail, tree, water	All	Reference map	-	-	The whole study area	The whole study area	Reference maps	-	-
	ROSIS (4 m)	Urban surface materials	All	Reference map	-	-	The whole study area	The whole study area	Reference maps	-	-
Xia et al. [128]	Landsat ETM+, OLI (30 m)	High albedo, vegetation low albedo, shadow	2	Google Earth images	Photointerpretation	-	100 polygons (30 × 30 m)	Random	Reference fractional abundance maps	Partial	-
Xu et al. [162]	Sentinel-2 (10–20–60 m)	Impervious surface, water body, vegetation, bare land	All	Google Earth images	Photointerpretation	In situ observations	-	-	Reference fractional abundance maps	Partial	-
Yang et al. [57]	AMMIS (0.5 m) AVIRIS ROSIS	-	All	Reference map	-	-	The whole study area	The whole study area	Reference maps	-	-
Yang [81]	AVIRIS (20 m)	Vegetation, water, soil	All	Reference map	-	-	The whole study area	The whole study area	Reference maps	-	-
	HYDICE (10 m)	Road, roof, soil, grass, trail, tree, water	All	Reference map	-	-	The whole study area	The whole study area	Reference maps	-	-
	ROSIS (4 m)	Urban surface materials	All	Reference map	-	-	The whole study area	The whole study area	Reference maps	-	-
Yang et al. [141]	Landsat-8 OLI (30 m)	Water, non-water	All	Google Earth images	-	-	The whole study area	The whole study area	Reference fractional abundance maps	Partial	-
Yi et al. [82]	AVIRIS (20 m)	Vegetation, water, soil	All	Reference map	-	-	The whole study area	The whole study area	Reference maps	-	-
	HYDICE (10 m)	Road, roof, soil, grass, trail, tree, water	All	Reference map	-	-	The whole study area	The whole study area	Reference maps	-	-
Yin et al. [82]	MODIS (0.250 km)	Water, soil	1	Landsat OLI image (30 m)	Modified normalized difference water index (MNDWI)	-	Landsat image	Representative area	Reference fractional abundance maps	Partial	Spatial resampling the reference maps
Zhang & Jiang [108]	Landsat (30 m) Sentinel-2 (20 m) MODIS (0.5 km)	Snow	1	GaoFen-2 image (3.2 m)	Supervised classification	-	-	-	Reference fractional abundance map	Partial	-
Zhang et al. [117]	HySpec (0.7 m)	Bitumen, red-painted metal sheets, blue fabric, red fabric, green fabric, grass	All	Reference map	-	-	-	-	Reference maps	Partial	-
Zhang et al. [83]	AVIRIS (20 m)	Andradite, chalcedony, kaolinite, jarosite, montmorillonite, nontronite	All	Reference map	-	-	The whole study area	The whole study area	Reference maps	-	-
Zhang et al. [93]	AVIRIS (10/20 m)	Dumortierite, muscovite, Alunite+muscovite, kaolinite, alunite, montmorillonite Tree, water, road, soil	All	Reference map	-	-	The whole study area	The whole study area	Reference maps	-	-
Zhang et al. [129]	Landsat-8 OLI (30 m)	Vegetation, impervious surfaces	All	GaoFen-1 image (2–8 m)	Photointerpretation	-	101 samples	Uniform	Reference fractional abundance maps	Partial	-
Zhang et al. [130]	Landsat-8 OLI (30 m)	Vegetation	All	GaoFen-1 image (2–8 m)	Object-based classification	-	101 samples	Uniform	Reference fractional abundance map	Partial	-

Table A1. *Cont.*

Paper	Remote Image	Determined Endmembers	Validated Endmembers	Sources of Reference Data	Method for Mapping the Endmembers	Validation of Reference Data with Other Reference Data	Sample Sizes and Number of Small Sample Sizes	Sampling Designs	Reference Data	Estimation of Fractional Abundances	Error in Co-Localization and Spatial Resampling
Zhang et al. [88]	AVIRIS (10/20 m)	Cuprite, road, trees, water, soil Asphalt, dirt, tree, roof	All	Reference map Reference map	-	-	The whole study area	The whole study area	Reference maps	-	-
Zhao et al. [84]	AVIRIS (10 m)	Road, trees, water, soil Asphalt, grass, tree, roof, metal, dirt	All	Reference map Reference map	-	-	The whole study area	The whole study area	Reference maps	-	-
Zhao et al. [96]	AVIRIS (10 m)	Road, trees, water, soil	All	Reference map	-	-	The whole study area	The whole study area	Reference maps	-	-
	HYDICE (10 m)	Road, roof, soil, grass, trail, tree, water	All	Reference map	-	-	The whole study area	The whole study area	Reference maps	-	-
Zhao et al. [94]	AVIRIS (10 m)	Road, trees, water, soil	All	Reference map	-	-	The whole study area	The whole study area	Reference maps	-	-
Zhao et al. [95]	AVIRIS (20 m)	Andradite, chalcedony, kaolinite, jarosite, montmorillonite, nontronite	All	Reference map	-	-	The whole study area	The whole study area	Reference maps	-	-
Zhao et al. [136]	Landsat-8 OLI (30 m) Sentinel-2 (10–20–60 m)	Impervious surfaces, vegetation, soil, water	2	WorldView-2 image (0.50–2 m)	-	-	172 polygons 480 × 480 m)	Random	Reference fractional abundance maps	Full	-
Zhao et al. [140]	Landsat (30 m) Spot (30 m)	Vegetation	1	Fractional vegetation cover reference maps (provided by VALERI project and the ImagineS)	-	In situ measurements of LAI (provided by VALERI project and the ImagineS)	445 squares (20 × 20 m or 30 × 30 m)	-	Reference fractional abundance map	Full	-
Zhao & Qin [168]	Sentinel-2 (10–20–60 m)	Vegetation, mineral area	All	In situ data	-	-	-	-	Reference fractional abundance maps	Partial	-
Zhu et al. [64]	AVHRR (1–5 km)	Snow, non-snow (bare land, vegetation, and water)	1	Landsat TM image (30 m)	Normalized difference snow index	-	Landsat image	Representative area	Reference fractional abundance map	Full	Spatial resolution variation
Zhu et al. [97]	AVIRIS (10 m)	Road, trees, water, soil	All	Reference map	-	-	The whole study area	The whole study area	Reference maps	-	-
	HYDICE (10 m)	Road, roof, soil, grass, trail, tree, water	All	Reference map	-	-	The whole study area	The whole study area	Reference maps	-	-
	Samson (3.2 m)	Soil, tree, water	All	Reference map	-	-	The whole study area	The whole study area	Reference maps	-	-

Table A2. Main characteristics of the eligible papers that were published in 2021.

Paper	Remote Image	Determined Endmembers	Validated Endmembers	Sources of Reference Data	Method for Mapping the Endmembers	Validation of Reference Data with Other Reference Data	Sample Sizes and Number of Small Sample Sizes	Sampling Designs	Reference Data	Estimation of Fractional Abundances	Error in Co-Localization and Spatial Resampling
Azar et al. [174]	AVIRIS CASI	Trees, Mostly Grass Ground Surface, Mixed Ground Surface, Dirt/Sand, Road	All All	Reference map CASI image	- Photointerpretation	-	The whole study area	The whole study area	Reference map Reference map	-	-
Badola et al. [226]	AVIRIS-NG (5 m) Sentinel-2 (10–20–60 m)	Black Spruce Birch Alder Gravel	All	In situ data	Photointerpretation	In situ observations	29 plots	Random	Reference map	-	-
Bai et al. [175]	AVIRIS	Asphalt, Grass, Tree, Roof, Metal, Dirt	All	Reference map	-	-	The whole study area	The whole study area	Reference map	-	-
Bair et al. [254]	Landsat MODIS	Snow, canopy	1	WorldView-2–3 images (0.34–0.55 m)	Photointerpretation	Airborne Snow Observatory (ASO) (3 m)	-	-	Reference fractional abundance map	Full	Spatial resampling the reference maps Evaluation of the errors in co-localization and spatial-resampling
Benhalouche et al. [230]	HYDICE (10 m) Samson (3.2)	Asphalt, grass, tree, roof Soil, tree, water	All	Reference map	-	-	The whole study area	The whole study area	Reference map	-	-
Benhalouche et al. [265]	PRISMA (30 m)	Mineral	All	Geological map	-	-	The whole study area	The whole study area	Reference map	-	-
Borsoi et al. [176]	AVIRIS	-	All	Reference map	-	-	The whole study area	The whole study area	Reference map	-	-
Cerra et al. [238]	HySpex	Target	All	In situ data	-	Reference targets and Aeronet data	-	-	Reference fractional abundance maps	-	-
Chang et al. [229]	GF-5 (30 m) Sentinel 2 (10–20–60 m) ZY-1-02D (30 m)	-	All	In situ data	-	-	-	-	Reference fractional abundance maps	-	-
Chen et al. [239]	Landsat	-	All	UAV images	-	Ground survey data	-	-	Reference fractional abundance maps	-	-
Chen et al. [245]	Landsat	Vegetation, impervious surface, bare soil, and water	All	Google Earth images	-	-	-	-	Reference fractional abundance maps	-	-

Table A2. *Cont.*

Paper	Remote Image	Determined Endmembers	Validated End-mem-bers	Sources of Reference Data	Method for Mapping the Endmembers	Validation of Reference Data with Other Reference Data	Sample Sizes and Number of Small Sample Sizes	Sampling Designs	Reference Data	Estimation of Fractional Abun-dances	Error in Co-Localization and Spatial Resampling
Chen et al. [246]	Landsat	-	All	Google Earth images	-	Field surveys	300 plots	Random	Reference fractional abundance maps	-	-
Converse et al. [247]	Landsat	Green vegetation, non-photosynthetic vegetation, soil	All	UAS images	-	Field surveys	Plots	-	Reference fractional abundance maps	Full	-
Di et al. [177]	AVIRIS	Cuprite	All	Reference map	-	-	The whole study area	The whole study area	Reference map	-	-
Dong & Yuan [178]	AVIRIS	-	All	Reference map	-	-	The whole study area	The whole study area	Reference map	-	-
Dong et al. [179]	AVIRIS	-	All	Reference map	-	-	The whole study area	The whole study area	Reference map	-	-
Dong et al. [180]	AVIRIS	-	All	Reference map	-	-	The whole study area	The whole study area	Reference map	-	-
Dutta et al. [248]	Landsat	Vegetation, impervious surface, bare soil,	1	In situ data	-	Built-up density, urban expansion and population density of the area	-	-	Reference fractional abundance maps	Full	-
Ekanayake et al. [181]	AVIRIS	-	All	Reference map	-	-	The whole study area	The whole study area	Reference map	-	-
Elrewainy & Sherif [182]	AVIRIS	-	All	Reference map	-	-	The whole study area	The whole study area	Reference map	-	-
Feng & Fan [255]	Landsat (30 m) Sentinel 2 (10–20–60 m)	Vegetation, high-albedo impervious surface, low-albedo impervious surface soil	All	In situ data	-	-	18000 testing areas	random	Reference fractional abundance maps	Full	-
Fernández-García et al. [256]	Landsat (30 m)	Arboreal vegetation, shrubby vegetation, herbaceous vegetation, rock and bare soil, water	All	Orthophotographs (0.25 m)	-	-	250 plots (30 × 30 m)	random	Reference fractional abundance maps	Full	Spatial resolution variation
Finger et al. [249]	Landsat (30 m)	-	All	California Department of Fish and Wildlife (CDFW) aerial survey canopy area product	-	-	-	-	Reference fractional abundance maps	Full	-
Gu et al. [183]	AVIRIS	-	All	Reference map	-	-	The whole study area	The whole study area	Reference map	-	-
Guo et al. [184]	AVIRIS	-	All	Reference map	-	-	The whole study area	The whole study area	Reference map	-	-
Gu et al. [185]	AVIRIS	-	All	Reference map	-	-	The whole study area	The whole study area	Reference map	-	-
Han et al. [186]	AVIRIS	Asphalt, grass, tree, roof	All	Reference map	-	-	The whole study area	The whole study area	Reference map	-	-
Han et al. [268]	AVIRIS	-	All	Reference map	-	-	The whole study area	The whole study area	Reference map	-	-
Haq et al. [234]	Hyperion (30 m)	Clean snow, blue ice, refreezing ice dirty snow, dirty glacier ice, firn, moraine, and glacier ice	All	In situ data	-	Sentinel-2 images	-	-	Reference fractional abundance maps	Full	-
He et al. [231]	HYDICE (10 m) MODIS (0.5–1 km)	-	All All	Reference map Finer Resolution Observation and Monitoring of Global Land Cov (30 m)	-	-	- 61 scenes	-	Reference fractional abundance maps	Full	-
He et al. [56]	ROSIS (4m)	Urban surface materials	All	Reference map	-	-	The whole study area	The whole study area	Reference map	-	-
Hua et al. [187]	AVIRIS	-	All	Reference map	-	-	The whole study area	The whole study area	Reference map	-	-
Hua et al. [188]	AVIRIS Samson (3.2)	- Soil, tree, water	All	Reference map	-	-	The whole study area	The whole study area	Reference map	-	-
Huang et al. [189]	AVIRIS	-	All	Reference map	-	-	The whole study area	The whole study area	Reference map	-	-
Jia et al. [190]	AVIRIS	Cuprite	All	Reference map	-	-	The whole study area	The whole study area	Reference map	-	-
Ji et al. [235]	Hyperion (30 m)	Photosynthetic vegetation, non-photosynthetic vegetation, bore soil	All	Reference map	-	-	The whole study area	The whole study area	Reference map	-	-
Jiji [250]	Landsat (30 m)	Heavy metals	All	In situ data	-	-	17 samples	Random	Reference fractional abundance maps	Full	-
Jin et al. [267]	ROSIS (4 m) Samson (3.2 m)	Urban surface materials Soil, tree, water	All	Reference map	-	-	The whole study area	The whole study area	Reference map	-	-
Kneib et al. [271]	Sentinel 2 (10–20–60 m)	-	all	Pleiades images (2 m)	Photointerpretation	-	-	-	Reference fractional abundance maps	Full	-
Kucuk & Yuksel [202]	AVIRIS	Cuprite	All	Reference map	-	-	The whole study area	The whole study area	Reference map	-	-
Kumar & Chakravortty [191]	AVIRIS ROSIS (4 m)	- Urban surface materials	All	Reference map	-	-	The whole study area	The whole study area	Reference map	-	-
Li et al. [203]	AVIRIS	Cuprite	All	Reference map	-	-	The whole study area	The whole study area	Reference map	-	-
Li et al. [192]	AVIRIS HYDICE (10 m)	Cuprite -	All	Reference map	-	-	The whole study area	The whole study area	Reference map	-	-
Li et al. [193]	AVIRIS		All	Reference map	-	-	The whole study area	The whole study area	Reference map	-	-
Li [194]	AVIRIS	Cuprite	All	Reference map	-	-	The whole study area	The whole study area	Reference map	-	-
Li et al. [195]	AVIRIS	Cuprite	All	Reference map	-	-	The whole study area	The whole study area	Reference map	-	-

Table A2. *Cont.*

Paper	Remote Image	Determined Endmembers	Validated End-members	Sources of Reference Data	Method for Mapping the Endmembers	Validation of Reference Data with Other Reference Data	Sample Sizes and Number of Small Sample Sizes	Sampling Designs	Reference Data	Estimation of Fractional Abun-dances	Error in Co-Localization and Spatial Resampling
Li et al. [196]	AVIRIS	Cuprite	All	Reference map	-	-	The whole study area	The whole study area	Reference map	-	-
Li et al. [197]	AVIRIS	Cuprite	All	Reference map	-	-	The whole study area	The whole study area	Reference map	-	-
Li et al. [251]	Landsat (30 m)	Impervious, vegetation, bare land, water	All	Google Earth images	-	Field surveys	4296 sampled points	Random	Reference fractional abundance maps	Full	-
Li [257]	Landsat (30 m)	Impervious, soil, vegetation	All	Images	-	-	200 sample points	Random	Reference fractional abundance maps	Full	-
Li et al. [204]	AVIRIS HYDICE (10 m)	-	All	Reference map	-	-	The whole study area	The whole study area	Reference map	-	-
Li et al. [205]	AVIRIS	Cuprite	All	Reference map	-	-	The whole study area	The whole study area	Reference map	-	-
Liu et al. [206]	AVIRIS	-	All	Reference map	-	-	The whole study area	The whole study area	Reference map	-	-
Lui & Zhu [207]	AVIRIS Samson (3.2 m)	- Soil, tree, water	All	Reference map	-	-	The whole study area	The whole study area	Reference map	-	-
Lombard & Andrieu [240]	Landsat	-	3	Google Earth images	Phointerpretation	-	8490 sample points	Random	Reference fractional abundance maps	Full	-
Luo & Chen [260]	Landsat	Vegetation, impervious, soil	1	Gaofen-2 and WorldView-2 images	-	-	-	-	Reference fractional abundance maps	Full	Spatial resolution variation
Ma et al. [276]	WorldView-3	Vegetation	All	Digital cover photography	-	Vegetation spectra	30 sample points	-	Reference fractional abundance map	Full	-
Mudereri et al. [273]	Sentinel 2 (10–20–60 m)	-	All	Google Earth images	-	Field surveys	1370 pixels	Random	Reference fractional abundance maps	Full	-
Muhuri et al. [258]	Landsat Sentinel 2 (10–20–60 m)	Snow cover	All	In situ data	-	Airborne Snow Observatory (ASO) (2 m)	-	-	Reference fractional abundance maps	Full	-
Okujeni et al. [228]	Simulated EnMAP	-	All	Google Earth images	-	Landsat images	3183 sites	Random	Reference fractional abundance maps	Full	-
Ou et al. [233]	HyMap (4.5 m)	Soil organic matter, soil heavy meta	All	In situ data	-	-	95 soil samples	Random	Reference fractional abundance maps	Full	-
Pan et al. [261]	MODIS (0.5–1 km)	Snow	All	Landsat images	MESMA	-	The whole study area	The whole study area	Reference fractional abundance maps	Full	-
Patel et al. [208]	AVIRIS	-	All	Reference map	-	-	The whole study area	The whole study area	Reference map	-	-
Peng et al. [209]	AVIRIS	-	All	Reference map	-	-	The whole study area	The whole study area	Reference map	-	-
Qin et al. [210]	AVIRIS Samson (3.2 m)	- Soil, tree, water	All	Reference map	-	-	The whole study area	The whole study area	Reference map	-	-
Racoviteanu et al. [241]	Landsat	Debris-covered glaciers	All	Pléiades 1A image (2 m) RapidEye image (5 m) PlanetScope (3 m)	Phointerpretation	DEM	151 test pixels	Random	Reference fractional abundance maps	Full	-
Rittger et al. [262]	MODIS (0.5–1 km)	Snow	All	Landsat images	-	-	-	Random	Reference fractional abundance maps	Full	Spatial resolution variation
Sall et al. [252]	Landsat (30 m)	Waterbodies	All	DigitalGlobe WorldView-2 (0.46 m)	-	National Agriculturelmagery Program (NAIP)	-	-	Reference fractional abundance maps	Full	-
Sarkar & Sur [173]	ASTER (15–30–90 m)	Bauxite minerals	All	In situ data	-	Petrological, EPMA, SEM-EDS studies DEM	-	-	Reference fractional abundance maps	Full	-
Seydi & Hasanlou [236]	Hyperion (30 m)	-	All	In situ data	-	-	73505 samples	Random	Reference fractional abundance maps	Full	-
Seydi & Hasanlou [237]	Hyperion (30 m)	-	All	In situ data	-	-	-	-	Reference fractional abundance maps	Full	-
Shahid & Schizas [211]	AVIRIS Samson (3.2 m)	- Soil, tree, water	All	Reference map	-	-	The whole study area	The whole study area	Reference map	-	-
Shen et al. [242]	Landsat (30 m)	Impervious, non-impervious surface	All	Land use map by the National Basic Geographic Information Center	-	-	-	-	Reference map	-	-
Shen et al. [270]	Sentinel 2 (10–20–60 m)	-	All	Google Earth images	Phointerpretation	-	467 polygons	Random	Reference fractional abundance maps	Full	-
Shumack et al. [243]	Landsat (30 m) Sentinel 2 (10–20–60 m)	Bare soil, photosynthetic vegetation, non-photosynthetic vegetation	All	Orthorectified mosaic images (0.02 m)	Object based image analyses	SLATS dataset of fractional ground cover surveys	400 point per images	Random	Reference fractional abundance maps	Full	-
Song et al. [232]	HYDICE (10 m) Samson (3.2 m)	Road, trees, water, soil Soil, tree, water	All	Reference map	-	-	The whole study area	The whole study area	Reference map	-	-

Table A2. *Cont.*

Paper	Remote Image	Determined Endmembers	Validated End-members	Sources of Reference Data	Method for Mapping the Endmembers	Validation of Reference Data with Other Reference Data	Sample Sizes and Number of Small Sample Sizes	Sampling Designs	Reference Data	Estimation of Fractional Abundances	Error in Co-Localization and Spatial Resampling
Soydan et al. [272]	Sentinel 2 (10–20–60 m)	-	All	Laboratory analysis of field collected samples through Inductive Coupled Plasma	-	Laboratory analysis of field collected samples through X-Ray Diffraction, and ASD spectral analysis	-	-	Reference fractional abundance maps	Full	-
Su et al. [212]	AVIRIS HYDICE (10 m) Hyperion (30 m)	- Road, trees, water, soil	All	Reference map	-	-	The whole study area	The whole study area	Reference map	-	-
Sun et al. [263]	MODIS (0.5–1 km)	Green vegetation, sand, saline, and dark surface	All	Google Earth images Field observations	-	-	89 samples 10 plots (1 × 1 km)	Random	Reference fractional abundance maps	Full	Spatial resolution variation
Sun et al. [275]	WorldView-2	Mosses, lichens, rock, water, snow		In situ data	-	Photos and spectra	32 plots (2 × 2 m)	Random	Reference fractional abundance maps	-	-
Tan et al. [198]	AVIRIS	Cuprite	All	Reference map	-	-	The whole study area	The whole study area	Reference map	-	-
Vibhute et al. [213]	AVIRIS	Tree, soil, water, road	All	Reference map	-	-	The whole study area	The whole study area	Reference map	-	-
Wan et al. [214]	AVIRIS HYDICE (10 m) Samson (3.2 m)	- Soil, tree, water	All	Reference map	-	-	The whole study area	The whole study area	Reference map	-	-
Wang et al. [215]	AVIRIS	-	All	Reference map	-	-	The whole study area	The whole study area	Reference map	-	-
Vermeulen et al. [244]	Landsat Sentinel 2 (10–20–60 m)	Soil, Photosynthetic Vegetation, Non-Photosynthetic Vegetation	All	Images, field data	-	-	(10 × 10 m) plots	-	Reference fractional abundance maps	-	-
Wang et al. [199]	AVIRIS	-	All	Reference map	-	-	The whole study area	The whole study area	Reference map	-	-
Wang et al. [216]	AVIRIS HYDICE (10 m)	-	All	Reference map	-	-	The whole study area	The whole study area	Reference map	-	-
Wang [217]	AVIRIS ROSIS (4 m)	- Urban surface materials	All	Reference map	-	-	The whole study area	The whole study area	Reference map	-	-
Wang et al. [200]	AVIRIS ROSIS (4 m)	- Urban surface materials	All	Reference map	-	-	The whole study area	The whole study area	Reference map	-	-
Wu et al. [253]	Landsat Sentinel 2 (10–20–60 m)	Bare soil, agricultural crop Water, vegetation, urban	All	Google Maps	Photointerpretation	-	-	-	Reference fractional abundance maps	Full	-
Xiong et al. [201]	AVIRIS	-	All	Reference map	-	-	The whole study area	The whole study area	Reference map	-	-
Xiong et al. [218]	AVIRIS HYDICE (10 m)	- Road, trees, water, soil	All	Reference map	-	-	The whole study area	The whole study area	Reference map	-	-
Xu et al. [219]	AVIRIS	Cuprite	All	Reference map	-	-	The whole study area	The whole study area	Reference map	-	-
Xu & Somers [269]	Sentinel 2 (10–20–60 m)	Vegetation, soil, impervious surface	All	Google Earth images	Object-oriented classification	-	-	-	Reference fractional abundance maps	Full	-
Yang et al. [264]	MODIS (0.5–1 km)	Vegetation, soil	All	GF-1, Google Earth images	-	-	2044 samples (0.5 × 0.5 km)	Random	Reference fractional abundance maps	Full	-
Ye et al. [220]	AVIRIS	-	All	Reference map	-	-	The whole study area	The whole study area	Reference map	-	-
Yu et al. [227]	Landsat (30 m) CASI	-	All	GF-1 image (2 m) GeoEye image (2 m) Reference map	Classification	-	The whole study area	The whole study area	Reference fractional abundance maps	Partial	-
Yuan et al. [274]	UAV multispectral image	-	All	In situ data	-	-	67 samples	-	Reference fractional abundance maps	Full	-
Yuan & Dong [221]	AVIRIS	-	All	Reference map	-	-	The whole study area	The whole study area	Reference map	-	-
Yuan et al. [222]	AVIRIS	-	All	Reference map	-	-	The whole study area	The whole study area	Reference map	-	-
Zang et al. [259]	Landsat	Vegetation, soil, impervious surface	All	Google Earth Pro image		Night light data, population data at township scale, administrative data	120 samples	Random	Reference fractional abundance maps	Full	-
Zhang & Pezeril [223]	AVIRIS	-	All	Reference map	-	-	The whole study area	The whole study area	Reference map	-	-
Zhao et al. [266]	ROSIS (4 m)	Urban surface materials	All	Reference map	-	-	The whole study area	The whole study area	Reference map	-	-
Zheng et al. [224]	AVIRIS Samson (3.2 m)	- Soil, tree, water	All	Reference map	-	-	The whole study area	The whole study area	Reference map	-	-
Zhu et al. [225]	AVIRIS Samson (3.2 m)	- Soil, tree, water	All	Reference map	-	-	The whole study area	The whole study area	Reference map	-	-

Table A3. Main characteristics of the eligible papers that were published in 2020.

Paper	Remote Image	Determined Endmembers	Validated End-members	Sources of Reference Data	Method for Mapping the Endmembers	Validation of Reference Data with Other Reference Data	Sample Sizes and Number of Small Sample Sizes	Sampling Designs	Reference Data	Estimation of Fractional Abun-dances	Error in Co-Localization and Spatial Resampling
Aalstad et al. [340]	Landsat MODIS Sentinel2	Shadow, cloudy, snow, snow-free	All	305 terrestrial images	Classification	DEM	-	-	Reference fractional abundance maps	Full	-
Aldeghlawi et al. [334]	HYDICE	Urban surface materials	All	Reference maps	-	-	The whole study area	The whole study area	Reference map	-	-
Arai et al. [368]	PROBA-V	Vegetation, soil, shade	All	Landsat images (30 m)	Calculate Geometry function	Land use and land cover map produced by the MapBiomas Project and the Agricultural Census	298 sampling units	Uniform	Reference fractional abundance maps	Full	Spatial resampling the reference maps
Bai et al. [281]	AVIRIS	-	All	Reference map	-	-	The whole study area	The whole study area	Reference map	-	-
Benhalouche et al. [278]	ASTER	-	All	In situ data	-	-	2 samples	-	Reference fractional abundance maps	Full	-
Binh et al. [341]	Landsat	-	All	Google Earth images	Phointerpretation	Field surveys	-	-	Reference fractional abundance maps	Full	Evaluation of the errors in co-localization and spatial-resampling
Borsoi et al. [283]	AVIRIS	-	All	Reference map	-	-	The whole study area	The whole study area	Reference map	-	-
Borsoi et al. [282]	AVIRIS	-	All	Reference map	-	-	The whole study area	The whole study area	Reference map	-	-
Borsoi et al. [176]	AVIRIS	-	All	Reference map	-	-	The whole study area	The whole study area	Reference map	-	-
Bullock et al. [349]	Landsat	-	All	In situ data	-	-	500 samples	Random	Reference fractional abundance maps	Full	-
Carlson et al. [377]	Sentinel (10–20–60 m)	-	All	In situ data	-	Aerial photograhs	-	Random	Reference fractional abundance maps	Full	-
Chen et al. [299]	AVIRIS HYDICE	Road, trees, water, soil	All	Reference map	-	-	The whole study area	The whole study area	Reference map	-	-
Cheng et al. [543]	Hyperspectral	-	All	In situ data	-	-	-	Random	Reference fractional abundance maps	Full	Evaluation of the errors in co-localization and spatial-resampling
Cooper et al. [330]	Simulated EnMAP (30 m)	-	All	Google Earth images	Phointerpretation	-	260 polygons (90 × 90 m)	Random	Reference fractional abundance maps	Full	-
Czekajlo et al. [350]	Landsat	-	All	Google Earth images	Phointerpretation	-	1085 grids (6 × 6 m)	Random	Reference fractional abundance maps	Full	-
Dai et al. [351]	Landsat	-	All	In situ data		DEM	2223 samples sites	Random			
Das et al. [300]	AVIRIS	-	All	Reference map	-	-	The whole study area	The whole study area	Reference map	-	-
Dou et al. [301]	AVIRIS Samson	Soil, tree, water	All	Reference map	-	-	The whole study area	The whole study area	Reference map	-	-
Drumetz et al. [329]	CASI	-	All	Reference map	-	-	The whole study area	The whole study area	Reference map	-	-
Elkholy et al. [284]	AVIRIS Samson	Soil, tree, water	All	Reference map	-	-	The whole study area	The whole study area	Reference map	-	-
Fang et al. [285]	AVIRIS ROSIS	Urban surface materials	All	Reference map	-	-	The whole study area	The whole study area	Reference map	-	-
Fathy et al. [286]	AVIRIS	-	All	Reference map	-	-	The whole study area	The whole study area	Reference map	-	-
Fernández-Guisuraga et al. [342]	Landsat WorldView-2	Photosynthetic vegetation, non-photosynthetic vegetation, soil and shade	All	In situ data	-	-	85 (30 × 30 m) field plots 360 (2 × 2 m) field plots	Random	Reference fractional abundance maps	Full	Co-localization the maps
Firozjaei et al. [364]	MODIS	-	All	Landsat images	-	Annual primary energy consumption, Global gridded population density, Population size data, Normalized difference vegetation index (NDVI) Data, CO and NOx emissions	The whole study area	The whole study area	Reference fractional abundance maps	Full	-
Fraga et al. [378]	Sentinel-2 (10–20–60 m)	-	All	In situ data	-		15 sampling points	Random	Reference fractional abundance maps	Full	-
Gharbi et al. [545]	Hyperspectral	-	All	Reference map	-	-	The whole study area	The whole study area	Reference map	-	-
Girolamo-Neto et al. [379]	Sentinel-2 (10–20–60 m)	-	All	In situ data			461 field observations	Random	Reference fractional abundance maps	Full	-
Godinho Cassol et al. [369]	PROBA-V	Vegetation, soil, shade	All	Landsat images (30 m)	-	-	622 sampling units	Uniform	Reference fractional abundance maps	Full	-
Han et al. [287]	AVIRIS	-	All	Reference map	-	-	The whole study area	The whole study area	Reference map	-	-
He et al. [356]	Landsat	-	All	In situ data	-	Photos	118 field sites	Random	Reference fractional abundance maps	Full	-
Holland & Du [288]	AVIRIS	-	All	Reference map	-	-	The whole study area	The whole study area	Reference map	-	-
Hua et al. [289]	AVIRIS	-	All	Reference map	-	-	The whole study area	The whole study area	Reference map	-	-
Huang et al. [302]	AVIRIS	-	All	Reference map	-	-	The whole study area	The whole study area	Reference map	-	-

Table A3. *Cont.*

Paper	Remote Image	Determined Endmembers	Validated End-members	Sources of Reference Data	Method for Mapping the Endmembers	Validation of Reference Data with Other Reference Data	Sample Sizes and Number of Small Sample Sizes	Sampling Designs	Reference Data	Estimation of Fractional Abundances	Error in Co-Localization and Spatial Resampling
Huechacona-Ruiz et al. [380]	Sentinel-2 (10–20–60 m)	-	All	In situ data	-	GPS	288 sampling units	Random	Reference fractional abundance maps	Full	-
Imbiriba et al. [303]	AVIRIS Samson	- Soil, tree, water	All	Reference map	-	-	The whole study area	The whole study area	Reference map	-	-
Jarchow et al. [358]	Landsat	-	All	WorldView-2 (0.5 m)	-	National Agriculture Imagery Program (NAIP) scene	154 pods	Random	Reference fractional abundance maps	Full	-
Ji et al. [333]	GF1 Landsat Sentinel-2 (10–20–60 m)	-	All	In situ data	-	GPS	111 surveyed fractional-cover sites	Random	Reference fractional abundance maps	Full	-
Jiang et al. [304]	AVIRIS	-	All	Reference map	-	-	The whole study area	The whole study area	Reference map	-	-
Karoui et al. [290]	AVIRIS	-	All	Reference map	-	-	The whole study area	The whole study area	Reference map	-	-
Khan et al. [352]	Landsat	-	All	In situ data	-	GPS, "Land Use, Land Use Change and Forestry Projects"	108 circular sample plots	Random	Reference fractional abundance maps	Full	-
Kompella et al. [328]	AWiFS Sentienl-2 (10–20–60 m)	-	All	In situ data	-	GPS	2 sampling areas	-	Reference fractional abundance maps	Partial	Co-localization the maps
Laamrani et al. [343]	Landsat	-	All	Photographs	-	Field surveys, GPS	70 (30 × 30 m) sampling area	-	Reference fractional abundance maps	Full	Co-localization the maps
Lewińska et al. [359]	MODIS	Soil, green vegetation, non-photosynthetic vegetation shade		Land cover classifications (30 m), Map of the Natural Vegetation of Europe	-	-	The whole study area	The whole study area	Reference fractional abundance maps	Full	-
Li et al. [305]	AVIRIS Samson	- Soil, tree, water	All	Reference map	-	-	The whole study area	The whole study area	Reference map	-	-
Li [360]	Landsat	Vegetation, high albedo, low albedo, soil	All	Orthophotography images, Google Earth images	-	-	The whole study area	The whole study area	Reference fractional abundance maps	Full	-
Ling et al. [365]	MODIS	water and land	All	Radar altimetry water levels	-	-	The whole study area	The whole study area	Reference fractional abundance maps	Full	-
Liu et al. [332]	GF1 GF2 Landsat Sentinel-2 (10–20–60 m)	Water, vegetation, soil	All	Google Earth images		Meteorological data	129 sample points		Reference fractional abundance maps	Full	-
Lu et al. [306]	AVIRIS	-	All	Reference map	-	-	The whole study area	The whole study area	Reference map	-	-
Lymburner et al. [348]	Landsat	-	All	LIDAR survey	-	-	100 (10 × 10 km) tiles	Random	Reference fractional abundance maps	Full	-
Lyu et al. [338]	Hyperion (30 m)	-	All	In situ data	-	Land use data	36 plots	Random	Reference fractional abundance maps	Full	-
Markiet & Mõttus [277]	AISA Eagle II airborne hyperspectral scanner	-	-	In situ data	-	Site fertility class, tree species composition, diameter at breast height, median tree height, effective leaf area index calculated from canopy gap fraction	250 plots	Random	Reference fractional abundance maps	Full	-
Mei et al. [307]	AVIRIS HYDICE	- Road, trees, water, soil	All	Reference map	-	-	The whole study area	The whole study area	Reference map	-	-
Moghadam et al. [336]	HyMap Hyperion (30 m)	-	All	Geological map	-	-	The whole study area	The whole study area	Reference fractional abundance maps	Partial	-
Montorio et al. [339]	Landsat Sentinel-2 (10–20–60 m)	-	All	Pléiades-1A orthoimage	-	-	275/280 plots	Random	Reference fractional abundance maps	Full	-
Park et al. [546]	Hyperspectral	-	All	In situ data	-	-	-	-	Reference fractional abundance maps	Full	-
Patel et al. [372]	ROSIS	Urban surface materials	All	Reference map	-	-	The whole study area	The whole study area	Reference map	-	-
Peng et al. [297]	AVIRIS	-	All	Reference map	-	-	The whole study area	The whole study area	Reference map	-	-
Peroni Venancio et al. [347]	Landsat	photosynthetic vegetation, soil/non-photosynthetic vegetation	All	In situ data	-	-	-	Random	Reference fractional abundance maps	Full	-
Qi et al. [312]	AVIRIS	-	All	Reference map	-	-	The whole study area	The whole study area	Reference map	-	-
Qi et al. [308]	AVIRIS	-	All	Reference map	-	-	The whole study area	The whole study area	Reference map	-	-
Qian et al. [309]	AVIRIS HYDICE	- Road, trees, water, soil	All	Reference map	-	-	The whole study area	The whole study area	Reference map	-	-
Qu & Bao [321]	AVIRIS HYDICE	-	All	Reference map	-	-	The whole study area	The whole study area	Reference map	-	-

Table A3. *Cont.*

Paper	Remote Image	Determined Endmembers	Validated End-mem-bers	Sources of Reference Data	Method for Mapping the Endmembers	Validation of Reference Data with Other Reference Data	Sample Sizes and Number of Small Sample Sizes	Sampling Designs	Reference Data	Estimation of Fractional Abun-dances	Error in Co-Localization and Spatial Resampling
Quintano et al. [381]	Sentinel-2 (10–20–60 m)	Char, green vegetation, non-photosynthetic vegetation, soil, shade	All	Official burn severity (three severity levels) and fire perimeter maps provided by Portuguese Study Center of Forest Fires	-	-	The whole study area	The whole study area	Reference map	-	-
Rasti et al. [320]	AVIRIS Samson	- Trees, water, soil	All	Reference map	-	-	The whole study area	The whole study area	Reference map	-	-
Redowan et al. [371]	Landsat	-	All	Google Earth images	-	-	Representative areas	Representative areas	Reference fractional abundance maps	Full	-
Rathnayake et al. [293]	AVIRIS HYDICE	- Road, trees, water, soil	All	Reference map	-	-	The whole study area	The whole study area	Reference map	-	-
Salvatore et al. [385]	WorldView-2 WorldView-3	-	All	In situ data	-	-	-	-	Reference fractional abundance maps	Full	-
Sall et al. [252]	Landsat	-	All	WorldView-2 (0.46 m)		National Agriculture Imagery Program (NAIP	89 waterbodies	The whole study area	Reference fractional abundance maps	Full	-
Salehi et al. [280]	HyMap ASTER Landsat Sentinel-2	-	All	In situ data	-	Geological map, X-ray fluorescence analysis	-	-	Reference fractional abundance maps	Full	-
Senf et al. [345]	Landsat	-	All	Aerial images	-	-	360 sample areas	Random	Reference fractional abundance maps	Full	-
Shah et al. [313]	AVIRIS	-	All	Reference map	-	-	The whole study area	The whole study area	Reference map	-	-
Shih et al. [354]	Landsat	Vegetation, Impervious, Soil	All	Google Earth VHR images	-	-	107 (90 × 90 m) samples	Random	Reference fractional abundance maps	Partial	
Shimabukuro et al. [370]	PROBA-V	-	All	Sentinel-2	-	-	Representative areas	Representative areas	Reference fractional abundance maps	Full	-
Shimabukuro et al. [353]	Landsat Suomi NPP-VIIRS ROBA-V	-	All	Sentinel-2 MODIS	-	Annual classifications of the Program for Monitoring Deforestation in the Brazilian Amazon (PRODES), Global Burned Area Products (Fire CCI, MCD45A1,MCD64A1)	-	-	Reference fractional abundance maps	Partial	-
Siebels et al. [319]	AVIRIS	-	All	Reference map	-	-	The whole study area	The whole study area	Reference map	-	-
Sing & Gray [363]	Landsat	-	All	In situ data	-	-	346 field plots	Random	Reference fractional abundance maps	Full	-
Sun et al. [331]	GF-1	-	All	Google Earth images	-	-	4500 pixels	Random	Reference fractional abundance maps	Full	-
Takodjou Wambo et al. [279]	ASTER Landsat	-	All	In situ data	-	Geological map, X-ray diffraction analysis	7 outcrops, 53 rock samples	-	Reference fractional abundance maps	Full	-
Tao et al. [315]	AVIRIS Samson	- Soil, tree, water	All	Reference map	-	-	The whole study area	The whole study area	Reference map	-	-
Thayn et al. [357]	Landsat	-	All	Low-altitude aerial imagery collected from a DJI Mavic Pro drone	-	-	Representative areas	Representative areas	Reference fractional abundance maps	Full	-
Tong et al. [311]	AVIRIS	-	All	Reference map	-	-	The whole study area	The whole study area	Reference map	-	-
Topouzelis et al. [382]	Sentinel-2 (10–20–60 m)	-	All	Unmanned Aerial System images	-	-	Representative areas	Representative areas	Reference fractional abundance maps	Full	-
Topouzelis et al. [383]	Sentinel-2 (10–20–60 m)	-	All	Unmanned Aerial System images	-	-	Representative areas	Representative areas	Reference fractional abundance maps	Full	-
Trinder & Liu [344]	Landsat	-	All	Ziyuan-3 image, Gaofen-1 satellite image,	-	-	-	-	Reference fractional abundance maps	Full	-
Uezato et al. [325]	AVIRIS	-	All	Reference map	-	-	The whole study area	The whole study area	Reference map	-	-
Vijayashekhar et al. [292]	AVIRIS HYDICE	- Road, trees, water, soil	All	Reference map	-	-	The whole study area	The whole study area	Reference map	-	-
Wang et al. [375]	Samson	Soil, tree, water	All	Reference map	-	-	The whole study area	The whole study area	Reference map	-	-
Wang et al. [366]	PlanetScope (3 m)	Green vegetation Non-photosynthetic vegetation	All	In situ data		Field measurements of LAI, phenocam-based leafless tree-crown fraction, phenocam-based leafy tree-crown fraction	no	no	Reference fractional abundance maps	Full	Expansion of the windows of field sample size

Table A3. *Cont.*

Paper	Remote Image	Determined Endmembers	Validated End-members	Sources of Reference Data	Method for Mapping the Endmembers	Validation of Reference Data with Other Reference Data	Sample Sizes and Number of Small Sample Sizes	Sampling Designs	Reference Data	Estimation of Fractional Abun-dances	Error in Co-Localization and Spatial Resampling
Wang et al. [346]	Landsat	Water, urban, agriculture, forest	All	Reference map	-	-	The whole study area	The whole study area	Reference map	-	-
Wang et al. [373]	ROSIS (4 m)	Urban surface materials	All	Reference map	-	-	The whole study area	The whole study area	Reference map	-	-
Wang et al. [322]	AVIRIS HYDICE	-	All	Reference map	-	-	The whole study area	The whole study area	Reference map	-	-
Wright & Polashenski [362]	MODIS (0.5 m)	-	All	WorldView-2 (0.46 m) WorldView-3 (0.31 m)	-		Representative areas	Representative areas	Reference fractional abundance maps	Full	-
Xiong et al. [323]	AVIRIS Samson	- Soil, tree, water	All	Reference map	-	-	The whole study area	The whole study area	Reference map	-	-
Xu et al. [295]	AVIRIS	-	All	Reference map	-	-	The whole study area	The whole study area	Reference map	-	-
Xu et al. [296]	AVIRIS	-	All	Reference map	-	-	The whole study area	The whole study area	Reference map	-	-
Xu et al. [316]	AVIRIS HYDICE	- Road, trees, water, soil	All	Reference map	-	-	The whole study area	The whole study area	Reference map	-	-
Xu et al. [318]	AVIRIS HYDICE	- Asphalt, trees, water, soil	All	Reference map	-	-	The whole study area	The whole study area	Reference map	-	-
Yang & Chen [294]	AVIRIS	-	All	Reference map	-	-	The whole study area	The whole study area	Reference map	-	-
Yang et al. [327]	AVIRIS	-	All	Reference map	-	-	The whole study area	The whole study area	Reference map	-	-
Yang et al. [298]	AVIRIS HYDICE	- Asphalt, trees, water, soil	All	Reference map	-	-	The whole study area	The whole study area	Reference map	-	-
Yang et al. [374]	Samson	Soil, tree, water	All	Reference map	-	-	The whole study area	The whole study area	Reference map	-	-
Yin et al. [355]	Landsat	-	All	Google Earth images	-	-	500 samples	Random	Reference fractional abundance maps	Full	-
Yuan et al. [314]	AVIRIS	-	All	Reference map	-	-	The whole study area	The whole study area	Reference map	-	-
Yue et al. [376]	Sentinel-2 (10–20–60 m)	-	All	Digital photos	-	-	The whole study area	The whole study area	Reference fractional abundance maps	Full	-
Zeng et al. [317]	AVIRIS	-	All	Reference map	-	-	The whole study area	The whole study area	Reference map	-	-
Zhang et al. [337]	HySpex (0.7 m)			Google Earth images	-	-	-	-	Reference fractional abundance maps	Full	-
Zhang et al. [384]	UAV hyperspectral data	-	All	In situ data	-	Laboratory analysis	35 samples	-	Reference fractional abundance maps	Full	-
Zhang et al. [326]	AVIRIS	Cuprite	All	Reference map	-	-	The whole study area	The whole study area	Reference map	-	-
Zhou et al. [310]	AVIRIS HYDICE	- Asphalt, trees, water, soil	All	Reference map	-	-	The whole study area	The whole study area	Reference map	-	-
Zhou et al. [324]	AVIRIS HYDICE Samson	- Soil, tree, water	All	Reference map	-	-	The whole study area	The whole study area	Reference map	-	-
Zhou et al. [291]	AVIRIS (16 m) AVIRIS NG (4 m)	Turfgrass, non-photosynthetic vegetation (NPV), paved, roof, soil, and tree	All	NAIP high-resolution images (1 m)	-	-	64 regions of interest (180 × 180 m)	Random	Reference fractional abundance maps	Partial	-
Zhu et al. [335]	HYDICE	Asphalt, trees, water, soil	All	Reference map	-	-	The whole study area	The whole study area	Reference map	-	-

Table A4. Main characteristics of the eligible papers that were published in 2011.

Paper	Remote Image	Determined Endmembers	Validated End-members	Sources of Reference Data	Method for Mapping the Endmembers	Validation of Reference Data with Other Reference Data	Sample Sizes and Number of Small Sample Sizes	Sampling Designs	Reference Data	Estimation of Fractional Abun-dances	Error in Co-Localization and Spatial Resampling
Altmann et al. [404]	AVIRIS	-	All	Reference map	-	-	The whole study area	The whole study area	Reference map	-	-
Ambikapathi et al. [405]	AVIRIS	Cuprite	All	Reference map	-	-	The whole study area	The whole study area	Reference map	-	-
Bartholomeus et al. [386]	AHS	Maize	All	In situ data	-	-	14 samples	Random	Reference fractional abundance maps	Partial	-
Bouaziz et al. [420]	MODIS	-	All	In situ data	-	-	102 samples	Random	Reference fractional abundance maps	Partial	-
Canham et al. [406]	AVIRIS	Cuprite	All	Reference map	-	-	The whole study area	The whole study area	Reference map	-	-
Cao et al. [429]	HJ-1 (30 m)	-	All	In situ data	-	-	13 sample plots	Random	Reference fractional abundance maps	-	-
Castrodad et al. [392]	AVIRIS HYDICE HyMAP	- Asphalt, trees, water, soil	All	Reference map	-	-	The whole study area	The whole study area	Reference map	-	-
Chen et al. [430]	HJ-1 (30 m)	-	All	In situ data	-	-	13 sample plots	Random	Reference fractional abundance maps	-	-

Table A4. *Cont.*

Paper	Remote Image	Determined Endmembers	Validated End-members	Sources of Reference Data	Method for Mapping the Endmembers	Validation of Reference Data with Other Reference Data	Sample Sizes and Number of Small Sample Sizes	Sampling Designs	Reference Data	Estimation of Fractional Abun-dances	Error in Co-Localization and Spatial Resampling
Chudnovsky et al. [428]	Hyperion (30 m)	-	All	In situ data	-	Bulk mineral, geo-chemical composition	8 samples	-	Reference fractional abundance maps	-	-
Cui et al. [421]	MODIS (0.5–1 km)	-	All	Landsat image	-	-	Landsat image	Representative area	Reference fractional abundance maps	Partial	-
de Jong et al. [427]	HyMAP (5 m)	-	All	In situ data	-	Physical characterization, infiltration measurements	107 plots	Random	Reference fractional abundance maps	-	-
Dopido et al. [393]	AVIRIS	Cuprite	All	Reference map	-	-	The whole study area	The whole study area	Reference map	-	-
Eches et al. [407]	AVIRIS	Cuprite	All	Reference map	-	-	The whole study area	The whole study area	Reference map	-	-
Ghrefat & Goodell [387]	ASTER AVIRIS Hyperion	-	All	In situ data	-	-	-	-	Reference fractional abundance maps	-	-
Gilichinsky et al. [439]	Landsat Landsat SPOT	-	-	In situ data	-	-	229 validation areas	Random	Reference fractional abundance maps	-	-
Gillis & Plemmons [424]	HYDICE	Asphalt, trees, water, soil	All	Reference map	-	-	The whole study area	The whole study area	Reference map	-	-
Griffin et al. [431]	Landsat	-	All	In situ data	-	-	304 samples	Random	Reference fractional abundance maps	Full	-
Halimi et al. [394]	AVIRIS	-	All	Reference map	-	-	The whole study area	The whole study area	Reference map	-	-
Halimi et al. [408]	AVIRIS	-	All	Reference map	-	-	The whole study area	The whole study area	Reference map	-	-
Hamada et al. [441]	QuickBird (0.6–2.4 m) SPOT (10 m)	-	All	Infrared aerial photography (0.15 m)	Phointerpretation	-	30 samples	Random	Reference fractional abundance maps	Full	Spatial resolution variation
Heylen et al. [395]	AVIRIS	Cuprite	All	Reference map	-	-	The whole study area	The whole study area	Reference map	-	-
Heylen et al. [396]	AVIRIS	Cuprite	All	Reference map	-	-	The whole study area	The whole study area	Reference map	-	-
Heylen & Scheunders [397]	AVIRIS	Cuprite	All	Reference map	-	-	The whole study area	The whole study area	Reference map	-	-
Hosseinjani & Tangestani [388]	ASTER	-	All	In situ data	-	Geological map, X-ray diffraction analysis	8 samples	Random	Reference fractional abundance maps	Full	-
Hu & Weng [390]	ASTER	-	All	Images	-	-	Representative area	Representative area	Reference fractional abundance maps	Full	-
Iordache et al. [398]	AVIRIS	Cuprite	All	Reference map	-	-	The whole study area	The whole study area	Reference map	-	-
Iordache et al. [409]	AVIRIS	Cuprite	All	Reference map	-	-	The whole study area	The whole study area	Reference map	-	-
Ji & Feng [442]	QuickBird (2.4 m)	-	All	QuickBird (0.6 m)	-	-	The whole study area	The whole study area	Reference fractional abundance maps	Partial	-
Jiao et al. [434]	Landsat	-	All	Airborne images	-	-	Representative area	Representative area	Reference fractional abundance maps	Full	-
Kamal & Phinn [418]	CASI	-	All	Map of the mangrove speciesderived from aerial photographic interpretation at scale of 1:25,000 Provided by Queensland Herbar-ium/Environmental Protection Agency (EPA)	-	-	400 samples	Random	Reference fractional abundance maps	Partial	-
Knight & Voth [422]	MODIS	-	All	Landsat image	-	-	The whole study area	The whole study area	Reference fractional abundance maps	Full	-
Liu et al. [399]	AVIRIS	-	All	Reference map	-	-	The whole study area	The whole study area	Reference map	-	-
Lu et al. [435]	Landsat	High-albedo, low-albedo, vegetation, soil	All	QuickBird	Hybrid method	-	250 points	Random	Reference fractional abundance maps	Partial	Spatial resolution variation
Lu et al. [432]	Landsat	High-albedo, low-albedo, vegetation, soil	All	QuickBird	Hybrid method	-	1512 samples	Random	Reference fractional abundance maps	Partial	-
Lu et al. [423]	Landsat MODIS	Forest and non-forest Vegetation, shade and soil	All	Annual classifications of the Program for Monitoring Deforestation in the Brazilian Amazon (PRODES)	-	Official truth-terrain data from deforested and non-deforested areas prepared by PRODES	-	-	Reference fractional abundance maps	Full	-
Martin & Plaza [410]	AVIRIS	Cuprite	All	Reference map	-	-	The whole study area	The whole study area	Reference map	-	-
Martin et al. [411]	AVIRIS	Cuprite	All	Reference map	-	-	The whole study area	The whole study area	Reference map	-	-
Mei al. [307]	AVIRIS	-	All	Reference map	-	-	The whole study area	The whole study area	Reference map	-	-
Mei & He [412]	AVIRIS	Cuprite	All	Reference map	-	-	The whole study area	The whole study area	Reference map	-	-
Mianji et al. [400]	AVIRIS	-	All	Reference map	-	-	The whole study area	The whole study area	Reference map	-	-
Negrón-Juárez et al. [433]	Landsat	Photosynthetic vegetation, non-photosynthetic vegetation	All	In situ data	-	-	30 pixel	random	Reference fractional abundance maps	Partial	-
Qian et al. [425]	HYDICE	Asphalt, trees, water, soil	All	Reference map	-	-	The whole study area	The whole study area	Reference map	-	-

Table A4. *Cont.*

Paper	Remote Image	Determined Endmembers	Validated End-members	Sources of Reference Data	Method for Mapping the Endmembers	Validation of Reference Data with Other Reference Data	Sample Sizes and Number of Small Sample Sizes	Sampling Designs	Reference Data	Estimation of Fractional Abundances	Error in Co-Localization and Spatial Resampling
Reno et al. [436]	Landsat	Vegetation, soil, water	All	In situ data	-	Photos, botanical observations	168 ground points	-	Reference fractional abundance maps	Full	-
Sankey & Glenn [437]	Landsat	-	All	In situ data	-	-	100 plots (30 × 30 m)	Random	Reference fractional abundance maps	Full	-
Sunderman & Weisberg [438]	Landsat	-	All	In situ data	-	-	400 plots	Random	Reference fractional abundance maps	Full	-
Swatantran et al. [401]	AVIRIS	-	All	In situ data	-	Laser Vegetation Imaging Sensor	125 field plots classified based on WHR type for analysis by species/vegetation type	Random	Reference fractional abundance maps	Full	-
Vicente & de Souza Filho [389]	ASTER	-	All	In situ data	-	X-ray diffraction analysis on the same samples	42 soil samples	Random	Reference fractional abundance maps	Full	-
Villa et al. [413]	AVIRIS	-	All	Reference map	-	-	The whole study area	The whole study area	Reference map	-	-
Weng et al. [391]	ASTER	Green vegetation, soils low-albedo surfaces and high-albedo surface	All	Other ASTER images	Same procedures	-	The whole study area	The whole study area	Reference fractional abundance maps	Full	-
Xia et al. [414]	AVIRIS HYDICE	- Asphalt, trees, water, soil	All	Reference map	-	-	The whole study area	The whole study area	Reference map	-	-
Xia et al. [402]	AVIRIS	-	All	Reference map	-	-	The whole study area	The whole study area	Reference map	-	-
Yang et al. [415]	AVIRIS HYDICE	-	All	Reference map	-	-	The whole study area	The whole study area	Reference map	-	-
Youngentob et al. [426]	HyMap (3.5 m)	-	All	In situ data	-	-	9* isolated eucalypt paddock trees	Random	Reference fractional abundance maps	Full	-
Zare [403]	AVIRIS	-	All	Reference map	-	-	The whole study area	The whole study area	Reference map	-	-
Zhan et al. [416]	AVIRIS	-	All	Reference map	-	-	The whole study area	The whole study area	Reference map	-	
Zhao et al. [417]	AVIRIS	-	All	Reference map	-	-	The whole study area	The whole study area	Reference map	-	
Zurita-Milla et al. [419]	MERIS	-	All	High-spatial-resolution land-cover dataset (Dutch land-use database) (25 m)	-	-	The whole study area	The whole study area	Reference fractional abundance maps	Full	Spatial resampling the reference maps

Table A5. Main characteristics of the eligible papers that were published in 2010.

Paper	Remote Image	Determined Endmembers	Validated End-members	Sources of Reference Data	Method for Mapping the Endmembers	Validation of Reference Data with Other Reference Data	Sample Sizes and Number of Small Sample Sizes	Sampling Designs	Reference Data	Estimation of Fractional Abundances	Error in Co-Localization and Spatial Resampling
Alves Aguilar et al. [496]	MODIS (0.5–1 km)	Vegetation, soil	1	Landsat TM image (30 m)	NDVI	In situ observations	Landsat image	Representative area	Reference fractional abundance map	Partial	-
Biggs et al. [477]	Landsat (30 m)	Green vegetation, nonphotosynthetic vegetation, impervious surfaces, soil, shade	All	High resolution imagery	Photointerpretation	-	38 squares	Random	Reference fractional abundance maps	Full	-
Bolman [478]	Landsat (30 m)	Deciduous crowns, fully leaved crowns, shade	2	In situ data		-	17 plots	Uniform	Reference fractional abundance maps	Full	-
Borfecchia et al. [489]	Landsat (30 m)	-	-	QuickBird image (2.8 m)	Maximum Likelihood classification	Aerial photos	The whole study area	The whole study area	Reference fractional abundance maps	Full	-
Castrodad et al. [471]	HYDICE	Trees, grass, road	All	Reference map	-	-	The whole study area	The whole study area	Reference maps	-	-
	HyMAP	Coniferous trees, deciduous trees, grass, water, crop, road, concrete, gravel	All	Reference map	-	-	The whole study area	The whole study area	Reference maps	-	-
Cavalli et al. [494]	MIVIS (3 m)	Vegetation, soil	1	Land cover map	-	In situ observations	-	Random	Reference maps	-	-
Chang et al. [458]	AVIRIS (20 m)	Cuprite, vegetation, soil	All	Reference map	-	-	The whole study area	The whole study area	Reference maps	-	-
	HYDICE (1.5 m)	-	-	Reference map	-	-	The whole study area	The whole study area	Reference maps	-	-
Chen et al. [475]	HJ-1 CCD (30 m)	Vegetation	All	In situ data	-	Land-use, land-cover, vegetation maps	-	-	Reference fractional abundance map	Full	-
Eches et al. [457]	AVIRIS (20 m)	Cuprite, vegetation, soil	All	Reference map	-	-	The whole study area	The whole study area	Reference maps	-	-
Eckmann et al. [496]	MODIS (0.5–1 km)	Fire	1	Band 9 of ASTER image (30 m)	-	GLC 2000 land-cover	Aster image	Representative area	Reference map	-	-
Elatawneh et al. [473]	Hyperion (30 m)	Land-cover classes	All	QuickBird image	-	In situ observations	The whole study area	The whole study area	Reference fractional abundance maps	Full	-

Table A5. *Cont.*

Paper	Remote Image	Determined Endmembers	Validated Endmembers	Sources of Reference Data	Method for Mapping the Endmembers	Validation of Reference Data with Other Reference Data	Sample Sizes and Number of Small Sample Sizes	Sampling Designs	Reference Data	Estimation of Fractional Abundances	Error in Co-Localization and Spatial Resampling
Elmore & Guin [484]	Landsat (30 m)	Vegetation, substrate, and shade	All	Aerial photographs	Photointerpretation	Land cover based on aerial photography called GIRAS	-	Random	Reference fractional abundance maps	Full	-
Estes et al. [447]	ASTER (15–30–90 m)	-	-	In situ data	-	-	127 circles (11.3 m radius)	-	Reference fractional abundance maps	Full	Change the windows of pixels
Gilichinsky et al. [492]	Landsat (30 m) SPOT (10 m)	Lichen classes	1	In situ data	-	-	229 plots	Uniform	Reference fractional abundance maps	Full	-
Golubiewski & Wessman [456]	AVIRIS (20 m)	Vegetation, soil, manmade materials	All	In situ data	-	-	-	-	Reference fractional abundance maps	-	-
He et al. [485]	Landsat (30 m)	2 vegetations, water	All	QuickBird image	Classification	-	The whole study area	The whole study area	Reference fractional abundance maps	Full	-
Hendrix et al. [464]	CASI	-	-	In situ data	-	-	The whole study area	The whole study area	Reference maps	-	-
Hu & Weng [445]	ASTER (15–30–90 m)	-	-	QuickBird image (0.61 m)	Classification	-	The whole study area	The whole study area	Reference fractional abundance maps	Full	-
Huang et al. [479]	Landsat (30 m)	Fractional vegetation cover	All	In situ data	-	-	12 polygons (45 × 30 m)	Random	Reference fractional abundance map	Full	-
Huang et al. [449]	AVIRIS (20 m)	Road, trees, lawn, path, roof, shadow	All	Reference map	-	-	The whole study area	The whole study area	Reference maps	-	-
Huck et al. [459]	AVIRIS (20 m)	Minerals	All	Reference map	-	-	The whole study area	The whole study area	Reference maps	-	-
Iordarche et al. [460]	AVIRIS (20 m)	Minerals	All	Reference map	-	-	The whole study area	The whole study area	Reference maps	-	-
Jin et al. [450]	AVIRIS (20 m) AVIRIS (20 m)	Minerals -	All	Reference map	-	-	The whole study area	The whole study area	Reference maps	-	-
Li et al. [482]	Landsat (30 m)	Low albedo, high albedo, soil, vegetation	All	In situ data	-	-	400 samples	Random	Reference fractional abundance map	Full	-
Liu et al. [491]	Landsat (30 m)	Urban, forest, water, cropland, grass, developing land	All	QuickBird image (0.61 m)	Photointerpretation	In situ observations	3000 samples	Uniform	Reference fractional abundance map	Full	-
Liu & Yue [486]	Landsat TM (30 m) SPOT (10–20 m)	Urban vegetation fraction	All	In situ data	-	-	samples	Random	Reference fractional abundance map	Full	-
Luo et al. [451]	AVIRIS (20 m)	-	-	Reference map	-	-	The whole study area	The whole study area	Reference maps	-	-
Luo et al. [452]	AVIRIS (20 m)	-	-	Reference map	-	-	The whole study area	The whole study area	Reference maps	-	-
Martin et al. [461]	AVIRIS (20 m)	Alunite, buddingtonite, calcite, kaolinite and muscovite	All	Reference map	-	-	The whole study area	The whole study area	Reference maps	-	-
Martin & Plaza [462]	AVIRIS (20 m)	Minerals	All	Reference map	-	-	The whole study area	The whole study area	Reference maps	-	-
Martin & Plaza [462]	AVIRIS (20 m)	Minerals	Field reference data	-	-	The whole of study area		Reference maps	-	-	
Mei et al. [453]	AVIRIS (20 m)	Vegetation	All	Reference map	-	-	The whole study area	The whole study area	Reference maps	-	-
Mei et al. [454]	AVIRIS (20 m)	Mineral	All	Reference map	-	-	The whole study area	The whole study area	Reference maps	-	-
Meng et al. [476]	HJ-1A/1B (30 m)	Road, vegetation, Building	All	Aerial photo	Photointerpretation Classification	-	The whole study area	The whole study area	Reference fractional abundance maps	Full	-
Meusburger et al. [497]	QuickBird (2.4 m)	Vegetations Soil	-	In situ data	-	-	43 plots (10 × 10 m)	Random	Reference fractional abundance map	Full	-
Meusburger et al. [498]	QuickBird (2.4 m)	Vegetations Soil	All	In situ data	-	-	63 samples	Random	Reference fractional abundance map	Full	-
Mezned et al. [446]	ASTER (30 m) Landsat ETM+ (15 m)	Calcite, clays, gypsum, oxyhydroxides, pyrite	All	In situ data	-	-	-	Random	Reference fractional abundance maps	Partial	-
Mucher et al. [444]	AHS (2.4 m)	Heathland vegetation	All	In situ data	-	Aerial photos	104 circles (3 m radius)	-	Reference fractional abundance maps	Full	-
Pacheco & McNairn [480]	Landsat (30 m) SPOT (20 m)	Vegetation, soil and residue	All	Digital photographs	-	Soil Landscapes of Canada Working Group, 2007	Digital images	Representative area	Reference fractional abundance maps	Full	Size and spatial resolution of the reference maps
Pascucci et al. [101]	ATM (2 m) CASI (2 m)	Soil, vegetation	All	Land cover map		In situ observations	25 samples	Random	Reference fractional abundance maps	Full	-
Plaza & Plaza [465]	DAIS (6 m)	Cork-oak trees, pasture, bare soil	All	ROSIS image (1.2 m)	Maximum-likelihood supervised classification	-	The whole study area	The whole study area	Reference fractional abundance maps	Full	Co-localization the maps
Powell & Roberts [483]	Landsat (30 m)	Vegetation, impervious soil	All	Aerial photos	-	-	41 samples	-	Reference fractional abundance maps	Full	-
Raksuntorn et al. [463]	AVIRIS (10 m) HYDICE (10 m)	Minerals -	All -	Reference map Reference map	-	-	The whole study area The whole study area	The whole study area The whole study area	Reference maps Reference maps	-	-
Ruescas et al. [448]	AVHRR (1 km)	Vegetation, burnt area, rocks, soil	All	AHS image (6 m)	Maximum likelihood classification	Statistic reports provided by the Environmental Ministry of Spain	AHS image	Representative area	Reference fractional abundance maps	Full	Evaluation of the errors in co-localization and spatial-resampling

Table A5. *Cont.*

Paper	Remote Image	Determined Endmembers	Validated End-members	Sources of Reference Data	Method for Mapping the Endmembers	Validation of Reference Data with Other Reference Data	Sample Sizes and Number of Small Sample Sizes	Sampling Designs	Reference Data	Estimation of Fractional Abundances	Error in Co-Localization and Spatial Resampling
Sarapirome & Kulrat [493]	Landsat (30 m)	Vegetation, impervious soil; vegetation, soil, shade	All	Air photos	-	In situ observations	-	-	Reference fractional abundance maps	Full	-
Schmidt & Witte [499]	SPOT (2.5–10 m)	Water, soil, vegetation	All	In situ data	-	-	Polygons	Random	Reference maps	-	-
Silván-Cárdenas & Wang [490]	Landsat (30 m)	Vegetations	All	AISA image (1 m)	Spectral angle mapper classification	In situ observations	300 points (30 × 30 m)	Random	Reference fractional abundance maps	Full	-
Soenen et al. [500]	SPOT (10–25 m)	Sunlit canopy, sunlit background, shadow	All	In situ data	-	-	36 plots (400 m^2)	Random	Reference fractional abundance maps	Full	The size of reference maps
Solans Vila & Barbosa [481]	Landsat (15 m)	Green vegetation, soil, shade, non-photosynthetic vegetation	All	In situ data	-	-	-	-	Reference fractional abundance maps	Full	-
Somers et al. [472]	Landsat (30 m) Hyperion (30 m)	Eucalyptus trees, soil, litter and grass	All	In situ data	-	-	46 plots	Stratified random	Reference fractional abundance map	Full	-
Tommervik et al. [487]	Landsat (30 m)	Vegetations	All	Aerial photographs and QuickBird-2 image	Photointerpretation	-	10 plots	Random	Reference fractional abundance map	Full	-
Verrelst et al. [467]	CHRIS (17 m)	Vegetation, snow	All	Aerial photographs	-	-	Aerial photographs	Representative area	Reference fractional abundance map	Full	-
Villa et al. [455]	AVIRIS (10 m) HYDICE (10 m)	- Asphalt, trees, water, soil	- -	Reference map Reference map	-	-	The whole study area The whole study area	The whole study area The whole study area	Reference maps Reference maps	-	-
Xiong et al. [470]	HYDICE (10 m)	-	-	Reference map	-	-	The whole study area	The whole study area	Reference maps	-	-
Yang & Everitt [443]	Airborne hyperspectral image (about 1.5 m)	Invasive weeds	All	In situ data	-	-	425 circular areas (diameter of 3 m)	Stratified random	Reference fractional abundance map	Full	-
Yang et al. [488]	Landsat TM (30 m)	2Vegetation, impervious surfaces (low and high albedo), soil	All	Aerial photographs	Photointerpretation	-	138 samples	Random	Reference fractional abundance maps	Full	-

Table A6. Main characteristics of the eligible papers that were published in 1996.

Paper	Remote Image	Determined Endmembers	Validated End-members	Sources of Reference Data	Method for Mapping the Endmembers	Validation of Reference Data with Other Reference Data	Sample Sizes and Number of Small Sample Sizes	Sampling Designs	Reference Data	Estimation of Fractional Abundances	Error in Co-Localization and Spatial Resampling
Ben-dor et al. [507]	SPOT	Mineral	All	Geological map	-	GER scanner data	The whole study area	The whole study area	Reference fractional abundance map	Partial	Co-localization the maps
Bowers & Rowan [503]	AVIRIS	Mineral	All	Geological map	-	-	The whole study area	The whole study area	Reference fractional abundance map	Partial	-
Hunt et al. [502]	AVIRIS	-	All	Landsat image	Unconstrained linear spectral unmixing	-	The whole study area	The whole study area	Reference fractional abundance map	Partial	-
Rosenthal et al. [505]	Landsat	-	All	High resolution aerial photographs	-	-	The whole study area	The whole study area	Reference fractional abundance map	Full	-
Thomas et al. [14]	Landsat	-	All	Images	Photointerpretation	-	The whole study area	The whole study area	Reference fractional abundance map	Full	-
Ustin et al. [501]	AVIRIS	-	All	Aerial photograph	-	Field based vegetation map	The whole study area	The whole study area	Reference fractional abundance map	Full	-
Van der Meer [504]	GERIS	-	All	Map	-	-	The whole study area	The whole study area	Reference fractional abundance map	Partial	-
Van der Meer [506]	Landsat	-	All	Map	-	-	The whole study area	The whole study area	Reference fractional abundance map	Partial	-

Table A7. Main characteristics of the eligible papers that were published in 1995.

Paper	Remote Image	Determined Endmembers	Validated Endmembers	Sources of Reference Data	Method for Mapping the Endmembers	Validation of Reference Data with Other Reference Data	Sample Sizes and Number of Small Sample Sizes	Sampling Designs	Reference Data	Estimation of Fractional Abundances	Error in Co-Localization and Spatial Resampling
Bianchi et al. [514]	MIVIS (4 m)	Oil, water, wood, cultivated field, smooth and grooved surface soil, rice field	1	In situ data	-	-	200 samples	Uniform	Reference fractional abundance map	Full	-
Dwyer et al. [509]	AVIRIS (20 m)	Minerals	All	Geological map	-	Remotely sensed and ground-based data	The whole study area	The whole study area	Reference maps	-	
Hall et al. [515]	MMR	Canopy, canopy plus background, background	All	In situ data	-	-	-	-	Reference fractional abundance map	Full	-
Kerdiles & Grondona [508]	AVHRR (1 km)	Vegetation, soil	All	Landsat TM image (30 m)	classification	-	-	-	Reference fractional abundance maps	Full	-
Lacaze et al. [510]	AVIRIS (20 m)	Vegetation, soil, rock	All	Landsat TM image (30 m)	classification	-	-	-	Reference fractional abundance maps	Full	-
Lavreau et al. [512]	Landsat (30 m)	Vegetation	All	Land cover map	-	-	-		Reference maps	-	-
Rowan et al. [511]	AVIRIS (20 m)	Minerals	All	Geological map	-	-	The whole study area	The whole study area	Reference maps	-	-
Van Der Meer [513]	Landsat (30 m)	Minerals	All	Geological map	-	In situ observations	The whole study area	The whole study area	Reference fractional abundance maps	Full	-

References

1. Ichoku, C.; Karnieli, A. A Review of Mixture Modeling Techniques for Sub-Pixel Land Cover Estimation. *Remote Sens. Rev.* **1996**, *13*, 161–186. [CrossRef]
2. Plaza, A.; Martinez, P.; Perez, R.; Plaza, J. A New Approach to Mixed Pixel Classification of Hyperspectral Imagery Based on Extended Morphological Profiles. *Pattern Recognit.* **2004**, *37*, 1097–1116. [CrossRef]
3. Mei, S.; Feng, D.; He, M. Hopfield Neural Network Based Mixed Pixel Unmixing for Multispectral Data. In Proceedings of the Satellite Data Compression, Communication, and Processing IV, San Diego, CA, USA, 28 August 2008; Volume 7084, pp. 88–95.
4. Borsoi, R.A.; Imbiriba, T.; Bermudez, J.C.M.; Richard, C.; Chanussot, J.; Drumetz, L.; Tourneret, J.-Y.; Zare, A.; Jutten, C. Spectral Variability in Hyperspectral Data Unmixing: A Comprehensive Review. *IEEE Geosci. Remote Sens. Mag.* **2021**, *9*, 223–270. [CrossRef]
5. Wei, J.; Wang, X. An Overview on Linear Unmixing of Hyperspectral Data. *Math. Probl. Eng.* **2020**, *2020*, 1–12. [CrossRef]
6. Keshava, N.; Mustard, J.F. Spectral Unmixing. *IEEE Signal Process. Mag.* **2002**, *19*, 44–57. [CrossRef]
7. Bioucas-Dias, J.M.; Plaza, A. Hyperspectral Unmixing: Geometrical, Statistical, and Sparse Regression-Based Approaches. In Proceedings of the Mage and Signal Processing for Remote Sensing XVI, SPIE, Toulouse, France, 7 October 2010; Bruzzone, L., Ed.; p. 78300A.
8. Heylen, R.; Parente, M.; Gader, P. A Review of Nonlinear Hyperspectral Unmixing Methods. *IEEE J. Sel. Top. Appl. Earth Obs. Remote Sens.* **2014**, *7*, 1844–1868. [CrossRef]
9. Settle, J.; Drake, N. Linear Mixing and the Estimation of Ground Cover Proportions. *Int. J. Remote Sens.* **1993**, *14*, 1159–1177. [CrossRef]
10. Borel, C.C.; Gerstl, S.A.W. Nonlinear Spectral Mixing Models for Vegetative and Soil Surfaces. *Remote Sens. Environ.* **1994**, *47*, 403–416. [CrossRef]
11. Ray, T.W.; Murray, B.C. Nonlinear Spectral Mixing in Desert Vegetation. *Remote Sens. Environ.* **1996**, *55*, 59–64. [CrossRef]
12. Johnson, P.E.; Smith, M.O.; Taylor-George, S.; Adams, J.B. A Semiempirical Method for Analysis of the Reflectance Spectra of Binary Mineral Mixtures. *J. Geophys. Res. Solid Earth* **1983**, *88*, 3557–3561. [CrossRef]
13. Smith, M.O.; Johnson, P.E.; Adams, J.B. Quantitative Determination of Mineral Types and Abundances from Reflectance Spectra Using Principal Components Analysis. *J. Geophys. Res. Solid Earth* **1985**, *90*, C797–C804. [CrossRef]
14. Thomas, G.; Hobbs, S.E.; Dufour, M. Woodland Area Estimation by Spectral Mixing: Applying a Goodness-of-Fit Solution Method. *Int. J. Remote Sens.* **1996**, *17*, 291–301. [CrossRef]
15. Goetz, A.F.H.; Boardman, J.W. Quantitative Determination of Imaging Spectrometer Specifications Based on Spectral Mixing Models. In Proceedings of the 12th Canadian Symposium on Remote Sensing Geoscience and Remote Sensing Symposium, Vancouver, BC, Canada, 10–14 July 1989; Volume 2, pp. 1036–1039.
16. Adams, J.B.; Smith, M.O.; Gillespie, A.R. Simple Models for Complex Natural Surfaces: A Strategy for The Hyperspectral Era of Remote Sensing. In Proceedings of the 12th Canadian Symposium on Remote Sensing Geoscience and Remote Sensing Symposium, Vancouver, BC, Canada, 10–14 July 1989; Volume 1, pp. 16–21.
17. Gillespie, A. Interpretation of Residual Images: Spectral Mixture Analysis of AVIRIS Images, Owens Valley, California. Jet Propulsion Laboratory. In Proceedings of the Second Airborne Visible/Infrared Imaging Spectrometer (AVIRIS) Workshop, Owens Valley, CA, USA, 4–5 June 1990; pp. 243–270.
18. Gillespie, A.R. Spectral Mixture Analysis of Multispectral Thermal Infrared Images. *Remote Sens. Environ.* **1992**, *42*, 137–145. [CrossRef]
19. Sabol, D.E.; Adams, J.B.; Smith, M.O. Quantitative Subpixel Spectral Detection of Targets in Multispectral Images. *J. Geophys. Res.* **1992**, *97*, 2659. [CrossRef]
20. Farrand, W.H.; Harsanyi, J.C. Discrimination of Poorly Exposed Lithologies in AVIRIS Data. In Proceedings of the JPL, Summaries of the 4th Annual JPL Airborne Geoscience Workshop, Washington, DC, USA, 28–29 October 1993; Volume 1. AVIRIS Workshop.
21. Adams, J.B.; Sabol, D.E.; Kapos, V.; Almeida Filho, R.; Roberts, D.A.; Smith, M.O.; Gillespie, A.R. Classification of Multispectral Images Based on Fractions of Endmembers: Application to Land-Cover Change in the Brazilian Amazon. *Remote Sens. Environ.* **1995**, *52*, 137–154. [CrossRef]
22. Tompkins, S. Optimization of Endmembers for Spectral Mixture Analysis. *Remote Sens. Environ.* **1997**, *59*, 472–489. [CrossRef]
23. Adams, J.B.; Smith, M.O.; Johnson, P.E. Spectral Mixture Modeling: A New Analysis of Rock and Soil Types at the Viking Lander 1 Site. *J. Geophys. Res.* **1986**, *91*, 8098. [CrossRef]
24. Huete, A.; Escadafal, R. Assessment of Soil-Vegetation-Senesced Materials with Spectral Mixture Modeling: Preliminary Analysis. In Proceedings of the 10th Annual International Symposium on Geoscience and Remote Sensing, College Park, MD, USA, 20–24 May 1990; pp. 1621–1624.
25. Sohn, Y.; McCoy, R.M. Mapping Desert Shrub Rangeland Using Spectral Unmixing and Modeling Spectral Mixtures with TM Data. *Photogramm. Eng. Remote Sens.* **1997**, *63*, 707–716.
26. Boardman, J.W.; Kruse, F.A.; Green, R.O. Mapping Target Signatures via Partial Unmixing of AVIRIS Data. In Proceedings of the Summaries of the Fifth Annual JPL Airborne Earth Science Workshop, Pasadena, CA, USA, 23–26 January 1995.
27. Adams, J.B.; McCord, T.B. Optical Properties of Mineral Separates, Glass, and Anorthositic Fragments from Apollo Mare Samples. In Proceedings of the Lunar and Planetary Science Conference Proceedings, Woodlands, TX, USA, 11–14 January 1971; Volume 2, p. 2183.

28. Singer, R.B.; McCord, T.B. Mars-Large Scale Mixing of Bright and Dark Surface Materials and Implications for Analysis of Spectral Reflectance. In Proceedings of the Lunar and Planetary Science Conference Proceedings, Huston, TX, USA, 19–23 March 1979; Volume 10, pp. 1835–1848.
29. Hapke, B. Bidirectional Reflectance Spectroscopy: 1. Theory. *J. Geophys. Res. Solid Earth* **1981**, *86*, 3039–3054. [CrossRef]
30. Shimabukuro, Y.; Carvalho, V.; Rudorff, B. NOAA-AVHRR Data Processing for the Mapping of Vegetation Cover. *Int. J. Remote Sens.* **1997**, *18*, 671–677. [CrossRef]
31. Shimabukuro, Y.E.; Smith, J.A. The Least-Squares Mixing Models to Generate Fraction Images Derived from Remote Sensing Multispectral Data. *IEEE Trans. Geosci. Remote Sens.* **1991**, *29*, 16–20. [CrossRef]
32. Boardman, J.W. Inversion of Imaging Spectrometry Data Using Singular Value Decomposition. In Proceedings of the 12th Canadian Symposium on Remote Sensing Geoscience and Remote Sensing Symposium, Vancouver, BC, Canada, 10–14 July 1989; Volume 4, pp. 2069–2072.
33. Heinz, D.C. Chein-I-Chang Fully Constrained Least Squares Linear Spectral Mixture Analysis Method for Material Quantification in Hyperspectral Imagery. *IEEE Trans. Geosci. Remote Sens.* **2001**, *39*, 529–545. [CrossRef]
34. Keshava, N. A Survey of Spectral Unmixing Algorithms. *Linc. Lab. J.* **2003**, 55–78.
35. Martínez, P.J.; Pérez, R.M.; Plaza, A.; Aguilar, P.L.; Cantero, M.C.; Plaza, J. Endmember Extraction Algorithms from Hyperspectral Images. *Ann. Geophys.* **2006**, *49*, 93–101. [CrossRef]
36. Veganzones, M.A.; Grana, M. Endmember Extraction Methods: A Short Review. In Proceedings of the International Conference on Knowledge-Based and Intelligent Information and Engineering Systems, Zagreb, Croatia, 3–5 September 2008; pp. 400–407.
37. Parente, M.; Plaza, A. Survey of Geometric and Statistical Unmixing Algorithms for Hyperspectral Images. In Proceedings of the 2010 2nd Workshop on Hyperspectral Image and Signal Processing: Evolution in Remote Sensing, Reykjavik, Iceland, 14–16 June 2010; pp. 1–4.
38. Bioucas-Dias, J.M.; Plaza, A. An Overview on Hyperspectral Unmixing: Geometrical, Statistical, and Sparse Regression Based Approaches. In Proceedings of the 2011 IEEE International Geoscience and Remote Sensing Symposium, Vancouver, BC, Canada, 24–29 July 2011; pp. 1135–1138.
39. Somers, B.; Asner, G.P.; Tits, L.; Coppin, P. Endmember Variability in Spectral Mixture Analysis: A Review. *Remote Sens. Environ.* **2011**, *115*, 1603–1616. [CrossRef]
40. Bioucas-Dias, J.M.; Plaza, A.; Dobigeon, N.; Parente, M.; Du, Q.; Gader, P.; Chanussot, J. Hyperspectral Unmixing Overview: Geometrical, Statistical, and Sparse Regression-Based Approaches. *IEEE J. Sel. Top. Appl. Earth Obs. Remote Sens.* **2012**, *5*, 354–379. [CrossRef]
41. Quintano, C.; Fernández-Manso, A.; Shimabukuro, Y.E.; Pereira, G. Spectral Unmixing. *Int. J. Remote Sens.* **2012**, *33*, 5307–5340. [CrossRef]
42. Ismail, M.M.B.; Bchir, O. Survey on Number of Endmembers Estimation Techniques for Hyperspectral Data Unmixing. In Proceedings of the 2014 International Conference on Audio, Language and Image Processing, Shanghai, China, 7–9 July 2014; pp. 651–655.
43. Shi, C.; Wang, L. Incorporating Spatial Information in Spectral Unmixing: A Review. *Remote Sens. Environ.* **2014**, *149*, 70–87. [CrossRef]
44. Drumetz, L.; Chanussot, J.; Jutten, C. Variability of the Endmembers in Spectral Unmixing: Recent Advances. In Proceedings of the 2016 8th Workshop on Hyperspectral Image and Signal Processing: Evolution in Remote Sensing (WHISPERS), Los Angeles, CA, USA, 21–24 August 2016; pp. 1–5.
45. Wang, L.; Shi, C.; Diao, C.; Ji, W.; Yin, D. A Survey of Methods Incorporating Spatial Information in Image Classification and Spectral Unmixing. *Int. J. Remote Sens.* **2016**, *37*, 3870–3910. [CrossRef]
46. Bassani, C.; Cavalli, R.M.; Antonelli, P. Influence of Aerosol and Surface Reflectance Variability on Hyperspectral Observed Radiance. *Atmos. Meas. Tech.* **2012**, *5*, 1193–1203. [CrossRef]
47. Abbate, G.; Cavalli, R.M.; Pascucci, S.; Pignatti, S.; Poscolieri, M. Others Relations between Morphological Settings and Vegetation Covers in a Medium Relief Landscape of Central Italy. *Ann. Geophys.* **2006**, *49*, 153–166.
48. CEOS Working Group on Calibration & Validation (WGCV). Available online: https://ceos.org/ourwork/workinggroups/wgcv/ (accessed on 22 March 2023).
49. Moher, D.; Liberati, A.; Tetzlaff, J.; Altman, D.G. The PRISMA Group Preferred Reporting Items for Systematic Reviews and Meta-Analyses: The PRISMA Statement. *PLoS Med.* **2009**, *6*, e1000097. [CrossRef]
50. Page, M.J.; McKenzie, J.E.; Bossuyt, P.M.; Boutron, I.; Hoffmann, T.C.; Mulrow, C.D.; Shamseer, L.; Tetzlaff, J.M.; Akl, E.A.; Brennan, S.E.; et al. The PRISMA 2020 Statement: An Updated Guideline for Reporting Systematic Reviews. *Int. J. Surg.* **2021**, *88*, 105906. [CrossRef]
51. Foody, G.; Cox, D. Sub-Pixel Land Cover Composition Estimation Using a Linear Mixture Model and Fuzzy Membership Functions. *Remote Sens.* **1994**, *15*, 619–631. [CrossRef]
52. Jasinski, M.F.; Eagleson, P.S. Estimation of Subpixel Vegetation Cover Using Red-Infrared Scattergrams. *IEEE Trans. Geosci. Remote Sens.* **1990**, *28*, 253–267. [CrossRef]
53. Macomber, S.A.; Woodcock, C.E. Mapping and Monitoring Conifer Mortality Using Remote Sensing in the Lake Tahoe Basin. *Remote Sens. Environ.* **1994**, *50*, 255–266. [CrossRef]

54. Marsh, Switzer, Kowalik And Lyon Resolving the Percentage of Component Terrains within Single Resolution Elements. *Photogramm. Eng. Remote Sens.* **1980**, *46*, 10791086.
55. Cen, Y.; Zhang, L.; Zhang, X.; Wang, Y.; Qi, W.; Tang, S.; Zhang, P. Aerial Hyperspectral Remote Sensing Classification Dataset of Xiongan New Area (Matiwan Village). *J. Remote Sens* **2020**, *24*, 1299–1306.
56. He, D.; Shi, Q.; Liu, X.; Zhong, Y.; Liu, X. Spectral–Spatial Fusion Sub-Pixel Mapping Based on Deep Neural Network. *IEEE Geosci. Remote Sens. Lett.* **2022**, *19*, 1–5. [CrossRef]
57. Yang, X.; Cao, W.; Lu, Y.; Zhou, Y. Hyperspectral Image Transformer Classification Networks. *IEEE Trans. Geosci. Remote Sens.* **2022**, *60*, 1–15. [CrossRef]
58. Schaepman, M.E.; Jehle, M.; Hueni, A.; D'Odorico, P.; Damm, A.; Weyermann, J.; Schneider, F.D.; Laurent, V.; Popp, C.; Seidel, F.C.; et al. Advanced Radiometry Measurements and Earth Science Applications with the Airborne Prism Experiment (APEX). *Remote Sens. Environ.* **2015**, *158*, 207–219. [CrossRef]
59. Palsson, B.; Sveinsson, J.R.; Ulfarsson, M.O. Blind Hyperspectral Unmixing Using Autoencoders: A Critical Comparison. *IEEE J. Sel. Top. Appl. Earth Obs. Remote Sens.* **2022**, *15*, 1340–1372. [CrossRef]
60. Advanced Spaceborne Thermal Emission and Reflection Radiometer (ASTER). Available online: https://terra.nasa.gov/about/terra-instruments/aster (accessed on 15 May 2023).
61. Roy, P. Detection of Iron-Bearing Mineral Assemblages in Nainarmalai Granulite Region, South India, Based on Satellite Image Processing and Geochemical Anomalies. *Environ. Monit. Assess.* **2022**, *194*, 866. [CrossRef]
62. Abay, H.H.; Legesse, D.; Venkata Suryabhagavan, K.; Atnafu, B. Mapping of Ferric (Fe^{3+}) and Ferrous (Fe^{2+}) Iron Oxides Distribution Using ASTER and Landsat 8 OLI Data, in Negash Lateritic Iron Deposit, Northern Ethiopia. *Geol. Ecol. Landsc.* **2022**, 1–18. [CrossRef]
63. Advanced Very High Resolution Radiometer (AVHRR). Available online: https://www.earthdata.nasa.gov/sensors/avhrr (accessed on 15 May 2023).
64. Zhu, J.; Cao, S.; Shang, G.; Shi, J.; Wang, X.; Zheng, Z.; Liu, C.; Yang, H.; Xie, B. Subpixel Snow Mapping Using Daily AVHRR/2 Data over Qinghai–Tibet Plateau. *Remote Sens.* **2022**, *14*, 2899. [CrossRef]
65. Pan, F.; Jiang, L. Accuracy Evaluation of Several AVHRR Fractional Snow Cover Retrieval Algorithms in Asia Water Tower Region. In Proceedings of the IGARSS 2022–2022 IEEE International Geoscience and Remote Sensing Symposium, Kuala Lumpur, Malaysia, 17–22 July 2022; pp. 3860–3863.
66. Pan, F.; Jiang, L.; Zheng, Z.; Wang, G.; Cui, H.; Zhou, X.; Huang, J. Retrieval of Fractional Snow Cover over High Mountain Asia Using 1 Km and 5 Km AVHRR/2 with Simulated Mid-Infrared Reflective Band. *Remote Sens.* **2022**, *14*, 3303. [CrossRef]
67. Airborne Visible/Infrared Imaging Spectrometer (AVIRIS). Available online: https://aviris.jpl.nasa.gov/ (accessed on 15 May 2023).
68. Hadi, F.; Yang, J.; Ullah, M.; Ahmad, I.; Farooque, G.; Xiao, L. DHCAE: Deep Hybrid Convolutional Autoencoder Approach for Robust Supervised Hyperspectral Unmixing. *Remote Sens.* **2022**, *14*, 4433. [CrossRef]
69. Hong, D.; Gao, L.; Yao, J.; Yokoya, N.; Chanussot, J.; Heiden, U.; Zhang, B. Endmember-Guided Unmixing Network (EGU-Net): A General Deep Learning Framework for Self-Supervised Hyperspectral Unmixing. *IEEE Trans. Neural Netw. Learn. Syst.* **2022**, *33*, 6518–6531. [CrossRef]
70. Dhaini, M.; Berar, M.; Honeine, P.; Van Exem, A. End-to-End Convolutional Autoencoder for Nonlinear Hyperspectral Unmixing. *Remote Sens.* **2022**, *14*, 3341. [CrossRef]
71. Fang, Y.; Wang, Y.; Xu, L.; Zhuo, R.; Wong, A.; Clausi, D.A. BCUN: Bayesian Fully Convolutional Neural Network for Hyperspectral Spectral Unmixing. *IEEE Trans. Geosci. Remote Sens.* **2022**, *60*, 1–14. [CrossRef]
72. Hua, Z.; Li, X.; Feng, Y.; Zhao, L. Dual Branch Autoencoder Network for Spectral-Spatial Hyperspectral Unmixing. *IEEE Geosci. Remote Sens. Lett.* **2022**, *19*, 1–5. [CrossRef]
73. Jin, Q.; Ma, Y.; Mei, X.; Ma, J. TANet: An Unsupervised Two-Stream Autoencoder Network for Hyperspectral Unmixing. *IEEE Trans. Geosci. Remote Sens.* **2022**, *60*, 1–15. [CrossRef]
74. Li, M.; Yang, B.; Wang, B. Robust Nonlinear Unmixing for Hyperspectral Images Based on an Extended Multilinear Mixing Model. In Proceedings of the IGARSS 2022–2022 IEEE International Geoscience and Remote Sensing Symposium, Kuala Lumpur, Malaysia, 17 July 2022; pp. 1780–1783.
75. Li, H.; Wu, K.; Xu, Y. An Integrated Change Detection Method Based on Spectral Unmixing and the CNN for Hyperspectral Imagery. *Remote Sens.* **2022**, *14*, 2523. [CrossRef]
76. Li, Z.; Altmann, Y.; Chen, J.; Mclaughlin, S.; Rahardja, S. Sparse Linear Spectral Unmixing of Hyperspectral Images Using Expectation-Propagation. *IEEE Trans. Geosci. Remote Sens.* **2022**, *60*, 1–13. [CrossRef]
77. Luo, W.; Gao, L.; Hong, D.; Chanussot, J. Endmember Purification with Affine Simplicial Cone Model. *IEEE Trans. Geosci. Remote Sens.* **2022**, *60*, 1–23. [CrossRef]
78. Ma, K.Y.; Chang, C.-I. Kernel-Based Constrained Energy Minimization for Hyperspectral Mixed Pixel Classification. *IEEE Trans. Geosci. Remote Sens.* **2022**, *60*, 1–23. [CrossRef]
79. Shi, S.; Zhao, M.; Zhang, L.; Altmann, Y.; Chen, J. Probabilistic Generative Model for Hyperspectral Unmixing Accounting for Endmember Variability. *IEEE Trans. Geosci. Remote Sens.* **2022**, *60*, 1–15. [CrossRef]
80. Sun, C.; Xing, F.; Liu, D.; Han, J.; Yang, B. Nonlinear Spectral Unmixing of Hyperspectral Imagery Based on Hapke Model and Relevance Vector Regression Algorithm. *J. Phys. Conf. Ser.* **2022**, *2219*, 012044. [CrossRef]

81. Yang, B. Supervised Nonlinear Hyperspectral Unmixing With Automatic Shadow Compensation Using Multiswarm Particle Swarm Optimization. *IEEE Trans. Geosci. Remote Sens.* **2022**, *60*, 1–18. [CrossRef]
82. Yi, C.; Liu, Y.; Zheng, L.; Gan, Y. Joint Processing of Spatial Resolution Enhancement and Spectral Unmixing for Hyperspectral Image. *IEEE Geosci. Remote Sens. Lett.* **2022**, *19*, 1–5. [CrossRef]
83. Zhang, H.; Lei, L.; Zhang, S.; Huang, M.; Li, F.; Deng, C.; Wang, S. Spatial Graph Regularized Nonnegative Matrix Factorization for Hyperspectral Unmixing. In Proceedings of the IGARSS 2022—2022 IEEE International Geoscience and Remote Sensing Symposium, Kuala Lumpur, Malaysia, 17 July 2022; pp. 1624–1627.
84. Zhao, M.; Wang, X.; Chen, J.; Chen, W. A Plug-and-Play Priors Framework for Hyperspectral Unmixing. *IEEE Trans. Geosci. Remote Sens.* **2022**, *60*, 1–13. [CrossRef]
85. Wu, Z.; Wang, B. Kernel-Based Decomposition Model with Total Variation and Sparsity Regularizations via Union Dictionary for Nonlinear Hyperspectral Anomaly Detection. *IEEE Trans. Geosci. Remote Sens.* **2022**, *60*, 1–16. [CrossRef]
86. Guan, Q.; Xu, T.; Feng, S.; Yu, F.; Song, K. Optimal Segmentation and Improved Abundance Estimation for Superpixel-Based Hyperspectral Unmixing. *Eur. J. Remote Sens.* **2022**, *55*, 485–506. [CrossRef]
87. Wang, G.; Zhang, Y.; Xie, W.-F.; Qu, Y.; Feng, L. Hyperspectral Linear Unmixing Based on Collaborative Sparsity and Multi-Band Non-Local Total Variation. *Int. J. Remote Sens.* **2022**, *43*, 1–26. [CrossRef]
88. Zhang, J.; Zhang, X.; Meng, H.; Sun, C.; Wang, L.; Cao, X. Nonlinear Unmixing via Deep Autoencoder Networks for Generalized Bilinear Model. *Remote Sens.* **2022**, *14*, 5167. [CrossRef]
89. Qi, L.; Gao, F.; Dong, J.; Gao, X.; Du, Q. SSCU-Net: Spatial–Spectral Collaborative Unmixing Network for Hyperspectral Images. *IEEE Trans. Geosci. Remote Sens.* **2022**, *60*, 1–15. [CrossRef]
90. Shi, S.; Zhang, L.; Altmann, Y.; Chen, J. Deep Generative Model for Spatial–Spectral Unmixing With Multiple Endmember Priors. *IEEE Trans. Geosci. Remote Sens.* **2022**, *60*, 1–14. [CrossRef]
91. Tao, X.; Paoletti, M.E.; Han, L.; Haut, J.M.; Ren, P.; Plaza, J.; Plaza, A. Fast Orthogonal Projection for Hyperspectral Unmixing. *IEEE Trans. Geosci. Remote Sens.* **2022**, *60*, 1–13. [CrossRef]
92. Wang, Z.; Li, J.; Liu, Y.; Xie, F.; Li, P. An Adaptive Surrogate-Assisted Endmember Extraction Framework Based on Intelligent Optimization Algorithms for Hyperspectral Remote Sensing Images. *Remote Sens.* **2022**, *14*, 892. [CrossRef]
93. Zhang, G.; Mei, S.; Xie, B.; Ma, M.; Zhang, Y.; Feng, Y.; Du, Q. Spectral Variability Augmented Sparse Unmixing of Hyperspectral Images. *IEEE Trans. Geosci. Remote Sens.* **2022**, *60*, 1–13. [CrossRef]
94. Zhao, M.; Wang, M.; Chen, J.; Rahardja, S. Perceptual Loss-Constrained Adversarial Autoencoder Networks for Hyperspectral Unmixing. *IEEE Geosci. Remote Sens. Lett.* **2022**, *19*, 1–5. [CrossRef]
95. Zhao, M.; Gao, T.; Chen, J.; Chen, W. Hyperspectral Unmixing via Nonnegative Matrix Factorization with Handcrafted and Learned Priors. *IEEE Geosci. Remote Sens. Lett.* **2022**, *19*, 1–5. [CrossRef]
96. Zhao, M.; Wang, M.; Chen, J.; Rahardja, S. Hyperspectral Unmixing for Additive Nonlinear Models With a 3-D-CNN Autoencoder Network. *IEEE Trans. Geosci. Remote Sens.* **2022**, *60*, 1–15. [CrossRef]
97. Zhu, Q.; Wang, L.; Chen, J.; Zeng, W.; Zhong, Y.; Guan, Q.; Yang, Z. S ^{3} TRM: Spectral-Spatial Unmixing of Hyperspectral Imagery Based on Sparse Topic Relaxation-Clustering Model. *IEEE Trans. Geosci. Remote Sens.* **2022**, *60*, 1–13. [CrossRef]
98. Gu, J.; Yang, B.; Wang, B. Nonlinear Unmixing for Hyperspectral Images via Kernel-Transformed Bilinear Mixing Models. *IEEE Trans. Geosci. Remote Sens.* **2022**, *60*, 1–13. [CrossRef]
99. Airborne Visible InfraRed Imaging Spectrometer—Next Generation (AVIRIS NG). Available online: https://www.jpl.nasa.gov/missions/airborne-visible-infrared-imaging-spectrometer-next-generation-aviris-ng (accessed on 15 May 2023).
100. Lyngdoh, R.B.; Dave, R.; Anand, S.S.; Ahmad, T.; Misra, A. Hyperspectral Unmixing with Spectral Variability Using Endmember Guided Probabilistic Generative Deep Learning. In Proceedings of the IGARSS 2022–2022 IEEE International Geoscience and Remote Sensing Symposium, Kuala Lumpur, Malaysia, 17 July 2022; pp. 1768–1771.
101. Pascucci, S.; Cavalli, R.; Palombo, A.; Pignatti, S. Suitability of CASI and ATM Airborne Remote Sensing Data for Archaeological Subsurface Structure Detection under Different Land Cover: The Arpi Case Study (Italy). *J. Geophys. Eng.* **2010**, *7*, 183–189. [CrossRef]
102. DLR Earth Sensing Imaging Spectrometer (DESIS). Available online: https://www.dlr.de/os/en/desktopdefault.aspx/tabid-12923/ (accessed on 15 May 2023).
103. Legleiter, C.J.; King, T.V.; Carpenter, K.D.; Hall, N.C.; Mumford, A.C.; Slonecker, T.; Graham, J.L.; Stengel, V.G.; Simon, N.; Rosen, B.H. Spectral Mixture Analysis for Surveillance of Harmful Algal Blooms (SMASH): A Field-, Laboratory-, and Satellite-Based Approach to Identifying Cyanobacteria Genera from Remotely Sensed Data. *Remote Sens. Environ.* **2022**, *279*, 113089. [CrossRef]
104. Cerra, D.; Ji, C.; Heiden, U. Solar Panels Area Estimation Using the Spaceborne Imaging Spectrometer Desis: Outperforming Multispectral Sensors. *ISPRS Ann. Photogramm. Remote Sens. Spat. Inf. Sci.* **2022**, *V-1–2022*, 9–14. [CrossRef]
105. EnMAP (Environmental Mapping and Analysis Program). Available online: https://www.dlr.de/eoc/en/desktopdefault.aspx/tabid-5514/20470_read-47899/ (accessed on 15 May 2023).
106. Gaofen (GF). Available online: https://space.skyrocket.de/doc_sdat/gf-6.htm (accessed on 15 May 2023).
107. Li, Y.; Sun, B.; Gao, Z.; Su, W.; Wang, B.; Yan, Z.; Gao, T. Extraction of Rocky Desertification Information in Karst Area by Using Different Multispectral Sensor Data and Multiple Endmember Spectral Mixture Analysis Method. *Front. Environ. Sci.* **2022**, *10*, 996708. [CrossRef]

108. Zhang, C.; Jiang, L. Fractional Snow Cover Mapping with High Spatiotemporal Resolution Based on Landsat, Sentinel-2 And Modis Observation. In Proceedings of the IGARSS 2022—2022 IEEE International Geoscience and Remote Sensing Symposium, Kuala Lumpur, Malaysia, 17 July 2022; pp. 3935–3938.
109. Shao, Z.; Zhang, Y.; Zhang, C.; Huang, X.; Cheng, T. Mapping Impervious Surfaces with a Hierarchical Spectral Mixture Analysis Incorporating Endmember Spatial Distribution. *Geo-Spat. Inf. Sci.* **2022**, *25*, 550–567. [CrossRef]
110. Rickard, L.J.; Basedow, R.W.; Zalewski, E.F.; Silverglate, P.R.; Landers, M. HYDICE: An Airborne System for Hyperspectral Imaging. In Proceedings of the Imaging Spectrometry of the Terrestrial Environment, SPIE, Orlando, FL, USA, 23 September 1993; Volume 1937, pp. 173–179.
111. Kuester, J.; Anastasiadis, J.; Middelmann, W.; Heizmann, M. Investigating the Influence of Hyperspectral Data Compression on Spectral Unmixing. In Proceedings of the Image and Signal Processing for Remote Sensing XXVIII, Edinburgh, UK, 26–28 September 2022; Pierdicca, N., Bruzzone, L., Bovolo, F., Eds.; SPIE: Berlin, Germany, 2022; p. 16.
112. Pearlman, J.S.; Barry, P.S.; Segal, C.C.; Shepanski, J.; Beiso, D.; Carman, S.L. Hyperion, a Space-Based Imaging Spectrometer. *IEEE Trans. Geosci. Remote Sens.* **2003**, *41*, 1160–1173. [CrossRef]
113. Kumar, V.; Pandey, K.; Panda, C.; Tiwari, V.; Agrawal, S. Assessment of Different Spectral Unmixing Techniques on Space Borne Hyperspectral Imagery. *Remote Sens. Earth Syst. Sci.* **2022**, 5. [CrossRef]
114. Cavalli, R.M. The Weight of Hyperion and PRISMA Hyperspectral Sensor Characteristics on Image Capability to Retrieve Urban Surface Materials in the City of Venice. *Sensors* **2023**, *23*, 454. [CrossRef] [PubMed]
115. Jamshid Moghadam, H.; Mohammady Oskouei, M.; Nouri, T. The Influence of Noise Intensity in the Nonlinear Spectral Unmixing of Hyperspectral Data. *PFG J. Photogramm. Remote Sens. Geoinf. Sci.* **2022**, *91*, 21–42. [CrossRef]
116. Rajendran, S.; Al-Naimi, N.; Al Khayat, J.A.; Sorino, C.F.; Sadooni, F.N.; Al Saad Al Kuwari, H. Chlorophyll-a Concentrations in the Arabian Gulf Waters of Arid Region: A Case Study from the Northern Coast of Qatar. *Reg. Stud. Mar. Sci.* **2022**, *56*, 102680. [CrossRef]
117. Zhang, G.; Scheunders, P.; Cerra, D.; Muller, R. Shadow-Aware Nonlinear Spectral Unmixing for Hyperspectral Imagery. *IEEE J. Sel. Top. Appl. Earth Obs. Remote Sens.* **2022**, *15*, 5514–5533. [CrossRef]
118. Landsat Satellite Missions. Available online: https://www.usgs.gov/landsat-missions/landsat-satellite-missions (accessed on 15 May 2023).
119. Sutton, A.; Fisher, A.; Metternicht, G. Assessing the Accuracy of Landsat Vegetation Fractional Cover for Monitoring Australian Drylands. *Remote Sens.* **2022**, *14*, 6322. [CrossRef]
120. Bera, D.; Kumar, P.; Siddiqui, A.; Majumdar, A. Assessing Impact of Urbanisation on Surface Runoff Using Vegetation-Impervious Surface-Soil (VIS) Fraction and NRCS Curve Number (CN) Model. *Model. Earth Syst. Environ.* **2022**, *8*, 309–322. [CrossRef]
121. Brice, E.M.; Halabisky, M.; Ray, A.M. Making the Leap from Ponds to Landscapes: Integrating Field-Based Monitoring of Amphibians and Wetlands with Satellite Observations. *Ecol. Indic.* **2022**, *135*, 108559. [CrossRef]
122. Ding, Q.; Pan, T.; Lin, T.; Zhang, C. Urban Land-Cover Changes in Major Cities in China from 1990 to 2015. *Int. J. Environ. Res. Public Health* **2022**, *19*, 16079. [CrossRef]
123. Halbgewachs, M.; Wegmann, M.; da Ponte, E. A Spectral Mixture Analysis and Landscape Metrics Based Framework for Monitoring Spatiotemporal Forest Cover Changes: A Case Study in Mato Grosso, Brazil. *Remote Sens.* **2022**, *14*, 1907. [CrossRef]
124. Lathrop, R.G.; Merchant, D.; Niles, L.; Paludo, D.; Santos, C.D.; Larrain, C.E.; Feigin, S.; Smith, J.; Dey, A. Multi-Sensor Remote Sensing of Intertidal Flat Habitats for Migratory Shorebird Conservation. *Remote Sens.* **2022**, *14*, 5016. [CrossRef]
125. Nill, L.; Grünberg, I.; Ullmann, T.; Gessner, M.; Boike, J.; Hostert, P. Arctic Shrub Expansion Revealed by Landsat-Derived Multitemporal Vegetation Cover Fractions in the Western Canadian Arctic. *Remote Sens. Environ.* **2022**, *281*, 113228. [CrossRef]
126. Ouyang, L.; Wu, C.; Li, J.; Liu, Y.; Wang, M.; Han, J.; Song, C.; Yu, Q.; Haase, D. Mapping Impervious Surface Using Phenology-Integrated and Fisher Transformed Linear Spectral Mixture Analysis. *Remote Sens.* **2022**, *14*, 1673. [CrossRef]
127. Tarazona Coronel, Y. Mapping Deforestation Using Fractions Indices and the Non-Seasonal PVts-β Detection Approach. *IEEE Geosci. Remote Sens. Lett.* **2022**, *19*, 1–5. [CrossRef]
128. Xia, Z.; Li, Y.; Zhang, W.; Chen, R.; Guo, S.; Zhang, P.; Du, P. Solar Photovoltaic Program Helps Turn Deserts Green in China: Evidence from Satellite Monitoring. *J. Environ. Manag.* **2022**, *324*, 116338. [CrossRef]
129. Zhang, Y.; Wang, Y.; Ding, N.; Yang, X. Spatial Pattern Impact of Impervious Surface Density on Urban Heat Island Effect: A Case Study in Xuzhou, China. *Land* **2022**, *11*, 2135. [CrossRef]
130. Zhang, Y.; Wang, Y.; Ding, N. Spatial Effects of Landscape Patterns of Urban Patches with Different Vegetation Fractions on Urban Thermal Environment. *Remote Sens.* **2022**, *14*, 5684. [CrossRef]
131. Santos, F.C.; da Silva Pinto Vieira, R.M.; Barbosa, A.A.; da Cruz Ferreira, Y.; Polizel, S.P.; Sestini, M.F.; Ometto, J.P.H.B. Application of Remote Sensing to Analyze the Loss of Natural Vegetation in the Jalapão Mosaic (Brazil) before and after the Creation of Protected Area (1970–2018). *Environ. Monit. Assess.* **2022**, *194*, 201. [CrossRef]
132. Shimabukuro, Y.E.; Arai, E.; da Silva, G.M.; Dutra, A.C.; Mataveli, G.; Duarte, V.; Martini, P.R.; Cassol, H.L.G.; Ferreira, D.S.; Junqueira, L.R. Mapping and Monitoring Forest Plantations in São Paulo State, Southeast Brazil, Using Fraction Images Derived from Multiannual Landsat Sensor Images. *Forests* **2022**, *13*, 1716. [CrossRef]
133. van Kuik, N.; de Vries, J.; Schwarz, C.; Ruessink, G. Surface-Area Development of Foredune Trough Blowouts and Associated Parabolic Dunes Quantified from Time Series of Satellite Imagery. *Aeolian Res.* **2022**, *57*, 100812. [CrossRef]

134. Compains Iso, L.; Fernández-Manso, A.; Fernández-García, V. Optimizing Spectral Libraries from Landsat Imagery for the Analysis of Habitat Richness Using MESMA. *Forests* **2022**, *13*, 1824. [CrossRef]
135. Sofan, P.; Chulafak, G.A.; Pambudi, A.I.; Yulianto, F. Assessment of Space-Based Tropical Smouldering Peatlands: Mixed Pixel Analysis. *IOP Conf. Ser. Earth Environ. Sci.* **2022**, *1109*, 012054. [CrossRef]
136. Zhao, Y.; Zhang, X.; Feng, W.; Xu, J. Deep Learning Classification by ResNet-18 Based on the Real Spectral Dataset from Multispectral Remote Sensing Images. *Remote Sens.* **2022**, *14*, 4883. [CrossRef]
137. Cipta, I.M.; Jaelani, L.M.; Sanjaya, H. Identification of Paddy Varieties from Landsat 8 Satellite Image Data Using Spectral Unmixing Method in Indramayu Regency, Indonesia. *ISPRS Int. J. Geo-Inf.* **2022**, *11*, 510. [CrossRef]
138. Viana-Soto, A.; Okujeni, A.; Pflugmacher, D.; García, M.; Aguado, I.; Hostert, P. Quantifying Post-Fire Shifts in Woody-Vegetation Cover Composition in Mediterranean Pine Forests Using Landsat Time Series and Regression-Based Unmixing. *Remote Sens. Environ.* **2022**, *281*, 113239. [CrossRef]
139. Silvan-Cardenas, J.L.; Wang, L. Fully Constrained Linear Spectral Unmixing: Analytic Solution Using Fuzzy Sets. *IEEE Trans. Geosci. Remote Sens.* **2010**, *48*, 3992–4002. [CrossRef]
140. Zhao, J.; Li, J.; Liu, Q.; Zhang, Z.; Dong, Y. Comparative Study of Fractional Vegetation Cover Estimation Methods Based on Fine Spatial Resolution Images for Three Vegetation Types. *IEEE Geosci. Remote Sens. Lett.* **2022**, *19*, 1–5. [CrossRef]
141. Yang, X.; Chu, Q.; Wang, L.; Yu, M. Water Body Super-Resolution Mapping Based on Multiple Endmember Spectral Mixture Analysis and Multiscale Spatio-Temporal Dependence. *Remote Sens.* **2022**, *14*, 2050. [CrossRef]
142. Wang, J.; Zhao, Y.; Fu, Y.; Xia, L.; Chen, J. Improving LSMA for Impervious Surface Estimation in an Urban Area. *Eur. J. Remote Sens.* **2022**, *55*, 37–51. [CrossRef]
143. Jin, M.; Ding, X.; Han, H.; Pang, J.; Wang, Y. An Improved Method Combining Fisher Transformation and Multiple Endmember Spectral Mixture Analysis for Lunar Mineral Abundance Quantification Using Spectral Data. *Icarus* **2022**, *380*, 115008. [CrossRef]
144. Bassani, C.; Cavalli, M.; Palombo, A.; Pignatti, S.; Madonna, F. Laboratory Activity for a New Procedure of MIVIS Calibration and Relative Validation with Test Dataa. *Ann. Geophys.* **2006**, *49*, 45–56.
145. Cavalli, R.M. Spatial Validation of Spectral Unmixing Results: A Case Study of Venice City. *Remote Sens.* **2022**, *14*, 5165. [CrossRef]
146. Medium Resolution Imaging Spectrometer (MERIS). Available online: https://earth.esa.int/eogateway/instruments/meris (accessed on 15 May 2023).
147. Ambarwulan, W.; Salama, M.S.; Verhoef, W.; Mannaerts, C.M. The Estimation of Total Suspended Matter from Satellite Imagery of Tropical Coastal Water Berau Estuary, Indonesia. *IOP Conf. Ser. Earth Environ. Sci.* **2022**, *950*, 012089. [CrossRef]
148. MODIS (Moderate Resolution Imaging Spectroradiomete. Available online: https://modis.gsfc.nasa.gov/about/ (accessed on 15 May 2023).
149. Hu, Z.; Kuipers Munneke, P.; Lhermitte, S.; Dirscherl, M.; Ji, C.; van den Broeke, M. FABIAN: A Daily Product of Fractional Austral-Summer Blue Ice over ANtarctica during 2000–2021 Based on MODIS Imagery Using Google Earth Engine. *Remote Sens. Environ.* **2022**, *280*, 113202. [CrossRef]
150. Wang, Q.; Ding, X.; Tong, X.; Atkinson, P.M. Real-Time Spatiotemporal Spectral Unmixing of MODIS Images. *IEEE Trans. Geosci. Remote Sens.* **2022**, *60*, 1–16. [CrossRef]
151. Yin, Z.; Ling, F.; Li, X.; Cai, X.; Chi, H.; Li, X.; Wang, L.; Zhang, Y.; Du, Y. A Cascaded Spectral–Spatial CNN Model for Super-Resolution River Mapping with MODIS Imagery. *IEEE Trans. Geosci. Remote Sens.* **2022**, *60*, 1–13. [CrossRef]
152. Ding, X.; Wang, Q.; Tong, X. Integrating 250 m MODIS Data in Spectral Unmixing for 500 m Fractional Vegetation Cover Estimation. *Int. J. Appl. Earth Obs. Geoinf.* **2022**, *111*, 102860. [CrossRef]
153. Song, M.; Zhong, Y.; Ma, A.; Xu, X.; Zhang, L. A Joint Spectral Unmixing and Subpixel Mapping Framework Based on Multiobjective Optimization. *IEEE Trans. Geosci. Remote Sens.* **2022**, *60*, 1–17. [CrossRef]
154. Pervin, R.; Robeson, S.M.; MacBean, N. Fusion of Airborne Hyperspectral and LiDAR Canopy-Height Data for Estimating Fractional Cover of Tall Woody Plants, Herbaceous Vegetation, and Other Soil Cover Types in a Semi-Arid Savanna Ecosystem. *Int. J. Remote Sens.* **2022**, *43*, 3890–3926. [CrossRef]
155. PRISMA (Hyperspectral Precursor of the Application Mission). Available online: https://www.asi.it/en/earth-science/prisma/ (accessed on 15 May 2023).
156. Benhalouche, F.Z.; Benabbou, O.; Karoui, M.S.; Kebir, L.W.; Bennia, A.; Deville, Y. Minerals Detection and Mapping in the Southwestern Algeria Gara-Djebilet Region with a Multistage Informed NMF-Based Unmixing Approach Using Prisma Remote Sensing Hyperspectral Data. In Proceedings of the IGARSS 2022—2022 IEEE International Geoscience and Remote Sensing Symposium, Kuala Lumpur, Malaysia, 17 July 2022; pp. 6422–6425.
157. Damarjati, S.; Nugraha, W.A.; Arjasakusuma, S. Mapping the Invasive Palm Species *Arenga Obtusifolia* Using Multiple Endmember Spectral Mixture Analysis (MESMA) and PRISMA Hyperspectral Data in Ujung Kulon National Park, Indonesia. *Geocarto Int.* **2022**, 1–21. [CrossRef]
158. Shaik, R.U.; Laneve, G.; Fusilli, L. An Automatic Procedure for Forest Fire Fuel Mapping Using Hyperspectral (PRISMA) Imagery: A Semi-Supervised Classification Approach. *Remote Sens.* **2022**, *14*, 1264. [CrossRef]
159. Bigdeli, B.; Samadzadegan, F.; Reinartz, P. A Multiple SVM System for Classification of Hyperspectral Remote Sensing Data. *J. Indian Soc. Remote Sens.* **2013**, *41*, 763–776. [CrossRef]
160. SENTINEL-2. Available online: https://sentinel.esa.int/web/sentinel/missions/sentinel-2 (accessed on 15 May 2023).

161. Fernández-Guisuraga, J.M.; Suárez-Seoane, S.; Quintano, C.; Fernández-Manso, A.; Calvo, L. Comparison of Physical-Based Models to Measure Forest Resilience to Fire as a Function of Burn Severity. *Remote Sens.* **2022**, *14*, 5138. [CrossRef]
162. Xu, Y.; Jiang, C.; Li, X.; Ji, E.; Ban, S. Research on the Extraction Method of Impervious Surface in Nanjing Based on Random Forest Classification. In Proceedings of the 2022 29th International Conference on Geoinformatics, Beijing, China, 15–18 August 2022; pp. 1–10.
163. Meng, R.; Xu, B.; Zhao, F.; Dong, Y.; Wang, C.; Sun, R.; Zhou, Y.; Zhou, L.; Gong, S.; Zhang, D. Characterizing the Provision and Inequality of Primary School Greenspaces in China's Major Cities Based on Multi-Sensor Remote Sensing. *Urban For. Urban Green.* **2022**, *75*, 127670. [CrossRef]
164. Cao, S.; Feng, J.; Hu, Z.; Li, Q.; Wu, G. Improving Estimation of Urban Land Cover Fractions with Rigorous Spatial Endmember Modeling. *ISPRS J. Photogramm. Remote Sens.* **2022**, *189*, 36–49. [CrossRef]
165. Sun, Z.; Zhu, Q.; Deng, S.; Li, X.; Hu, X.; Chen, R.; Shao, G.; Yang, H.; Yang, G. Estimation of Crop Residue Cover in Rice Paddies by a Dynamic-Quadripartite Pixel Model Based on Sentinel-2A Data. *Int. J. Appl. Earth Obs. Geoinf.* **2022**, *106*, 102645. [CrossRef]
166. Kremezi, M.; Kristollari, V.; Karathanassi, V.; Topouzelis, K.; Kolokoussis, P.; Taggio, N.; Aiello, A.; Ceriola, G.; Barbone, E.; Corradi, P. Increasing the Sentinel-2 Potential for Marine Plastic Litter Monitoring through Image Fusion Techniques. *Mar. Pollut. Bull.* **2022**, *182*, 113974. [CrossRef]
167. Ozer, E.; Leloglu, U.M. Wetland Spectral Unmixing Using Multispectral Satellite Images. *Geocarto Int.* **2022**, 1–24. [CrossRef]
168. Zhao, S.; Qin, Q. Detection and Identification of Surface Cover in Coalbed Methane Enrichment Area Based on Spectral Unmixing. In Proceedings of the IGARSS 2022—2022 IEEE International Geoscience and Remote Sensing Symposium, Kuala Lumpur, Malaysia, 17 July 2022; pp. 3732–3735.
169. Hajnal, W.; Priem, F.; Canters, F. M-CORE: A Novel Approach for Land Cover Fraction Mapping Using Multisite Spectral Libraries. *IEEE Trans. Geosci. Remote Sens.* **2022**, *60*, 1–15. [CrossRef]
170. Ronay, I.; Kizel, F.; Lati, R. The effect of spectral mixtures on weed species classification. *ISPRS Ann. Photogramm. Remote Sens. Spat. Inf. Sci.* **2022**, *V-3–2022*, 477–484. [CrossRef]
171. Satellite Pour l'Observation de La Terre (SPOT). Available online: https://earth.esa.int/eogateway/missions/spot (accessed on 15 May 2023).
172. WorldView. Available online: https://earth.esa.int/eogateway/missions/worldview (accessed on 15 May 2023).
173. Sarkar, D.; Sur, P. Targeting the Bauxite Rich Pockets from Lateritic Terrain Utilizing ASTER Data: A Case Study from Kabirdham District, Chhattisgarh, India. *J. Earth Syst. Sci.* **2021**, *130*, 189. [CrossRef]
174. Ghanbari Azar, S.; Meshgini, S.; Beheshti, S.; Yousefi Rezaii, T. Linear Mixing Model with Scaled Bundle Dictionary for Hyperspectral Unmixing with Spectral Variability. *Signal Process.* **2021**, *188*, 108214. [CrossRef]
175. Bai, J.; Feng, R.; Wang, L.; Zhong, Y.; Zhang, L. Weakly Supervised Convolutional Neural Networks for Hyperspectral Unmixing. In Proceedings of the 2021 IEEE International Geoscience and Remote Sensing Symposium IGARSS, Brussels, Belgium, 11 July 2021; pp. 3857–3860.
176. Borsoi, R.A.; Imbiriba, T.; Bermudez, J.C.M.; Richard, C. Deep Generative Models for Library Augmentation in Multiple Endmember Spectral Mixture Analysis. *IEEE Geosci. Remote Sens. Lett.* **2021**, *18*, 1831–1835. [CrossRef]
177. Di, W.-C.; Huang, J.; Wang, J.-J.; Huang, T.-Z. Enhancing Reweighted Low-Rank Representation for Hyperspectral Image Unmixing. In Proceedings of the 2021 IEEE International Geoscience and Remote Sensing Symposium IGARSSk, Brussels, Belgium, 11 July 2021; pp. 3825–3828.
178. Dong, L.; Yuan, Y. Sparse Constrained Low Tensor Rank Representation Framework for Hyperspectral Unmixing. *Remote Sens. Remote Sens.* **2021**, *13*, 1473. [CrossRef]
179. Dong, L.; Yuan, Y.; Lu, X. Spectral–Spatial Joint Sparse NMF for Hyperspectral Unmixing. *IEEE Trans. Geosci. Remote Sens.* **2021**, *59*, 2391–2402. [CrossRef]
180. Dong, L.; Lu, X.; Liu, G.; Yuan, Y. A Novel NMF Guided for Hyperspectral Unmixing From Incomplete and Noisy Data. *IEEE Trans. Geosci. Remote Sens.* **2022**, *60*, 1–15. [CrossRef]
181. Ekanayake, E.M.M.B.; Weerasooriya, H.M.H.K.; Ranasinghe, D.Y.L.; Herath, S.; Rathnayake, B.; Godaliyadda, G.M.R.I.; Ekanayake, M.P.B.; Herath, H.M.V.R. Constrained Nonnegative Matrix Factorization for Blind Hyperspectral Unmixing Incorporating Endmember Independence. *IEEE J. Sel. Top. Appl. Earth Obs. Remote Sens.* **2021**, *14*, 11853–11869. [CrossRef]
182. Elrewainy, A.; Sherif, S. Robust Anomaly Detection Algorithm for Hyperspectral Images Using Spectral Unmixing. In *Proceedings of the Image and Signal Processing for Remote Sensing XXVII*; Bruzzone, L., Bovolo, F., Benediktsson, J.A., Eds.; SPIE: Bellingham, DC, USA, 2021; p. 38.
183. Gu, J.; Cheng, T.; Wang, B. Reweighted Kernel-Based Nonlinear Hyperspectral Unmixing With Regional ℓ_1-Norm Regularization. *IEEE Geosci. Remote Sens. Lett.* **2022**, *19*, 1–5. [CrossRef]
184. Guo, Z.; Min, A.; Yang, B.; Chen, J.; Li, H.; Gao, J. A Sparse Oblique-Manifold Nonnegative Matrix Factorization for Hyperspectral Unmixing. *IEEE Trans. Geosci. Remote Sens.* **2022**, *60*, 1–13. [CrossRef]
185. Guo, Z.; Min, A.; Yang, B.; Chen, J.; Li, H. A Modified Huber Nonnegative Matrix Factorization Algorithm for Hyperspectral Unmixing. *IEEE J. Sel. Top. Appl. Earth Obs. Remote Sens.* **2021**, *14*, 5559–5571. [CrossRef]
186. Han, Z.; Hong, D.; Gao, L.; Zhang, B.; Chanussot, J. Deep Half-Siamese Networks for Hyperspectral Unmixing. *IEEE Geosci. Remote Sens. Lett.* **2021**, *18*, 1996–2000. [CrossRef]

187. Hua, Z.; Li, X.; Qiu, Q.; Zhao, L. Autoencoder Network for Hyperspectral Unmixing With Adaptive Abundance Smoothing. *IEEE Geosci. Remote Sens. Lett.* **2021**, *18*, 1640–1644. [CrossRef]
188. Hua, Z.; Li, X.; Jiang, J.; Zhao, L. Gated Autoencoder Network for Spectral–Spatial Hyperspectral Unmixing. *Remote Sens.* **2021**, *13*, 3147. [CrossRef]
189. Huang, J.; Di, W.-C.; Wang, J.-J.; Lin, J.; Huang, T.-Z. Bilateral Joint-Sparse Regression for Hyperspectral Unmixing. *IEEE J. Sel. Top. Appl. Earth Obs. Remote Sens.* **2021**, *14*, 10147–10161. [CrossRef]
190. Jia, P.; Zhang, M.; Shen, Y. Deep Spectral Unmixing Framework via 3D Denoising Convolutional Autoencoder. *IET Image Process.* **2021**, *15*, 1399–1409. [CrossRef]
191. Kumar, P.; Chakravortty, S. Generation of Sub-Pixel-Level Maps for Mixed Pixels in Hyperspectral Image Data. *Curr. Sci.* **2021**, *120*, 166. [CrossRef]
192. Li, C.; Jiang, Y.; Chen, X. Hyperspectral Unmixing via Noise-Free Model. *IEEE Trans. Geosci. Remote Sens.* **2021**, *59*, 3277–3291. [CrossRef]
193. Li, C.; Gu, Y.; Chen, X.; Zhang, Y.; Ruan, L. Hyperspectral Unmixing via Latent Multiheterogeneous Subspace. *IEEE Trans. Geosci. Remote Sens.* **2021**, *59*, 563–577. [CrossRef]
194. Li, F. Low-Rank and Spectral-Spatial Sparse Unmixing for Hyperspectral Remote Sensing Imagery. *Wirel. Commun. Mob. Comput.* **2021**, *2021*, 1–14. [CrossRef]
195. Li, F.; Zhang, S.; Liang, B.; Deng, C.; Xu, C.; Wang, S. Hyperspectral Sparse Unmixing With Spectral-Spatial Low-Rank Constraint. *IEEE J. Sel. Top. Appl. Earth Obs. Remote Sens.* **2021**, *14*, 6119–6130. [CrossRef]
196. Li, F.; Zhang, S.; Deng, C.; Liang, B.; Cao, J.; Wang, S. Robust Double Spatial Regularization Sparse Hyperspectral Unmixing. *IEEE J. Sel. Top. Appl. Earth Obs. Remote Sens.* **2021**, *14*, 12569–12582. [CrossRef]
197. Li, H.; Borsoi, R.A.; Imbiriba, T.; Closas, P.; Bermudez, J.C.M.; Erdogmus, D. Model-Based Deep Autoencoder Networks for Nonlinear Hyperspectral Unmixing. *IEEE Geosci. Remote Sens. Lett.* **2022**, *19*, 1–5. [CrossRef]
198. Tan, X.; Yu, Q.; Wang, Z.; Zhu, J. Semi-Supervised Unmixing of Hyperspectral Data via Spectral-Spatial Factorization. *IEEE Sens. J.* **2021**, *21*, 25963–25972. [CrossRef]
199. Wang, J.-J.; Huang, T.-Z.; Huang, J.; Deng, L.-J. A Two-Step Iterative Algorithm for Sparse Hyperspectral Unmixing via Total Variation. *Appl. Math. Comput.* **2021**, *401*, 126059. [CrossRef]
200. Wang, L.; Wang, S.; Jia, X.; Bi, T. A Novel Hyperspectral Unmixing Method Based on Least Squares Twin Support Vector Machines. *Eur. J. Remote Sens.* **2021**, *54*, 72–85. [CrossRef]
201. Xiong, F.; Zhou, J.; Ye, M.; Lu, J.; Qian, Y. NMF-SAE: An Interpretable Sparse Autoencoder for Hyperspectral Unmixing. In Proceedings of the ICASSP 2021—2021 IEEE International Conference on Acoustics, Speech and Signal Processing (ICASSP), Toronto, ON, Canada, 6 June 2021; pp. 1865–1869.
202. Kucuk, S.; Yuksel, S.E. Total Utility Metric Based Dictionary Pruning for Sparse Hyperspectral Unmixing. *IEEE Trans. Comput. Imaging* **2021**, *7*, 562–572. [CrossRef]
203. Li, C.; Chen, X.; Jiang, Y.; Yang, L. Elastic Constraints on Split Hierarchical Abundances for Blind Hyperspectral Unmixing. *Signal Process.* **2021**, *188*, 108229. [CrossRef]
204. Li, X.; Huang, R.; Zhao, L. Correntropy-Based Spatial-Spectral Robust Sparsity-Regularized Hyperspectral Unmixing. *IEEE Trans. Geosci. Remote Sens.* **2021**, *59*, 1453–1471. [CrossRef]
205. Li, Y.; Bao, W.; Qu, K.; Shen, X. Nonlocal Weighted Sparse Unmixing Based on Global Search and Parallel Optimization. *J. Appl. Rem. Sens.* **2021**, *15*. [CrossRef]
206. Liu, J.; Zhang, Y.; Liu, Y.; Mu, C. Hyperspectral Images Unmixing Based on Abundance Constrained Multi-Layer KNMF. *IEEE Access* **2021**, *9*, 91080–91090. [CrossRef]
207. Liu, R.; Zhu, X. Endmember Bundle Extraction Based on Multiobjective Optimization. *IEEE Trans. Geosci. Remote Sens.* **2021**, *59*, 8630–8645. [CrossRef]
208. Patel, J.R.; Joshi, M.V.; Bhatt, J.S. Spectral Unmixing Using Autoencoder with Spatial and Spectral Regularizations. In Proceedings of the 2021 IEEE International Geoscience and Remote Sensing Symposium IGARSS, Brussels, Belgium, 11 July 2021; pp. 3321–3324.
209. Peng, J.; Zhou, Y.; Sun, W.; Du, Q.; Xia, L. Self-Paced Nonnegative Matrix Factorization for Hyperspectral Unmixing. *IEEE Trans. Geosci. Remote Sens.* **2021**, *59*, 1501–1515. [CrossRef]
210. Qin, J.; Lee, H.; Chi, J.T.; Drumetz, L.; Chanussot, J.; Lou, Y.; Bertozzi, A.L. Blind Hyperspectral Unmixing Based on Graph Total Variation Regularization. *IEEE Trans. Geosci. Remote Sens.* **2021**, *59*, 3338–3351. [CrossRef]
211. Shahid, K.T.; Schizas, I.D. Unsupervised Hyperspectral Unmixing via Nonlinear Autoencoders. *IEEE Trans. Geosci. Remote Sens.* **2022**, *60*, 1–13. [CrossRef]
212. Su, Y.; Xu, X.; Li, J.; Qi, H.; Gamba, P.; Plaza, A. Deep Autoencoders With Multitask Learning for Bilinear Hyperspectral Unmixing. *IEEE Trans. Geosci. Remote Sens.* **2021**, *59*, 8615–8629. [CrossRef]
213. Vibhute, A.D.; Gaikwad, S.V.; Kale, K.V.; Mane, A.V. Hyperspectral Image Unmixing for Land Cover Classification. In Proceedings of the 2021 IEEE India Council International Subsections Conference (INDISCON), Nagpur, India, 27 August 2021; pp. 1–5.
214. Wan, L.; Chen, T.; Plaza, A.; Cai, H. Hyperspectral Unmixing Based on Spectral and Sparse Deep Convolutional Neural Networks. *IEEE J. Sel. Top. Appl. Earth Obs. Remote Sens.* **2021**, *14*, 11669–11682. [CrossRef]

215. Wang, J.-J.; Wang, D.-C.; Huang, T.-Z.; Huang, J. Endmember Constraint Non-Negative Tensor Factorization Via Total Variation for Hyperspectral Unmixing. In Proceedings of the 2021 IEEE International Geoscience and Remote Sensing Symposium IGARSS, Brussels, Belgium, 11 July 2021; pp. 3313–3316.
216. Wang, J.-J.; Wang, D.-C.; Huang, T.-Z.; Huang, J.; Zhao, X.-L.; Deng, L.-J. Endmember Independence Constrained Hyperspectral Unmixing via Nonnegative Tensor Factorization. *Knowl. Based Syst.* **2021**, *216*, 106657. [CrossRef]
217. Wang, J. A Novel Collaborative Representation Algorithm for Spectral Unmixing of Hyperspectral Remotely Sensed Imagery. *IEEE Access* **2021**, *9*, 89243–89248. [CrossRef]
218. Xiong, F.; Zhou, J.; Tao, S.; Lu, J.; Qian, Y. SNMF-Net: Learning a Deep Alternating Neural Network for Hyperspectral Unmixing. *IEEE Trans. Geosci. Remote Sens.* **2022**, *60*, 1–16. [CrossRef]
219. Xu, C.; Wu, Z.; Li, F.; Zhang, S.; Deng, C.; Wang, Y. Spectral-Spatial Joint Sparsity Unmixing of Hyperspectral Images Based on Framelet Transform. *Infrared Phys. Technol.* **2021**, *112*, 103564. [CrossRef]
220. Ye, C.; Liu, S.; Xu, M.; Du, B.; Wan, J.; Sheng, H. An Endmember Bundle Extraction Method Based on Multiscale Sampling to Address Spectral Variability for Hyperspectral Unmixing. *Remote Sens.* **2021**, *13*, 3941. [CrossRef]
221. Yuan, Y.; Dong, L. Weighted Sparsity Constraint Tensor Factorization for Hyperspectral Unmixing. In Proceedings of the 2021 IEEE International Geoscience and Remote Sensing Symposium IGARSS, Brussels, Belgium, 11 July 2021; pp. 3333–3336.
222. Yuan, Y.; Dong, L.; Li, X. Hyperspectral Unmixing Using Nonlocal Similarity-Regularized Low-Rank Tensor Factorization. *IEEE Trans. Geosci. Remote Sens.* **2022**, *60*, 1–14. [CrossRef]
223. Zhang, M.; Pezeril, S. A Comparative Study of Recent Multi-Component Unmixing Algorithms. In Proceedings of the 2021 11th Workshop on Hyperspectral Imaging and Signal Processing: Evolution in Remote Sensing (WHISPERS), Amsterdam, The Netherlands, 24 March 2021; pp. 1–5.
224. Zheng, P.; Su, H.; Du, Q. Sparse and Low-Rank Constrained Tensor Factorization for Hyperspectral Image Unmixing. *IEEE J. Sel. Top. Appl. Earth Obs. Remote Sens.* **2021**, *14*, 1754–1767. [CrossRef]
225. Zhu, Q.; Wang, L.; Zeng, W.; Guan, Q.; Hu, Z. A Sparse Topic Relaxion and Group Clustering Model for Hyperspectral Unmixing. *IEEE J. Sel. Top. Appl. Earth Obs. Remote Sens.* **2021**, *14*, 4014–4027. [CrossRef]
226. Badola, A.; Panda, S.K.; Roberts, D.A.; Waigl, C.F.; Bhatt, U.S.; Smith, C.W.; Jandt, R.R. Hyperspectral Data Simulation (Sentinel-2 to AVIRIS-NG) for Improved Wildfire Fuel Mapping, Boreal Alaska. *Remote Sens.* **2021**, *13*, 1693. [CrossRef]
227. Yu, J.; Wang, B.; Lin, Y.; Li, F.; Cai, J. A Novel Inequality-Constrained Weighted Linear Mixture Model for Endmember Variability. *Remote Sens. Environ.* **2021**, *257*, 112359. [CrossRef]
228. Okujeni, A.; Jänicke, C.; Cooper, S.; Frantz, D.; Hostert, P.; Clark, M.; Segl, K.; Van Der Linden, S. Multi-Season Unmixing of Vegetation Class Fractions across Diverse Californian Ecoregions Using Simulated Spaceborne Imaging Spectroscopy Data. *Remote Sens. Environ.* **2021**, *264*, 112558. [CrossRef]
229. Chang, M.; Meng, X.; Sun, W.; Yang, G.; Peng, J. Collaborative Coupled Hyperspectral Unmixing Based Subpixel Change Detection for Analyzing Coastal Wetlands. *IEEE J. Sel. Top. Appl. Earth Obs. Remote Sens.* **2021**, *14*, 8208–8224. [CrossRef]
230. Benhalouche, F.Z.; Deville, Y.; Karoui, M.S.; Ouamri, A. Hyperspectral Unmixing Based on Constrained Bilinear or Linear-Quadratic Matrix Factorization. *Remote Sens.* **2021**, *13*, 2132. [CrossRef]
231. He, D.; Zhong, Y.; Wang, X.; Zhang, L. Deep Convolutional Neural Network Framework for Subpixel Mapping. *IEEE Trans. Geosci. Remote Sens.* **2021**, *59*, 9518–9539. [CrossRef]
232. Song, H.; Wu, X.; Zou, A.; Liu, Y.; Zou, Y. Weighted Total Variation Regularized Blind Unmixing for Hyperspectral Image. *IEEE Geosci. Remote Sens. Lett.* **2022**, *19*, 1–5. [CrossRef]
233. Ou, D.; Tan, K.; Lai, J.; Jia, X.; Wang, X.; Chen, Y.; Li, J. Semi-Supervised DNN Regression on Airborne Hyperspectral Imagery for Improved Spatial Soil Properties Prediction. *Geoderma* **2021**, *385*, 114875. [CrossRef]
234. Haq, M.A.; Alshehri, M.; Rahaman, G.; Ghosh, A.; Baral, P.; Shekhar, C. Snow and Glacial Feature Identification Using Hyperion Dataset and Machine Learning Algorithms. *Arab. J. Geosci.* **2021**, *14*, 1525. [CrossRef]
235. Ji, C.; Li, X.; Wang, J.; Chen, M.; Pan, J. A Proposed Fully Constrained Least Squares for Solving Sparse Endmember Fractions with Linear Spectral Mixture Model. In Proceedings of the 2021 IEEE International Geoscience and Remote Sensing Symposium IGARSS, Brussels, Belgium, 11 July 2021; pp. 4143–4146.
236. Seydi, S.T.; Hasanlou, M. A New Structure for Binary and Multiple Hyperspectral Change Detection Based on Spectral Unmixing and Convolutional Neural Network. *Measurement* **2021**, *186*, 110137. [CrossRef]
237. Seydi, S.T.; Shah-Hosseini, R.; Hasanlou, M. New Framework for Hyperspectral Change Detection Based on Multi-Level Spectral Unmixing. *Appl. Geomat.* **2021**, *13*, 763–780. [CrossRef]
238. Cerra, D.; Pato, M.; Alonso, K.; Köhler, C.; Schneider, M.; de los Reyes, R.; Carmona, E.; Richter, R.; Kurz, F.; Reinartz, P.; et al. Dlr Hysu—a Benchmark Dataset for Spectral Unmixing. *Remote Sens.* **2021**, *13*, 2559. [CrossRef]
239. Chen, A.; Yang, X.; Xu, B.; Jin, Y.; Guo, J.; Xing, X.; Yang, D.; Wang, P.; Zhu, L. Monitoring the Spatiotemporal Dynamics of Aeolian Desertification Using Google Earth Engine. *Remote Sens.* **2021**, *13*, 1730. [CrossRef]
240. Lombard, F.; Andrieu, J. Mapping Mangrove Zonation Changes in Senegal with Landsat Imagery Using an OBIA Approach Combined with Linear Spectral Unmixing. *Remote Sens.* **2021**, *13*, 1961. [CrossRef]
241. Racoviteanu, A.E.; Nicholson, L.; Glasser, N.F. Surface Composition of Debris-Covered Glaciers across the Himalaya Using Linear Spectral Unmixing of Landsat 8 OLI Imagery. *Cryosphere* **2021**, *15*, 4557–4588. [CrossRef]

242. Shen, J.; Shuai, Y.; Li, P.; Cao, Y.; Ma, X. Extraction and Spatio-Temporal Analysis of Impervious Surfaces over Dongying Based on Landsat Data. *Remote Sens.* **2021**, *13*, 3666. [CrossRef]
243. Shumack, S.; Fisher, A.; Hesse, P.P. Refining Medium Resolution Fractional Cover for Arid Australia to Detect Vegetation Dynamics and Wind Erosion Susceptibility on Longitudinal Dunes. *Remote Sens. Environ.* **2021**, *265*, 112647. [CrossRef]
244. Vermeulen, L.M.; Munch, Z.; Palmer, A. Fractional Vegetation Cover Estimation in Southern African Rangelands Using Spectral Mixture Analysis and Google Earth Engine. *Comput. Electron. Agric.* **2021**, *182*, 105980. [CrossRef]
245. Chen, R.; Li, X.; Zhang, Y.; Zhou, P.; Wang, Y.; Shi, L.; Jiang, L.; Ling, F.; Du, Y. Spatiotemporal Continuous Impervious Surface Mapping by Fusion of Landsat Time Series Data and Google Earth Imagery. *Remote Sens.* **2021**, *13*, 2409. [CrossRef]
246. Chen, Y.; Huang, X.; Huang, J.; Liu, S.; Lu, D.; Zhao, S. Fractional Monitoring of Desert Vegetation Degradation, Recovery, and Greening Using Optimized Multi-Endmembers Spectral Mixture Analysis in a Dryland Basin of Northwest China. *GIScience Remote Sens.* **2021**, *58*, 300–321. [CrossRef]
247. Converse, R.L.; Lippitt, C.D.; Lippitt, C.L. Assessing Drought Vegetation Dynamics in Semiarid Grass- and Shrubland Using MESMA. *Remote Sens.* **2021**, *13*, 3840. [CrossRef]
248. Dutta, D.; Rahman, A.; Paul, S.K.; Kundu, A. Impervious Surface Growth and Its Inter-Relationship with Vegetation Cover and Land Surface Temperature in Peri-Urban Areas of Delhi. *Urban Clim.* **2021**, *37*, 100799. [CrossRef]
249. Finger, D.J.I.; McPherson, M.L.; Houskeeper, H.F.; Kudela, R.M. Mapping Bull Kelp Canopy in Northern California Using Landsat to Enable Long-Term Monitoring. *Remote Sens. Environ.* **2021**, *254*, 112243. [CrossRef]
250. Jiji, G.W. A Study on the Analysis of Heavy Metal Concentration Using Spectral Mixture Modelling Approach and Regression in Tirupur, India. *Earth Sci. Inf.* **2021**, *14*, 2077–2086. [CrossRef]
251. Li, M.; Zheng, Z.; Zhu, M.; He, Y.; Xia, J.; Chen, X.; Peng, Q.; He, Y.; Zhang, X.; Li, P. The Spatiotemporal Evolution of Urban Impervious Surface for Chengdu, China. *Photogramm Eng. Remote Sens.* **2021**, *87*, 491–502. [CrossRef]
252. Sall, I.; Jarchow, C.J.; Sigafus, B.H.; Eby, L.A.; Forzley, M.J.; Hossack, B.R. Estimating Inundation of Small Waterbodies with Sub-pixel Analysis of Landsat Imagery: Long-term Trends in Surface Water Area and Evaluation of Common Drought Indices. *Remote Sens. Ecol. Conserv.* **2021**, *7*, 109–124. [CrossRef]
253. Wu, K.; Chen, T.; Xu, Y.; Song, D.; Li, H. A Novel Change Detection Approach Based on Spectral Unmixing from Stacked Multitemporal Remote Sensing Images with a Variability of Endmembers. *Remote Sens.* **2021**, *13*, 2550. [CrossRef]
254. Bair, E.H.; Stillinger, T.; Dozier, J. Snow Property Inversion from Remote Sensing (SPIReS): A Generalized Multispectral Unmixing Approach with Examples from MODIS and Landsat 8 OLI. *IEEE Trans. Geosci. Remote Sens.* **2021**, *59*, 7270–7284. [CrossRef]
255. Feng, S.; Fan, F. Impervious Surface Extraction Based on Different Methods from Multiple Spatial Resolution Images: A Comprehensive Comparison. *Int. J. Digit. Earth* **2021**, *14*, 1148–1174. [CrossRef]
256. Fernández-García, V.; Marcos, E.; Fernández-Guisuraga, J.M.; Fernández-Manso, A.; Quintano, C.; Suárez-Seoane, S.; Calvo, L. Multiple Endmember Spectral Mixture Analysis (MESMA) Applied to the Study of Habitat Diversity in the Fine-Grained Landscapes of the Cantabrian Mountains. *Remote Sens.* **2021**, *13*, 979. [CrossRef]
257. Li, W. Improving Urban Impervious Surfaces Mapping through Integrating Statistical Methods and Spectral Mixture Analysis. *Remote Sens.* **2021**, *13*, 2474. [CrossRef]
258. Muhuri, A.; Gascoin, S.; Menzel, L.; Kostadinov, T.S.; Harpold, A.A.; Sanmiguel-Vallelado, A.; Lopez-Moreno, J.I. Performance Assessment of Optical Satellite-Based Operational Snow Cover Monitoring Algorithms in Forested Landscapes. *IEEE J. Sel. Top. Appl. Earth Obs. Remote Sens.* **2021**, *14*, 7159–7178. [CrossRef]
259. Zang, J.; Zhang, T.; Chen, L.; Li, L.; Liu, W.; Yuan, L.; Zhang, Y.; Liu, R.; Wang, Z.; Yu, Z.; et al. Optimization of Modelling Population Density Estimation Based on Impervious Surfaces. *Land* **2021**, *10*, 791. [CrossRef]
260. Luo, H.; Chen, N. A Combined Unmixing Framework for Impervious Surface Mapping on Medium-Resolution Images with Visible Shadows. *Photogramm. Eng. Remote Sens.* **2021**, *87*, 431–443. [CrossRef]
261. Pan, F.; Jiang, L.; Wang, G.; Su, X.; Zhou, X. Estimating Cloud-Free Fractional Snow Cover from Himawari-8, FY-4A and Modis Observation. In Proceedings of the 2021 IEEE International Geoscience and Remote Sensing Symposium IGARSS, Brussels, Belgium, 11 July 2021; pp. 5566–5569.
262. Rittger, K.; Krock, M.; Kleiber, W.; Bair, E.H.; Brodzik, M.J.; Stephenson, T.R.; Rajagopalan, B.; Bormann, K.J.; Painter, T.H. Multi-Sensor Fusion Using Random Forests for Daily Fractional Snow Cover at 30 m. *Remote Sens. Environ.* **2021**, *264*, 112608. [CrossRef]
263. Sun, Q.; Zhang, P.; Jiao, X.; Han, W.; Sun, Y.; Sun, D. Identifying and Understanding Alternative States of Dryland Landscape: A Hierarchical Analysis of Time Series of Fractional Vegetation-Soil Nexuses in China's Hexi Corridor. *Landsc. Urban Plan.* **2021**, *215*, 104225. [CrossRef]
264. Yang, Y.; Wu, T.; Zeng, Y.; Wang, S. An Adaptive-Parameter Pixel Unmixing Method for Mapping Evergreen Forest Fractions Based on Time-Series NDVI: A Case Study of Southern China. *Remote Sens.* **2021**, *13*, 4678. [CrossRef]
265. Benhalouche, F.Z.; Benabbou, O.; Kebir, L.W.; Bennia, A.; Karoui, M.S.; Deville, Y. An Informed NMF-Based Unmixing Approach for Mineral Detection and Mapping in the Algerian Central Hoggar Using PRISMA Remote Sensing Hyperspectral Data. In Proceedings of the 2021 IEEE International Geoscience and Remote Sensing Symposium IGARSS, Brussels, Belgium, 11 July 2021; pp. 1863–1866.
266. Zhao, M.; Chen, J.; Rahardja, S. Hyperspectral Shadow Removal via Nonlinear Unmixing. *IEEE Geosci. Remote Sens. Lett.* **2021**, *18*, 881–885. [CrossRef]

267. Jin, Q.; Ma, Y.; Fan, F.; Huang, J.; Mei, X.; Ma, J. Adversarial Autoencoder Network for Hyperspectral Unmixing. *IEEE Trans. Neural Netw. Learn. Syst.* **2021**, 1–15. [CrossRef]
268. Han, Z.; Hong, D.; Gao, L.; Chanussot, J.; Zhang, B. EvoNAS: Evolvable Neural Architecture Search for Hyperspectral Unmixing. In Proceedings of the 2021 IEEE International Geoscience and Remote Sensing Symposium IGARSS, Brussels, Belgium, 11 July 2021; pp. 3325–3328.
269. Xu, F.; Somers, B. Unmixing-Based Sentinel-2 Downscaling for Urban Land Cover Mapping. *ISPRS J. Photogramm. Remote Sens.* **2021**, *171*, 133–154. [CrossRef]
270. Shen, M.; Tang, M.; Li, Y. Phenology and Spectral Unmixing-Based Invasive Kudzu Mapping: A Case Study in Knox County, Tennessee. *Remote Sens.* **2021**, *13*, 4551. [CrossRef]
271. Kneib, M.; Miles, E.S.; Jola, S.; Buri, P.; Herreid, S.; Bhattacharya, A.; Watson, C.S.; Bolch, T.; Quincey, D.; Pellicciotti, F. Mapping Ice Cliffs on Debris-Covered Glaciers Using Multispectral Satellite Images. *Remote Sens. Environ.* **2021**, *253*, 112201. [CrossRef]
272. Soydan, H.; Koz, A.; Düzgün, H.Ş. Secondary Iron Mineral Detection via Hyperspectral Unmixing Analysis with Sentinel-2 Imagery. *Int. J. Appl. Earth Obs. Geoinf.* **2021**, *101*, 102343. [CrossRef]
273. Mudereri, B.T.; Abdel-Rahman, E.M.; Dube, T.; Niassy, S.; Khan, Z.; Tonnang, H.E.Z.; Landmann, T. A Two-Step Approach for Detecting Striga in a Complex Agroecological System Using Sentinel-2 Data. *Sci. Total Environ.* **2021**, *762*, 143151. [CrossRef]
274. Yuan, N.; Gong, Y.; Fang, S.; Liu, Y.; Duan, B.; Yang, K.; Wu, X.; Zhu, R. UAV Remote Sensing Estimation of Rice Yield Based on Adaptive Spectral Endmembers and Bilinear Mixing Model. *Remote Sens.* **2021**, *13*, 2190. [CrossRef]
275. Sun, X.; Wu, W.; Li, X.; Xu, X.; Li, J. Vegetation Abundance and Health Mapping Over Southwestern Antarctica Based on WorldView-2 Data and a Modified Spectral Mixture Analysis. *Remote Sens.* **2021**, *13*, 166. [CrossRef]
276. Ma, X.; Lu, L.; Ding, J.; Zhang, F.; He, B. Estimating Fractional Vegetation Cover of Row Crops from High Spatial Resolution Image. *Remote Sens.* **2021**, *13*, 3874. [CrossRef]
277. Markiet, V.; Mõttus, M. Estimation of Boreal Forest Floor Reflectance from Airborne Hyperspectral Data of Coniferous Forests. *Remote Sens. Environ.* **2020**, *249*, 112018. [CrossRef]
278. Benhalouche, F.Z.; Benabbou, O.; Karoui, M.S., Kebir, L.W.; Deville, Y. Detecting and Mapping Kaolinite In The Algerian Central Hoggar With A Partial Linear Nmf-Based Unmixing Method. In Proceedings of the 2020 Mediterranean and Middle-East Geoscience and Remote Sensing Symposium (M2GARSS), Tunis, Tunisia, 11 March 2020; pp. 204–207.
279. Takodjou Wambo, J.D.; Pour, A.B.; Ganno, S.; Asimow, P.D.; Zoheir, B.; Salles, R.D.R.; Nzenti, J.P.; Pradhan, B.; Muslim, A.M. Identifying High Potential Zones of Gold Mineralization in a Sub-Tropical Region Using Landsat-8 and ASTER Remote Sensing Data: A Case Study of the Ngoura-Colomines Goldfield, Eastern Cameroon. *Ore. Geol. Rev.* **2020**, *122*, 103530. [CrossRef]
280. Salehi, S.; Mielke, C.; Rogass, C. Mapping Ultramafic Complexes Using Airborne Imaging Spectroscopy and Spaceborne Data in Arctic Regions with Abundant Lichen Cover, a Case Study from the Niaqornarssuit Complex in South West Greenland. *Eur. J. Remote Sens.* **2020**, *53*, 156–175. [CrossRef]
281. Bai, J.; Feng, R.; Wang, L.; Li, H.; Li, F.; Zhong, Y.; Zhang, L. Semi-Supervised Hyperspectral Unmixing with Very Deep Convolutional Neural Networks. In Proceedings of the IGARSS 2020—2020 IEEE International Geoscience and Remote Sensing Symposium, Waikoloa, HI, USA, 26 September 2020; pp. 2400–2403.
282. Borsoi, R.A.; Imbiriba, T.; Bermudez, J.C.M. A Data Dependent Multiscale Model for Hyperspectral Unmixing With Spectral Variability. *IEEE Trans. Image Process.* **2020**, *29*, 3638–3651. [CrossRef] [PubMed]
283. Borsoi, R.A.; Imbiriba, T.; Bermudez, J.C.M.; Richard, C. A Blind Multiscale Spatial Regularization Framework for Kernel-Based Spectral Unmixing. *IEEE Trans. Image Process.* **2020**, *29*, 4965–4979. [CrossRef] [PubMed]
284. Elkholy, M.M.; Mostafa, M.; Ebied, H.M.; Tolba, M.F. Hyperspectral Unmixing Using Deep Convolutional Autoencoder. *Int. J. Remote Sens.* **2020**, *41*, 4799–4819. [CrossRef]
285. Fang, B.; Bai, Y.; Li, Y. Combining Spectral Unmixing and 3D/2D Dense Networks with Early-Exiting Strategy for Hyperspectral Image Classification. *Remote Sens.* **2020**, *12*, 779. [CrossRef]
286. Fathy, G.M.; Hassan, H.A.; Rahwan, S.; Sheta, W.M. Parallel Implementation of Multiple Kernel Self-Organizing Maps for Spectral Unmixing. *J. Real-Time Image Proc.* **2020**, *17*, 1267–1284. [CrossRef]
287. Han, H.; Wang, G.; Wang, M.; Miao, J.; Guo, S.; Chen, L.; Zhang, M.; Guo, K. Hyperspectral Unmixing Via Nonconvex Sparse and Low-Rank Constraint. *IEEE J. Sel. Top. Appl. Earth Obs. Remote Sens.* **2020**, *13*, 5704–5718. [CrossRef]
288. Holland, W.; Du, Q. Adversarially Regularized Autoencoder for Hyperspectral Image Unmixing. In Proceedings of the Image and Signal Processing for Remote Sensing XXVI, SPIE, Online Only, 20 September 2020; p. 29.
289. Hua, Z.; Li, X.; Chen, S.; Zhao, L. Hyperspectral Unmixing with Scaled and Perturbed Linear Mixing Model to Address Spectral Variability. *J. Appl. Rem. Sens.* **2020**, *14*, 1. [CrossRef]
290. Karoui, M.S.; Djerriri, K.; Boukerch, I. Unsupervised Hyperspectral Band Selection by Sequentially Clustering A Mahalanobis-Based Dissimilarity Of Spectrally Variable Endmembers. In Proceedings of the 2020 Mediterranean and Middle-East Geoscience and Remote Sensing Symposium (M2GARSS), Tunis, Tunisia, 11 March 2020; pp. 33–36.
291. Zhou, Y.; Wetherley, E.B.; Gader, P.D. Unmixing Urban Hyperspectral Imagery Using Probability Distributions to Represent Endmember Variability. *Remote Sens. Environ.* **2020**, *246*, 111857. [CrossRef]
292. Vijayashekhar, S.; Bhatt, J.S.; Chattopadhyay, B. Virtual Dimensionality of Hyperspectral Data: Use of Multiple Hypothesis Testing for Controlling Type-I Error. *IEEE J. Sel. Top. Appl. Earth Obs. Remote Sens.* **2020**, *13*, 2974–2985.

293. Rathnayake, B.; Ekanayake, E.M.M.B.; Weerakoon, K.; Godaliyadda, G.M.R.I.; Ekanayake, M.P.B.; Herath, H.M.V.R. Graph-Based Blind Hyperspectral Unmixing via Nonnegative Matrix Factorization. *IEEE Trans. Geosci. Remote Sens.* **2020**, *58*, 6391–6409. [CrossRef]
294. Yang, B.; Chen, Z. An Improved Bilinear Mixture Model Considering Adjacency and Shade Effects. In Proceedings of the IGARSS 2020—2020 IEEE International Geoscience and Remote Sensing Symposium, Waikoloa, HI, USA, 26 September 2020; pp. 2161–2164.
295. Xu, M.; Zhang, Y.; Fan, Y.; Chen, Y.; Song, D. Linear Spectral Mixing Model-Guided Artificial Bee Colony Method for Endmember Generation. *IEEE Geosci. Remote Sens. Lett.* **2020**, *17*, 2145–2149. [CrossRef]
296. Xu, N.; Hu, Y.; Geng, X.; Wang, Y. A Geometric View of Fast Gram Determinant-Based Endmember Extraction Algorithm for Hyperspectral Imagery. In Proceedings of the IGARSS 2020—2020 IEEE International Geoscience and Remote Sensing Symposium, Waikoloa, HI, USA, 26 September 2020; pp. 2181–2184.
297. Peng, J.; Jiang, F.; Sun, W.; Zhou, Y. Cauchy NMF for Hyperspectral Unmixing. In Proceedings of the IGARSS 2020—2020 IEEE International Geoscience and Remote Sensing Symposium, Waikoloa, HI, USA, 26 September 2020; pp. 2384–2387.
298. Yang, J.; Jia, M.; Xu, C.; Li, S. Joint Hyperspectral Unmixing for Urban Computing. *Geoinformatica* **2020**, *24*, 247–265. [CrossRef]
299. Chen, S.; Cao, Y.; Chen, L.; Guo, X. Geometrical Constrained Independent Component Analysis for Hyperspectral Unmixing. *Int. J. Remote Sens.* **2020**, *41*, 6783–6804. [CrossRef]
300. Das, S.; Routray, A.; Deb, A.K. Efficient Tensor Decomposition Approach for Estimation of the Number of Endmembers in a Hyperspectral Image. *J. Appl. Rem. Sens.* **2020**, *14*, 1. [CrossRef]
301. Dou, Z.; Gao, K.; Zhang, X.; Wang, H.; Wang, J. Hyperspectral Unmixing Using Orthogonal Sparse Prior-Based Autoencoder With Hyper-Laplacian Loss and Data-Driven Outlier Detection. *IEEE Trans. Geosci. Remote Sens.* **2020**, *58*, 6550–6564. [CrossRef]
302. Huang, Y.; Li, J.; Qi, L.; Wang, Y.; Gao, X. Spatial-Spectral Autoencoder Networks for Hyperspectral Unmixing. In Proceedings of the IGARSS 2020—2020 IEEE International Geoscience and Remote Sensing Symposium, Waikoloa, HI, USA, 26 September 2020; pp. 2396–2399.
303. Imbiriba, T.; Borsoi, R.A.; Bermudez, J.C.M. Low-Rank Tensor Modeling for Hyperspectral Unmixing Accounting for Spectral Variability. *IEEE Trans. Geosci. Remote Sens.* **2020**, *58*, 1833–1842. [CrossRef]
304. Jiang, X.; Gong, M.; Zhan, T.; Zhang, M. Multiobjective Endmember Extraction Based on Bilinear Mixture Model. *IEEE Trans. Geosci. Remote Sens.* **2020**, *58*, 8192–8210. [CrossRef]
305. Li, H.-C.; Liu, S.; Feng, X.-R.; Zhang, S.-Q. Sparsity-Constrained Coupled Nonnegative Matrix–Tensor Factorization for Hyperspectral Unmixing. *IEEE J. Sel. Top. Appl. Earth Obs. Remote Sens.* **2020**, *13*, 5061–5073. [CrossRef]
306. Lu, X.; Dong, L.; Yuan, Y. Subspace Clustering Constrained Sparse NMF for Hyperspectral Unmixing. *IEEE Trans. Geosci. Remote Sens.* **2020**, *58*, 3007–3019. [CrossRef]
307. Mei, S.; He, M.; Zhang, Y.; Wang, Z.; Feng, D. Improving Spatial–Spectral Endmember Extraction in the Presence of Anomalous Ground Objects. *IEEE Trans. Geosci. Remote Sens.* **2011**, *49*, 4210–4222. [CrossRef]
308. Qi, L.; Li, J.; Wang, Y.; Huang, Y.; Gao, X. Spectral–Spatial-Weighted Multiview Collaborative Sparse Unmixing for Hyperspectral Images. *IEEE Trans. Geosci. Remote Sens.* **2020**, *58*, 8766–8779. [CrossRef]
309. Qian, Y.; Xiong, F.; Qian, Q.; Zhou, J. Spectral Mixture Model Inspired Network Architectures for Hyperspectral Unmixing. *IEEE Trans. Geosci. Remote Sens.* **2020**, *58*, 7418–7434. [CrossRef]
310. Zhou, L.; Zhang, X.; Wang, J.; Bai, X.; Tong, L.; Zhang, L.; Zhou, J.; Hancock, E. Subspace Structure Regularized Nonnegative Matrix Factorization for Hyperspectral Unmixing. *IEEE J. Sel. Top. Appl. Earth Obs. Remote Sens.* **2020**, *13*, 4257–4270. [CrossRef]
311. Tong, L.; Zhou, J.; Qian, B.; Yu, J.; Xiao, C. Adaptive Graph Regularized Multilayer Nonnegative Matrix Factorization for Hyperspectral Unmixing. *IEEE J. Sel. Top. Appl. Earth Obs. Remote Sens.* **2020**, *13*, 434–447. [CrossRef]
312. Qi, L.; Li, J.; Wang, Y.; Lei, M.; Gao, X. Deep Spectral Convolution Network for Hyperspectral Image Unmixing with Spectral Library. *Signal Process.* **2020**, *176*, 107672. [CrossRef]
313. Shah, D.; Zaveri, T.; Trivedi, Y.N.; Plaza, A. Entropy-Based Convex Set Optimization for Spatial–Spectral Endmember Extraction from Hyperspectral Images. *IEEE J. Sel. Top. Appl. Earth Obs. Remote Sens.* **2020**, *13*, 4200–4213. [CrossRef]
314. Yuan, Y.; Zhang, Z.; Wang, Q. Improved Collaborative Non-Negative Matrix Factorization and Total Variation for Hyperspectral Unmixing. *IEEE J. Sel. Top. Appl. Earth Obs. Remote Sens.* **2020**, *13*, 998–1010. [CrossRef]
315. Tao, X.; Cui, T.; Plaza, A.; Ren, P. Simultaneously Counting and Extracting Endmembers in a Hyperspectral Image Based on Divergent Subsets. *IEEE Trans. Geosci. Remote Sens.* **2020**, *58*, 8952–8966. [CrossRef]
316. Xu, X.; Li, J.; Li, S.; Plaza, A. Generalized Morphological Component Analysis for Hyperspectral Unmixing. *IEEE Trans. Geosci. Remote Sens.* **2020**, *58*, 2817–2832. [CrossRef]
317. Zeng; Ritz; Zhao; Lan Attention-Based Residual Network with Scattering Transform Features for Hyperspectral Unmixing with Limited Training Samples. *Remote Sens.* **2020**, *12*, 400. [CrossRef]
318. Xu, X.; Li, J.; Li, S.; Plaza, A. Curvelet Transform Domain-Based Sparse Nonnegative Matrix Factorization for Hyperspectral Unmixing. *IEEE J. Sel. Top. Appl. Earth Obs. Remote Sens.* **2020**, *13*, 4908–4924. [CrossRef]
319. Siebels, K.; Goita, K.; Germain, M. Estimation of Mineral Abundance from Hyperspectral Data Using a New Supervised Neighbor-Band Ratio Unmixing Approach. *IEEE Trans. Geosci. Remote Sens.* **2020**, *58*, 6754–6766. [CrossRef]
320. Rasti, B.; Koirala, B.; Scheunders, P.; Ghamisi, P. How Hyperspectral Image Unmixing and Denoising Can Boost Each Other. *Remote Sens.* **2020**, *12*, 1728. [CrossRef]

321. Qu, K.; Bao, W. Multiple-*Priors* Ensemble Constrained Nonnegative Matrix Factorization for Spectral Unmixing. *IEEE J. Sel. Top. Appl. Earth Obs. Remote Sens.* **2020**, *13*, 963–975. [CrossRef]
322. Wang, W.; Qian, Y.; Liu, H. Multiple Clustering Guided Nonnegative Matrix Factorization for Hyperspectral Unmixing. *IEEE J. Sel. Top. Appl. Earth Obs. Remote Sens.* **2020**, *13*, 5162–5179. [CrossRef]
323. Xiong, F.; Zhou, J.; Lu, J.; Qian, Y. Nonconvex Nonseparable Sparse Nonnegative Matrix Factorization for Hyperspectral Unmixing. *IEEE J. Sel. Top. Appl. Earth Obs. Remote Sens.* **2020**, *13*, 6088–6100. [CrossRef]
324. Zhou, X.; Zhang, Y.; Zhang, J.; Shi, S. Alternating Direction Iterative Nonnegative Matrix Factorization Unmixing for Multispectral and Hyperspectral Data Fusion. *IEEE J. Sel. Top. Appl. Earth Obs. Remote Sens.* **2020**, *13*, 5223–5232. [CrossRef]
325. Uezato, T.; Yokoya, N.; He, W. Illumination Invariant Hyperspectral Image Unmixing Based on a Digital Surface Model. *IEEE Trans. Image Process.* **2020**, *29*, 3652–3664. [CrossRef] [PubMed]
326. Zhang, J.; Zhang, X.; Tang, X.; Chen, P.; Jiao, L. Sketch-Based Region Adaptive Sparse Unmixing Applied to Hyperspectral Image. *IEEE Trans. Geosci. Remote Sens.* **2020**, *58*, 8840–8856. [CrossRef]
327. Yang, B.; Chen, Z.; Wang, B. Nonlinear Endmember Identification for Hyperspectral Imagery via Hyperpath-Based Simplex Growing and Fuzzy Assessment. *IEEE J. Sel. Top. Appl. Earth Obs. Remote Sens.* **2020**, *13*, 351–366. [CrossRef]
328. Kompella, S.S.; Kadapala, B.K.R.; Abdul Hakeem, K.; Issac, A.M.; Annamalai, L. Accuracy Assessment and Normalisation of Water Spread Area Estimate from Multi-Sensor Satellite Data. *J. Indian Soc. Remote Sens.* **2020**, *48*, 1601–1611. [CrossRef]
329. Drumetz, L.; Chanussot, J.; Jutten, C.; Ma, W.-K.; Iwasaki, A. Spectral Variability Aware Blind Hyperspectral Image Unmixing Based on Convex Geometry. *IEEE Trans. Image Process.* **2020**, *29*, 4568–4582. [CrossRef]
330. Cooper, S.; Okujeni, A.; Jänicke, C.; Clark, M.; Van Der Linden, S.; Hostert, P. Disentangling Fractional Vegetation Cover: Regression-Based Unmixing of Simulated Spaceborne Imaging Spectroscopy Data. *Remote Sens. Environ.* **2020**, *246*, 111856. [CrossRef]
331. Sun, Q.; Zhang, P.; Wei, H.; Liu, A.; You, S.; Sun, D. Improved Mapping and Understanding of Desert Vegetation-Habitat Complexes from Intraannual Series of Spectral Endmember Space Using Cross-Wavelet Transform and Logistic Regression. *Remote Sens. Environ.* **2020**, *236*, 111516. [CrossRef]
332. Liu, D.; Chen, W.; Menz, G.; Dubovyk, O. Development of Integrated Wetland Change Detection Approach: In Case of Erdos Larus Relictus National Nature Reserve, China. *Sci. Total Environ.* **2020**, *731*, 139166. [CrossRef]
333. Ji, C.; Li, X.; Wei, H.; Li, S. Comparison of Different Multispectral Sensors for Photosynthetic and Non-Photosynthetic Vegetation-Fraction Retrieval. *Remote Sens.* **2020**, *12*, 115. [CrossRef]
334. Aldeghlawi, M.; Alkhatib, M.Q.; Velez-Reyes, M. Evaluating Column Subset Selection Methods for Endmember Extraction in Hyperspectral Unmixing. In Proceedings of the Algorithms, Technologies, and Applications for Multispectral and Hyperspectral Imagery XXVI, SPIE, Online Only, 9 June 2020; p. 46.
335. Zhu, F.; Honeine, P.; Chen, J. Pixel-Wise Linear/Nonlinear Nonnegative Matrix Factorization for Unmixing of Hyperspectral Data. In Proceedings of the ICASSP 2020—2020 IEEE International Conference on Acoustics, Speech and Signal Processing (ICASSP), Barcelona, Spain, 4–8 May 2020; pp. 4737–4741.
336. Moghadam, H.J.; Oskouei, M.M.; Nouri, T. Unmixing of Hyperspectral Data for Mineral Detection Using a Hybrid Method, Sar Chah-e Shur, Iran. *Arab. J. Geosci.* **2020**, *13*, 1041. [CrossRef]
337. Zhang, G.; Cerra, D.; Müller, R. Shadow Detection and Restoration for Hyperspectral Images Based on Nonlinear Spectral Unmixing. *Remote Sens.* **2020**, *12*, 3985. [CrossRef]
338. Lyu, X.; Li, X.; Dang, D.; Dou, H.; Xuan, X.; Liu, S.; Li, M.; Gong, J. A New Method for Grassland Degradation Monitoring by Vegetation Species Composition Using Hyperspectral Remote Sensing. *Ecol. Indic.* **2020**, *114*, 106310. [CrossRef]
339. Montorio, R.; Pérez-Cabello, F.; Borini Alves, D.; García-Martín, A. Unitemporal Approach to Fire Severity Mapping Using Multispectral Synthetic Databases and Random Forests. *Remote Sens. Environ.* **2020**, *249*, 112025. [CrossRef]
340. Aalstad, K.; Westermann, S.; Bertino, L. Evaluating Satellite Retrieved Fractional Snow-Covered Area at a High-Arctic Site Using Terrestrial Photography. *Remote Sens. Environ.* **2020**, *239*, 111618. [CrossRef]
341. Binh, D.V.; Wietlisbach, B.; Kantoush, S.; Loc, H.H.; Park, E.; Cesare, G.D.; Cuong, D.H.; Tung, N.X.; Sumi, T. A Novel Method for River Bank Detection from Landsat Satellite Data: A Case Study in the Vietnamese Mekong Delta. *Remote Sens.* **2020**, *12*, 3298. [CrossRef]
342. Fernández-Guisuraga, J.M.; Calvo, L.; Suárez-Seoane, S. Comparison of Pixel Unmixing Models in the Evaluation of Post-Fire Forest Resilience Based on Temporal Series of Satellite Imagery at Moderate and Very High Spatial Resolution. *ISPRS J. Photogramm. Remote Sens.* **2020**, *164*, 217–228. [CrossRef]
343. Laamrani, A.; Joosse, P.; McNairn, H.; Berg, A.; Hagerman, J.; Powell, K.; Berry, M. Assessing Soil Cover Levels during the Non-Growing Season Using Multitemporal Satellite Imagery and Spectral Unmixing Techniques. *Remote Sens.* **2020**, *12*, 1397. [CrossRef]
344. Trinder, J.; Liu, Q. Assessing Environmental Impacts of Urban Growth Using Remote Sensing. *Geo-Spat. Inf. Sci.* **2020**, *23*, 20–39. [CrossRef]
345. Senf, C.; Laštovička, J.; Okujeni, A.; Heurich, M.; Van Der Linden, S. A Generalized Regression-Based Unmixing Model for Mapping Forest Cover Fractions throughout Three Decades of Landsat Data. *Remote Sens. Environ.* **2020**, *240*, 111691. [CrossRef]
346. Wang, Q.; Zhang, C.; Tong, X.; Atkinson, P.M. General Solution to Reduce the Point Spread Function Effect in Subpixel Mapping. *Remote Sens. Environ.* **2020**, *251*, 112054. [CrossRef]

347. Peroni Venancio, L.; Chartuni Mantovani, E.; Do Amaral, C.H.; Usher Neale, C.M.; Zution Gonçalves, I.; Filgueiras, R.; Coelho Eugenio, F. Potential of Using Spectral Vegetation Indices for Corn Green Biomass Estimation Based on Their Relationship with the Photosynthetic Vegetation Sub-Pixel Fraction. *Agric. Water Manag.* **2020**, *236*, 106155. [CrossRef]
348. Lymburner, L.; Bunting, P.; Lucas, R.; Scarth, P.; Alam, I.; Phillips, C.; Ticehurst, C.; Held, A. Mapping the Multi-Decadal Mangrove Dynamics of the Australian Coastline. *Remote Sens. Environ.* **2020**, *238*, 111185. [CrossRef]
349. Bullock, E.L.; Woodcock, C.E.; Olofsson, P. Monitoring Tropical Forest Degradation Using Spectral Unmixing and Landsat Time Series Analysis. *Remote Sens. Environ.* **2020**, *238*, 110968. [CrossRef]
350. Czekajlo, A.; Coops, N.C.; Wulder, M.A.; Hermosilla, T.; Lu, Y.; White, J.C.; Van Den Bosch, M. The Urban Greenness Score: A Satellite-Based Metric for Multi-Decadal Characterization of Urban Land Dynamics. *Int. J. Appl. Earth Obs. Geoinf.* **2020**, *93*, 102210. [CrossRef]
351. Dai, J.; Roberts, D.A.; Stow, D.A.; An, L.; Hall, S.J.; Yabiku, S.T.; Kyriakidis, P.C. Mapping Understory Invasive Plant Species with Field and Remotely Sensed Data in Chitwan, Nepal. *Remote Sens. Environ.* **2020**, *250*, 112037. [CrossRef]
352. Khan, I.A.; Khan, M.R.; Baig, M.H.A.; Hussain, Z.; Hameed, N.; Khan, J.A. Assessment of Forest Cover and Carbon Stock Changes in Sub-Tropical Pine Forest of Azad Jammu & Kashmir (AJK), Pakistan Using Multi-Temporal Landsat Satellite Data and Field Inventory. *PLoS ONE* **2020**, *15*, e0226341. [CrossRef]
353. Shimabukuro, Y.E.; Dutra, A.C.; Arai, E.; Duarte, V.; Cassol, H.L.G.; Pereira, G.; Cardozo, F.D.S. Mapping Burned Areas of Mato Grosso State Brazilian Amazon Using Multisensor Datasets. *Remote Sens.* **2020**, *12*, 3827. [CrossRef]
354. Shih, H.; Stow, D.A.; Tsai, Y.; Roberts, D.A. Estimating the Starting Time and Identifying the Type of Urbanization Based on Dense Time Series of Landsat-Derived Vegetation-Impervious-Soil (V-I-S) Maps—A Case Study of North Taiwan from 1990 to 2015. *Int. J. Appl. Earth Obs. Geoinf.* **2020**, *85*, 101987. [CrossRef]
355. Yin, C.L.; Meng, F.; Xu, Y.N.; Yang, X.Y.; Xing, H.Q.; Fu, P.J. Developing Urban Built-up Area Extraction Method Based on Land Surface Emissivity Differences. *Infrared Phys. Technol.* **2020**, *110*, 103475. [CrossRef]
356. He, Y.; Yang, J.; Guo, X. Green Vegetation Cover Dynamics in a Heterogeneous Grassland: Spectral Unmixing of Landsat Time Series from 1999 to 2014. *Remote Sens.* **2020**, *12*, 3826. [CrossRef]
357. Thayn, J.B. Monitoring Narrow Mangrove Stands in Baja California Sur, Mexico Using Linear Spectral Unmixing. *Mar. Geod.* **2020**, *43*, 493–508. [CrossRef]
358. Jarchow, C.J.; Sigafus, B.H.; Muths, E.; Hossack, B.R. Using Full and Partial Unmixing Algorithms to Estimate the Inundation Extent of Small, Isolated Stock Ponds in an Arid Landscape. *Wetlands* **2020**, *40*, 563–575. [CrossRef]
359. Lewińska, K.E.; Hostert, P.; Buchner, J.; Bleyhl, B.; Radeloff, V.C. Short-Term Vegetation Loss versus Decadal Degradation of Grasslands in the Caucasus Based on Cumulative Endmember Fractions. *Remote Sens. Environ.* **2020**, *248*, 111969. [CrossRef]
360. Li, W. Mapping Urban Impervious Surfaces by Using Spectral Mixture Analysis and Spectral Indices. *Remote Sens.* **2019**, *12*, 94. [CrossRef]
361. Cavalli, R.M. Local, Daily, and Total Bio-Optical Models of Coastal Waters of Manfredonia Gulf Applied to Simulated Data of CHRIS, Landsat TM, MIVIS, MODIS, and PRISMA Sensors for Evaluating the Error. *Remote Sens.* **2020**, *12*, 1428. [CrossRef]
362. Wright, N.C.; Polashenski, C.M. How Machine Learning and High-Resolution Imagery Can Improve Melt Pond Retrieval from MODIS Over Current Spectral Unmixing Techniques. *J. Geophys. Res. Ocean.* **2020**, *125*. [CrossRef]
363. Singh, K.K.; Gray, J. Mapping Understory Invasive Plants in Urban Forests with Spectral and Temporal Unmixing of Landsat Imagery. *Photogramm. Eng. Remote Sens.* **2020**, *86*, 509–518. [CrossRef]
364. Firozjaei, M.K.; Weng, Q.; Zhao, C.; Kiavarz, M.; Lu, L.; Alavipanah, S.K. Surface Anthropogenic Heat Islands in Six Megacities: An Assessment Based on a Triple-Source Surface Energy Balance Model. *Remote Sens. Environ.* **2020**, *242*, 111751. [CrossRef]
365. Ling, F.; Li, X.; Foody, G.M.; Boyd, D.; Ge, Y.; Li, X.; Du, Y. Monitoring Surface Water Area Variations of Reservoirs Using Daily MODIS Images by Exploring Sub-Pixel Information. *ISPRS J. Photogramm. Remote Sens.* **2020**, *168*, 141–152. [CrossRef]
366. Wang, J.; Yang, D.; Detto, M.; Nelson, B.W.; Chen, M.; Guan, K.; Wu, S.; Yan, Z.; Wu, J. Multi-Scale Integration of Satellite Remote Sensing Improves Characterization of Dry-Season Green-up in an Amazon Tropical Evergreen Forest. *Remote Sens. Environ.* **2020**, *246*, 111865. [CrossRef]
367. PROBA-V. Available online: https://earth.esa.int/eogateway/missions/proba-v (accessed on 15 May 2023).
368. Arai, E.; Eyji Sano, E.; Dutra, A.C.; Cassol, H.L.G.; Hoffmann, T.B.; Shimabukuro, Y.E. Vegetation Fraction Images Derived from PROBA-V Data for Rapid Assessment of Annual Croplands in Brazil. *Remote Sens.* **2020**, *12*, 1152. [CrossRef]
369. Godinho Cassol, H.L.; Arai, E.; Eyji Sano, E.; Dutra, A.C.; Hoffmann, T.B.; Shimabukuro, Y.E. Maximum Fraction Images Derived from Year-Based Project for On-Board Autonomy-Vegetation (PROBA-V) Data for the Rapid Assessment of Land Use and Land Cover Areas in Mato Grosso State, Brazil. *Land* **2020**, *9*, 139. [CrossRef]
370. Shimabukuro, Y.E.; Arai, E.; Duarte, V.; Dutra, A.C.; Cassol, H.L.G.; Sano, E.E.; Hoffmann, T.B. Discriminating Land Use and Land Cover Classes in Brazil Based on the Annual PROBA-V 100 m Time Series. *IEEE J. Sel. Top. Appl. Earth Obs. Remote Sens.* **2020**, *13*, 3409–3420. [CrossRef]
371. Redowan, M.; Phinn, S.; Roelfsema, C.; Aziz, A.A. CLASlite Unmixing of Landsat Images to Estimate REDD+ Activity Data for Deforestation in a Bangladesh Forest. *J. Appl. Rem. Sens.* **2020**, *14*, 1. [CrossRef]
372. Patel, J.R.; Joshi, M.V.; Bhatt, J.S. A Novel Approach for Hyperspectral Image Superresolution Using Spectral Unmixing and Transfer Learning. In Proceedings of the IGARSS 2020—2020 IEEE International Geoscience and Remote Sensing Symposium, Waikoloa, HI, USA, 26 September 2020; pp. 1512–1515.

373. Wang, K.; Wang, Y.; Zhao, X.-L.; Chan, J.C.-W.; Xu, Z.; Meng, D. Hyperspectral and Multispectral Image Fusion via Nonlocal Low-Rank Tensor Decomposition and Spectral Unmixing. *IEEE Trans. Geosci. Remote Sens.* **2020**, *58*, 7654–7671. [CrossRef]
374. Yang, L.; Peng, J.; Su, H.; Xu, L.; Wang, Y.; Yu, B. Combined Nonlocal Spatial Information and Spatial Group Sparsity in NMF for Hyperspectral Unmixing. *IEEE Geosci. Remote Sens. Lett.* **2020**, *17*, 1767–1771. [CrossRef]
375. Wang, L.; Zhu, Q.; Zeng, W.; Zhong, Y.; Guan, Q.; Zhang, L.; Li, D. Semi-Automatic Fully Sparse Semantic Modeling Framework for Hyperspectral Unmixing. In Proceedings of the IGARSS 2020—2020 IEEE International Geoscience and Remote Sensing Symposium, Waikoloa, HI, USA, 26 September 2020; pp. 2388–2391.
376. Yue, J.; Tian, Q.; Dong, X.; Xu, N. Using Broadband Crop Residue Angle Index to Estimate the Fractional Cover of Vegetation, Crop Residue, and Bare Soil in Cropland Systems. *Remote Sens. Environ.* **2020**, *237*, 111538. [CrossRef]
377. Carlson, B.Z.; Hébert, M.; Van Reeth, C.; Bison, M.; Laigle, I.; Delestrade, A. Monitoring the Seasonal Hydrology of Alpine Wetlands in Response to Snow Cover Dynamics and Summer Climate: A Novel Approach with Sentinel-2. *Remote Sens.* **2020**, *12*, 1959. [CrossRef]
378. Fraga, R.S.; Guedes, H.A.S.; Martins, V.S.; Caballero, C.B.; Mendes, K.G.P.; Monks, J.L.F.; Fassoni-Andrade, A.C. Empirical Modelling of Suspended Solids in a Subtropical Lagoon (Brazil) Using Linear Spectral Mixing Algorithm. *Remote Sens. Appl. Soc. Environ.* **2020**, *20*, 100380. [CrossRef]
379. Girolamo-Neto, C.D.; Sato, L.Y.; Sanches, I.D.; Silva, I.C.O.; Rocha, J.C.S.; Almeida, C.A. Object based image analysis and texture features for pasture classification in brazilian savannah. *ISPRS Ann. Photogramm. Remote Sens. Spat. Inf. Sci.* **2020**, *V-3-2020*, 453–460. [CrossRef]
380. Huechacona-Ruiz, A.H.; Dupuy, J.M.; Schwartz, N.B.; Powers, J.S.; Reyes-García, C.; Tun-Dzul, F.; Hernández-Stefanoni, J.L. Mapping Tree Species Deciduousness of Tropical Dry Forests Combining Reflectance, Spectral Unmixing, and Texture Data from High-Resolution Imagery. *Forests* **2020**, *11*, 1234. [CrossRef]
381. Quintano, C.; Fernández-Manso, A.; Roberts, D.A. Enhanced Burn Severity Estimation Using Fine Resolution ET and MESMA Fraction Images with Machine Learning Algorithm. *Remote Sens. Environ.* **2020**, *244*, 111815. [CrossRef]
382. Topouzelis, K.; Papageorgiou, D.; Karagaitanakis, A.; Papakonstantinou, A.; Ballesteros, M.A. Plastic Litter Project 2019: Exploring the Detection of Floating Plastic Litter Using Drones and Sentinel 2 Satellite Images. In Proceedings of the IGARSS 2020—2020 IEEE International Geoscience and Remote Sensing Symposium, Waikoloa, HI, USA, 26 September 2020; pp. 6329–6332.
383. Topouzelis, K.; Papageorgiou, D.; Karagaitanakis, A.; Papakonstantinou, A.; Arias Ballesteros, M. Remote Sensing of Sea Surface Artificial Floating Plastic Targets with Sentinel-2 and Unmanned Aerial Systems (Plastic Litter Project 2019). *Remote Sens.* **2020**, *12*, 2013. [CrossRef]
384. Zhang, Y.; Wu, L.; Ren, H.; Deng, L.; Zhang, P. Retrieval of Water Quality Parameters from Hyperspectral Images Using Hybrid Bayesian Probabilistic Neural Network. *Remote Sens.* **2020**, *12*, 1567. [CrossRef]
385. Salvatore, M.R.; Borges, S.R.; Barrett, J.E.; Sokol, E.R.; Stanish, L.F.; Power, S.N.; Morin, P. Remote Characterization of Photosynthetic Communities in the Fryxell Basin of Taylor Valley, Antarctica. *Antarct. Sci.* **2020**, *32*, 255–270. [CrossRef]
386. Bartholomeus, H.; Kooistra, L.; Stevens, A.; Van Leeuwen, M.; Van Wesemael, B.; Ben-Dor, E.; Tychon, B. Soil Organic Carbon Mapping of Partially Vegetated Agricultural Fields with Imaging Spectroscopy. *Int. J. Appl. Earth Obs. Geoinf.* **2011**, *13*, 81–88. [CrossRef]
387. Ghrefat, H.A.; Goodell, P.C. Land Cover Mapping at Alkali Flat and Lake Lucero, White Sands, New Mexico, USA Using Multi-Temporal and Multi-Spectral Remote Sensing Data. *Int. J. Appl. Earth Obs. Geoinf.* **2011**, *13*, 616–625. [CrossRef]
388. Hosseinjani, M.; Tangestani, M.H. Mapping Alteration Minerals Using Sub-Pixel Unmixing of ASTER Data in the Sarduiyeh Area, SE Kerman, Iran. *Int. J. Digit. Earth* **2011**, *4*, 487–504. [CrossRef]
389. Vicente, L.E.; De Souza Filho, C.R. Identification of Mineral Components in Tropical Soils Using Reflectance Spectroscopy and Advanced Spaceborne Thermal Emission and Reflection Radiometer (ASTER) Data. *Remote Sens. Environ.* **2011**, *115*, 1824–1836. [CrossRef]
390. Hu, X.; Weng, Q. Estimating Impervious Surfaces from Medium Spatial Resolution Imagery: A Comparison between Fuzzy Classification and LSMA. *Int. J. Remote Sens.* **2011**, *32*, 5645–5663. [CrossRef]
391. Weng, Q.; Rajasekar, U.; Hu, X. Modeling Urban Heat Islands and Their Relationship with Impervious Surface and Vegetation Abundance by Using ASTER Images. *IEEE Trans. Geosci. Remote Sens.* **2011**, *49*, 4080–4089. [CrossRef]
392. Castrodad, A.; Xing, Z.; Greer, J.B.; Bosch, E.; Carin, L.; Sapiro, G. Learning Discriminative Sparse Representations for Modeling, Source Separation, and Mapping of Hyperspectral Imagery. *IEEE Trans. Geosci. Remote Sens.* **2011**, *49*, 4263–4281. [CrossRef]
393. Dopido, I.; Zortea, M.; Villa, A.; Plaza, A.; Gamba, P. Unmixing Prior to Supervised Classification of Remotely Sensed Hyperspectral Images. *IEEE Geosci. Remote Sens. Lett.* **2011**, *8*, 760–764. [CrossRef]
394. Halimi, A.; Altmann, Y.; Dobigeon, N.; Tourneret, J.-Y. Nonlinear Unmixing of Hyperspectral Images Using a Generalized Bilinear Model. *IEEE Trans. Geosci. Remote Sens.* **2011**, *49*, 4153–4162. [CrossRef]
395. Heylen, R.; Burazerovic, D.; Scheunders, P. Fully Constrained Least Squares Spectral Unmixing by Simplex Projection. *IEEE Trans. Geosci. Remote Sens.* **2011**, *49*, 4112–4122. [CrossRef]
396. Heylen, R.; Burazerovic, D.; Scheunders, P. Non-Linear Spectral Unmixing by Geodesic Simplex Volume Maximization. *IEEE J. Sel. Top. Signal Process.* **2011**, *5*, 534–542. [CrossRef]
397. Heylen, R.; Scheunders, P. Non-Linear Fully-Constrained Spectral Unmixing. In Proceedings of the 2011 IEEE International Geoscience and Remote Sensing Symposium, Vancouver, BC, Canada, 24–29 July 2011; pp. 1295–1298.

398. Iordache, M.-D.; Bioucas-Dias, J.M.; Plaza, A. Hyperspectral Unmixingwith Sparse Group Lasso. In Proceedings of the 2011 IEEE International Geoscience and Remote Sensing Symposium, Vancouver, BC, Canada, 24–29 July 2011; pp. 3586–3589.
399. Liu, X.; Xia, W.; Wang, B.; Zhang, L. An Approach Based on Constrained Nonnegative Matrix Factorization to Unmix Hyperspectral Data. *IEEE Trans. Geosci. Remote Sens.* **2011**, *49*, 757–772. [CrossRef]
400. Mianji, F.A.; Zhou, S.; Zhang, Y. Hyperspectral Unmixing Using a Novel Conversion Model. In Proceedings of the 2011 IEEE International Geoscience and Remote Sensing Symposium, Vancouver, BC, Canada, 24–29 July 2011; pp. 2527–2530.
401. Swatantran, A.; Dubayah, R.; Roberts, D.; Hofton, M.; Blair, J.B. Mapping Biomass and Stress in the Sierra Nevada Using Lidar and Hyperspectral Data Fusion. *Remote Sens. Environ.* **2011**, *115*, 2917–2930. [CrossRef]
402. Xia, W.; Wang, B.; Zhang, L.; Lu, Q. Simplex Volume Analysis Based on Triangular Factorization: A Framework for Hyperspectral Unmixing. In Proceedings of the 2011 IEEE International Geoscience and Remote Sensing Symposium, Vancouver, BC, Canada, 24–29 July 2011; pp. 1147–1150.
403. Zare, A. Spatial-Spectral Unmixing Using Fuzzy Local Information. In Proceedings of the 2011 IEEE International Geoscience and Remote Sensing Symposium, Vancouver, BC, Canada, 24–29 July 2011; pp. 1139–1142.
404. Altmann, Y.; Dobigeon, N.; Tourneret, J.-Y.; McLaughlin, S. Nonlinear Unmixing of Hyperspectral Images Using Radial Basis Functions and Orthogonal Least Squares. In Proceedings of the 2011 IEEE International Geoscience and Remote Sensing Symposium, Vancouver, BC, Canada, 24–29 July 2011; pp. 1151–1154.
405. Ambikapathi, A.; Chan, T.-H.; Ma, W.-K.; Chi, C.-Y. Chance-Constrained Robust Minimum-Volume Enclosing Simplex Algorithm for Hyperspectral Unmixing. *IEEE Trans. Geosci. Remote Sens.* **2011**, *49*, 4194–4209. [CrossRef]
406. Canham, K.; Schlamm, A.; Ziemann, A.; Basener, B.; Messinger, D. Spatially Adaptive Hyperspectral Unmixing. *IEEE Trans. Geosci. Remote Sens.* **2011**, *49*, 4248–4262. [CrossRef]
407. Eches, O.; Dobigeon, N.; Tourneret, J.-Y. Enhancing Hyperspectral Image Unmixing With Spatial Correlations. *IEEE Trans. Geosci. Remote Sens.* **2011**, *49*, 4239–4247. [CrossRef]
408. Halimi, A.; Altmann, Y.; Dobigeon, N.; Tourneret, J.-Y. Unmixing Hyperspectral Images Using the Generalized Bilinear Model. In Proceedings of the 2011 IEEE International Geoscience and Remote Sensing Symposium, Vancouver, BC, Canada, 24–29 July 2011; pp. 1886–1889.
409. Iordache, M.-D.; Bioucas-Dias, J.M.; Plaza, A. Sparse Unmixing of Hyperspectral Data. *IEEE Trans. Geosci. Remote Sens.* **2011**, *49*, 2014–2039. [CrossRef]
410. Martin, G.; Plaza, A. Region-Based Spatial Preprocessing for Endmember Extraction and Spectral Unmixing. *IEEE Geosci. Remote Sens. Lett.* **2011**, *8*, 745–749. [CrossRef]
411. Martin, G.; Plaza, A.; Zortea, M. Noise-Robust Spatial Preprocessing Prior to Endmember Extraction from Hyperspectral Data. In Proceedings of the 2011 IEEE International Geoscience and Remote Sensing Symposium, Vancouver, BC, Canada, 24–29 July 2011; pp. 1287–1290.
412. Mei, S.; He, M. Minimum Endmember-Wise Distance Constrained Nonnegative Matrix Factorization for Spectral Mixture Analysis of Hyperspectral Images. In Proceedings of the 2011 IEEE International Geoscience and Remote Sensing Symposium, Vancouver, BC, Canada, 24–29 July 2011; pp. 1299–1302.
413. Villa, A.; Chanussot, J.; Benediktsson, J.A.; Jutten, C. Spectral Unmixing for the Classification of Hyperspectral Images at a Finer Spatial Resolution. *IEEE J. Sel. Top. Signal Process.* **2011**, *5*, 521–533. [CrossRef]
414. Xia, W.; Liu, X.; Wang, B.; Zhang, L. Independent Component Analysis for Blind Unmixing of Hyperspectral Imagery with Additional Constraints. *IEEE Trans. Geosci. Remote Sens.* **2011**, *49*, 2165–2179. [CrossRef]
415. Zuyuan Yang; Guoxu Zhou; Shengli Xie; Shuxue Ding; Jun-Mei Yang; Jun Zhang Blind Spectral Unmixing Based on Sparse Nonnegative Matrix Factorization. *IEEE Trans. Image Process.* **2011**, *20*, 1112–1125. [CrossRef]
416. Zhang, B.; Sun, X.; Gao, L.; Yang, L. Endmember Extraction of Hyperspectral Remote Sensing Images Based on the Ant Colony Optimization (ACO) Algorithm. *IEEE Trans. Geosci. Remote Sens.* **2011**, *49*, 2635–2646. [CrossRef]
417. Zhao, Y.; Yang, J.; Zhang, Q.; Song, L.; Cheng, Y.; Pan, Q. Hyperspectral Imagery Super-Resolution by Sparse Representation and Spectral Regularization. *EURASIP J. Adv. Signal Process.* **2011**, *2011*, 87. [CrossRef]
418. Kamal, M.; Phinn, S. Hyperspectral Data for Mangrove Species Mapping: A Comparison of Pixel-Based and Object-Based Approach. *Remote Sens.* **2011**, *3*, 2222–2242. [CrossRef]
419. Zurita-Milla, R.; Gomez-Chova, L.; Guanter, L.; Clevers, J.G.P.W.; Camps-Valls, G. Multitemporal Unmixing of Medium-Spatial-Resolution Satellite Images: A Case Study Using MERIS Images for Land-Cover Mapping. *IEEE Trans. Geosci. Remote Sens.* **2011**, *49*, 4308–4317. [CrossRef]
420. Bouaziz, M.; Matschullat, J.; Gloaguen, R. Improved Remote Sensing Detection of Soil Salinity from a Semi-Arid Climate in Northeast Brazil. *Comptes Rendus Geosci.* **2011**, *343*, 795–803. [CrossRef]
421. Cui, Q.; Shi, J.; Xu, Y. Estimation of Sub-Pixel Water Area on Tibet Plateau Using Multiple Endmembers Spectral Mixture Spectral Analysis from MODIS Data. In Proceedings of the MIPPR 2011: Remote Sensing Image Processing, Geographic Information Systems, and Other Applications, Guilin, China, 20 November 2011; p. 80061T.
422. Knight, J.; Voth, M. Mapping Impervious Cover Using Multi-Temporal MODIS NDVI Data. *IEEE J. Sel. Top. Appl. Earth Obs. Remote Sens.* **2011**, *4*, 303–309. [CrossRef]
423. Lu, D.; Batistella, M.; Moran, E.; Hetrick, S.; Alves, D.; Brondizio, E. Fractional Forest Cover Mapping in the Brazilian Amazon with a Combination of MODIS and TM Images. *Int. J. Remote Sens.* **2011**, *32*, 7131–7149. [CrossRef]

424. Plemmons, R.J. Dimensionality Reduction, Classification, and Spectral Mixture Analysis Using Non-Negative Underapproximation. *Opt. Eng.* **2011**, *50*, 027001. [CrossRef]
425. Qian, Y.; Jia, S.; Zhou, J.; Robles-Kelly, A. Hyperspectral Unmixing via $L_{1/2}$ Sparsity-Constrained Nonnegative Matrix Factorization. *IEEE Trans. Geosci. Remote Sens.* **2011**, *49*, 4282–4297. [CrossRef]
426. Youngentob, K.N.; Roberts, D.A.; Held, A.A.; Dennison, P.E.; Jia, X.; Lindenmayer, D.B. Mapping Two Eucalyptus Subgenera Using Multiple Endmember Spectral Mixture Analysis and Continuum-Removed Imaging Spectrometry Data. *Remote Sens. Environ.* **2011**, *115*, 1115–1128. [CrossRef]
427. De Jong, S.M.; Addink, E.A.; Van Beek, L.P.H.; Duijsings, D. Physical Characterization, Spectral Response and Remotely Sensed Mapping of Mediterranean Soil Surface Crusts. *Catena* **2011**, *86*, 24–35. [CrossRef]
428. Chudnovsky, A.; Kostinski, A.; Herrmann, L.; Koren, I.; Nutesku, G.; Ben-Dor, E. Hyperspectral Spaceborne Imaging of Dust-Laden Flows: Anatomy of Saharan Dust Storm from the Bodélé Depression. *Remote Sens. Environ.* **2011**, *115*, 1013–1024. [CrossRef]
429. Cao, C.; Chen, W.; Li, G.; Jia, H.; Ji, W.; Xu, M.; Gao, M.; Ni, X.; Zhao, J.; Zheng, S.; et al. The Retrieval of Shrub Fractional Cover Based on a Geometric-Optical Model in Combination with Linear Spectral Mixture Analysis. *Can. J. Remote Sens.* **2011**, *37*, 348–358. [CrossRef]
430. Chen, W.; Cao, C.; Zhang, H.; Jia, H.; Ji, W.; Xu, M.; Gao, M.; Ni, X.; Zhao, J.; Zheng, S.; et al. Estimation of Shrub Canopy Cover Based on a Geometric-Optical Model Using HJ-1 Data. In Proceedings of the 2011 IEEE International Geoscience and Remote Sensing Symposium, Vancouver, BC, Canada, 1–5 August 2011; pp. 1922–1925.
431. Griffin, S.; Rogan, J.; Runfola, D.M. Application of Spectral and Environmental Variables to Map the Kissimmee Prairie Ecosystem Using Classification Trees. *GIScience Remote Sens.* **2011**, *48*, 299–323. [CrossRef]
432. Lu, D.; Moran, E.; Hetrick, S. Detection of Impervious Surface Change with Multitemporal Landsat Images in an Urban–Rural Frontier. *ISPRS J. Photogramm. Remote Sens.* **2011**, *66*, 298–306. [CrossRef]
433. Negrón-Juárez, R.I.; Chambers, J.Q.; Marra, D.M.; Ribeiro, G.H.P.M.; Rifai, S.W.; Higuchi, N.; Roberts, D. Detection of Subpixel Treefall Gaps with Landsat Imagery in Central Amazon Forests. *Remote Sens. Environ.* **2011**, *115*, 3322–3328. [CrossRef]
434. Jiao, Q.; Zhang, B.; Liu, L.; Hu, Y. Estimating Fractional Vegetation Cover in the Wenchuan Earthquake Disaster Area Using High-Resolution Airborne Image and Landsat TM Image. In Proceedings of the MIPPR 2011: Remote Sensing Image Processing, Geographic Information Systems, and Other Applications, SPIE, Guilin, China, 20 November 2011; p. 80062G.
435. Lu, D.; Li, G.; Moran, E.; Batistella, M.; Freitas, C.C. Mapping Impervious Surfaces with the Integrated Use of Landsat Thematic Mapper and Radar Data: A Case Study in an Urban–Rural Landscape in the Brazilian Amazon. *ISPRS J. Photogramm. Remote Sens.* **2011**, *66*, 798–808. [CrossRef]
436. Renó, V.F.; Novo, E.M.L.M.; Suemitsu, C.; Rennó, C.D.; Silva, T.S.F. Assessment of Deforestation in the Lower Amazon Floodplain Using Historical Landsat MSS/TM Imagery. *Remote Sens. Environ.* **2011**, *115*, 3446–3456. [CrossRef]
437. Sankey, T.; Glenn, N. Landsat-5 TM and Lidar Fusion for Sub-Pixel Juniper Tree Cover Estimates in a Western Rangeland. *Photogramm. Eng. Remote Sens.* **2011**, *77*, 1241–1248. [CrossRef]
438. Sunderman, S.O.; Weisberg, P.J. Remote Sensing Approaches for Reconstructing Fire Perimeters and Burn Severity Mosaics in Desert Spring Ecosystems. *Remote Sens. Environ.* **2011**, *115*, 2384–2389. [CrossRef]
439. Gilichinsky, M.; Sandström, P.; Reese, H.; Kivinen, S.; Moen, J.; Nilsson, M. Mapping Ground Lichens Using Forest Inventory and Optical Satellite Data. *Int. J. Remote Sens.* **2011**, *32*, 455–472. [CrossRef]
440. QuickBird. Available online: https://earth.esa.int/eogateway/missions/quickbird-2 (accessed on 15 May 2023).
441. Hamada, Y.; Stow, D.A.; Roberts, D.A. Estimating Life-Form Cover Fractions in California Sage Scrub Communities Using Multispectral Remote Sensing. *Remote Sens. Environ.* **2011**, *115*, 3056–3068. [CrossRef]
442. Ji, M.; Feng, J. Subpixel Measurement of Mangrove Canopy Closure via Spectral Mixture Analysis. *Front. Earth Sci.* **2011**, *5*, 130–137. [CrossRef]
443. Yang, C.; Everitt, J.H. Mapping Three Invasive Weeds Using Airborne Hyperspectral Imagery. *Ecol. Inform.* **2010**, *5*, 429–439. [CrossRef]
444. Mücher, C.; Kooistra, L.; Vermeulen, M.; Haest, B.; Spanhove, T.; Delalieux, S.; Borre, J.V.; Schmidt, A. Object Identification and Characterization with Hyperspectral Imagery to Identify Structure and Function of Natura 2000 Habitats. In Proceedings of the Proceedings Third GEOgraphic Object-Based Image Analysis Conference 2010, Ghent, Belgium, 29 June–2 July 2010; p. 5.
445. Hu, X.; Weng, Q. Estimation of Impervious Surfaces of Beijing, China, with Spectral Normalized Images Using Linear Spectral Mixture Analysis and Artificial Neural Network. *Geocarto Int.* **2010**, *25*, 231–253. [CrossRef]
446. Mezned, N.; Abdeljaoued, S.; Boussema, M.R. A Comparative Study for Unmixing Based Landsat ETM+ and ASTER Image Fusion. *Int. J. Appl. Earth Obs. Geoinf.* **2010**, *12*, S131–S137. [CrossRef]
447. Estes, L.D.; Reillo, P.R.; Mwangi, A.G.; Okin, G.S.; Shugart, H.H. Remote Sensing of Structural Complexity Indices for Habitat and Species Distribution Modeling. *Remote Sens. Environ.* **2010**, *114*, 792–804. [CrossRef]
448. Ruescas, A.B.; Sobrino, J.A.; Julien, Y.; Jiménez-Muñoz, J.C.; Sòria, G.; Hidalgo, V.; Atitar, M.; Franch, B.; Cuenca, J.; Mattar, C. Mapping Sub-Pixel Burnt Percentage Using AVHRR Data. Application to the Alcalaten Area in Spain. *Int. J. Remote Sens.* **2010**, *31*, 5315–5330. [CrossRef]
449. Huang, Y.; Zhang, L.; Li, P.; Zhong, Y. High-Resolution Hyper-Spectral Image Classification with Parts-Based Feature and Morphology Profile in Urban Area. *Geo-Spat. Inf. Sci.* **2010**, *13*, 111–122. [CrossRef]

450. Jin, J.; Wang, B.; Zhang, L. A Novel Approach Based on Fisher Discriminant Null Space for Decomposition of Mixed Pixels in Hyperspectral Imagery. *IEEE Geosci. Remote Sens. Lett.* **2010**, *7*, 699–703. [CrossRef]
451. Wen-Fei, L.; Liang, Z.; Bing, Z.; Lian-Ru, G. Null space spectral projection algorithm for hyperspectral image endmember extraction. *J. Infrared Millim. Waves* **2010**, *29*, 307.
452. Luo, W.-F.; Zhong, L.; Zhang, B.; Gao, L.-R. Independent Component Analysis for Spectral Unmixing in Hyperspectral Remote Sensing Image. *Spectrosc. Spectr. Anal.* **2010**, *30*, 1628–1633.
453. Mei, S.; He, M.; Dai, Y. Virtual Dimensionality Estimation by Double Subspace Projection for Hyperspectral Images. In Proceedings of the 2010 Second IITA International Conference on Geoscience and Remote Sensing, Qingdao, China, 28–31 August 2010; Volume 2, pp. 234–237.
454. Mei, S.; He, M.; Wang, Z.; Feng, D. Spatial Purity Based Endmember Extraction for Spectral Mixture Analysis. *IEEE Trans. Geosci. Remote Sens.* **2010**, *48*, 3434–3445. [CrossRef]
455. Villa, A.; Chanussot, J.; Benediktsson, J.A.; Jutten, C. Supervised Super-Resolution to Improve the Resolution of Hyperspectral Images Classification Maps. In Proceedings of the Image and Signal Processing for Remote Sensing XVI, SPIE, Toulouse, France, 20–22 September 2010; Volume 7830, pp. 168–175.
456. Golubiewski, N.E.; Wessman, C.A. Discriminating Urban Vegetation from a Metropolitan Matrix through Partial Unmixing with Hyperspectral AVIRIS Data. *Can. J. Remote Sens.* **2010**, *36*, 261–275. [CrossRef]
457. Eches, O.; Dobigeon, N.; Tourneret, J.-Y. Estimating the Number of Endmembers in Hyperspectral Images Using the Normal Compositional Model and a Hierarchical Bayesian Algorithm. *IEEE J. Sel. Top. Signal Process.* **2010**, *4*, 582–591. [CrossRef]
458. Chang, C.-I.; Xiong, W.; Liu, W.; Chang, M.-L.; Wu, C.-C.; Chen, C.C.-C. Linear Spectral Mixture Analysis Based Approaches to Estimation of Virtual Dimensionality in Hyperspectral Imagery. *IEEE Trans. Geosci. Remote Sens.* **2010**, 5595092. [CrossRef]
459. Huck, A.; Guillaume, M.; Blanc-Talon, J. Minimum Dispersion Constrained Nonnegative Matrix Factorization to Unmix Hyperspectral Data. *IEEE Trans. Geosci. Remote Sens.* **2010**, *48*, 2590–2602. [CrossRef]
460. Iordache, M.-D.; Plaza, A.; Bioucas-Dias, J. Recent Developments in Sparse Hyperspectral Unmixing. In Proceedings of the 2010 IEEE International Geoscience and Remote Sensing Symposium, Honolulu, HI, USA, 25–30 July 2010; pp. 1281–1284.
461. Martin, G.; Ruiz, V.G.; Plaza, A.J.; Ortiz, J.P.; Fernández, I.G. Impact of JPEG2000 Compression on Endmember Extraction and Unmixing of Remotely Sensed Hyperspectral Data. *J. Appl. Remote Sens.* **2010**, *4*, 041796.
462. Martin, G.; Plaza, A. Spatial Preprocessing for Endmember Extraction Using Unsupervised Clustering and Orthogonal Subspace Projection Concepts. In Proceedings of the 2010 IEEE International Geoscience and Remote Sensing Symposium, Honolulu, HI, USA, 25–30 July 2010; pp. 959–962.
463. Raksuntorn, N.; Du, Q. Nonlinear Spectral Mixture Analysis for Hyperspectral Imagery in an Unknown Environment. *IEEE Geosci. Remote Sens. Lett.* **2010**, *7*, 836–840. [CrossRef]
464. Hendrix, E.M.T.; Garcia, I.; Plaza, J.; Plaza, A. Minimum Volume Simplicial Enclosure for Spectral Unmixing of Remotely Sensed Hyperspectral Data. In Proceedings of the 2010 IEEE International Geoscience and Remote Sensing Symposium, Honolulu, HI, USA, 25–30 July 2010; pp. 193–196.
465. Plaza, J.; Plaza, A. Spectral Mixture Analysis of Hyperspectral Scenes Using Intelligently Selected Training Samples. *IEEE Geosci. Remote Sens. Lett.* **2010**, *7*, 371–375. [CrossRef]
466. Barnsley, M.J.; Settle, J.J.; Cutter, M.A.; Lobb, D.R.; Teston, F. The PROBA/CHRIS Mission: A Low-Cost Smallsat for Hyperspectral Multiangle Observations of the Earth Surface and Atmosphere. *IEEE Trans. Geosci. Remote Sens.* **2004**, *42*, 1512–1520. [CrossRef]
467. Verrelst, J.; Clevers, J.G.P.W.; Schaepman, M.E. Merging the Minnaert-k Parameter with Spectral Unmixing to Map Forest Heterogeneity With CHRIS/PROBA Data. *IEEE Trans. Geosci. Remote Sens.* **2010**, 5466035. [CrossRef]
468. Matabishi, J.G.; Braun, A.; Warth, G. Multiple endmember spectral mixture analysis of desis image to identify rooftops in kigali. *Int. Arch. Photogramm. Remote Sens. Spat. Inf. Sci.* **2022**, *XLVI-1/W1-2021*, 39–47. [CrossRef]
469. Paul, A.; Dutta, D.; Jha, C.S. Target detection using dlr earth sensing imaging spectrometer (desis) data. *Int. Arch. Photogramm. Remote Sens. Spat. Inf. Sci.* **2022**, *XLVI-1/W1-2021*, 57–64. [CrossRef]
470. Xiong, W.; Chang, C.-I.; Tsai, C.-T. Estimation of Virtual Dimensionality in Hyperspectral Imagery by Linear Spectral Mixture Analysis. In Proceedings of the 2010 IEEE International Geoscience and Remote Sensing Symposium, Honolulu, HI, USA, 25–30 July 2010; pp. 979–982.
471. Castrodad, A.; Xing, Z.; Greer, J.; Bosch, E.; Carin, L.; Sapiro, G. Discriminative Sparse Representations in Hyperspectral Imagery. In Proceedings of the 2010 IEEE International Conference on Image Processing, Hong Kong, 26–29 September 2010; pp. 1313–1316.
472. Somers, B.; Verbesselt, J.; Ampe, E.M.; Sims, N.; Verstraeten, W.W.; Coppin, P. Spectral Mixture Analysis to Monitor Defoliation in Mixed-Aged Eucalyptus Globulus Labill Plantations in Southern Australia Using Landsat 5-TM and EO-1 Hyperion Data. *Int. J. Appl. Earth Obs. Geoinf.* **2010**, *12*, 270–277. [CrossRef]
473. Elatawneh, A.; Manakos, I.; Kalaitzidis, C.; Schneider, T. Land-Cover Classification and Unmixing of Hyperion Image in Area of Anopoli. In *Imagin [e, g] Europe*; IOS Press: Amsterdam, The Netherlands, 2010; pp. 111–121.
474. Cavalli, R.M. Comparison of Split Window Algorithms for Retrieving Measurements of Sea Surface Temperature from MODIS Data in Near-Land Coastal Waters. *ISPRS Int. J. Geo-Inf.* **2018**, *7*, 30. [CrossRef]
475. Chen, W.; Cao, C.; He, Q.; Guo, H.; Zhang, H.; Li, R.; Zheng, S.; Xu, M.; Gao, M.; Zhao, J.; et al. Quantitative Estimation of the Shrub Canopy LAI from Atmosphere-Corrected HJ-1 CCD Data in Mu Us Sandland. *Sci. China Earth Sci.* **2010**, *53*, 26–33. [CrossRef]

476. Meng, D.; Gong, H.; Li, X.; Zhao, W.; Li, Y. Impervious Surface Coverage and Their Impact on Other Components of the Urban Ecosystem in Beijing. In Proceedings of the 2010 IEEE International Geoscience and Remote Sensing Symposium, Honolulu, HI, USA, 20–24 July 2010; pp. 2731–2734.
477. Biggs, T.W.; Atkinson, E.; Powell, R.; Ojeda-Revah, L. Land Cover Following Rapid Urbanization on the US–Mexico Border: Implications for Conceptual Models of Urban Watershed Processes. *Landsc. Urban Plan.* **2010**, *96*, 78–87. [CrossRef]
478. Bohlman, S.A. Landscape Patterns and Environmental Controls of Deciduousness in Forests of Central Panama: Patterns/Controls of Tropical Deciduousness. *Glob. Ecol. Biogeogr.* **2010**, *19*, 376–385. [CrossRef]
479. Huang, C.; Asner, G.P.; Barger, N.N.; Neff, J.C.; Floyd, M.L. Regional Aboveground Live Carbon Losses Due to Drought-Induced Tree Dieback in Piñon–Juniper Ecosystems. *Remote Sens. Environ.* **2010**, *114*, 1471–1479. [CrossRef]
480. Pacheco, A.; McNairn, H. Evaluating Multispectral Remote Sensing and Spectral Unmixing Analysis for Crop Residue Mapping. *Remote Sens. Environ.* **2010**, *114*, 2219–2228. [CrossRef]
481. Solans Vila, J.P.; Barbosa, P. Post-Fire Vegetation Regrowth Detection in the Deiva Marina Region (Liguria-Italy) Using Landsat TM and ETM+ Data. *Ecol. Model.* **2010**, *221*, 75–84. [CrossRef]
482. Li, C.; Du, J.; Su, Y.; Li, Q.; Chen, L. Extraction of Impervious Surface Based on Multi-Source Satellite Data of Qinhuai River Basin from 1979–2009. In Proceedings of the 2010 18th International Conference on Geoinformatics, Beijing, China, 18–20 June 2010; pp. 1–6.
483. Powell, R.L.; Roberts, D.A. Characterizing Urban Land-Cover Change in Rondônia, Brazil: 1985 to 2000. *J. Lat. Am. Geogr.* **2010**, *9*, 183–211. [CrossRef]
484. Elmore, A.J.; Guinn, S.M. Synergistic Use of Landsat Multispectral Scanner with GIRAS Land-Cover Data to Retrieve Impervious Surface Area for the Potomac River Basin in 1975. *Remote Sens. Environ.* **2010**, *114*, 2384–2391. [CrossRef]
485. He, M.; Zhao, B.; Ouyang, Z.; Yan, Y.; Li, B. Linear Spectral Mixture Analysis of Landsat TM Data for Monitoring Invasive Exotic Plants in Estuarine Wetlands. *Int. J. Remote Sens.* **2010**, *31*, 4319–4333. [CrossRef]
486. Liu, Y.; Yue, W. Estimation of Urban Vegetation Fraction by Image Fusion and Spectral Unmixing. *Acta Ecol. Sin.* **2010**, *30*, 93–99.
487. Tømmervik, H.; Dunfjeld, S.; Olsson, G.A.; Nilsen, M.Ø. Detection of Ancient Reindeer Pens, Cultural Remains and Anthropogenic Influenced Vegetation in Byrkije (Børgefjell) Mountains, Fennoscandia. *Landsc. Urban Plan.* **2010**, *98*, 56–71. [CrossRef]
488. Yang, F.; Matsushita, B.; Fukushima, T. A Pre-Screened and Normalized Multiple Endmember Spectral Mixture Analysis for Mapping Impervious Surface Area in Lake Kasumigaura Basin, Japan. *ISPRS J. Photogramm. Remote Sens.* **2010**, *65*, 479–490. [CrossRef]
489. Borfecchia, F.; De Cecco, L.; Pollino, M.; La Porta, L.; Lugari, A.; Martini, S.; Ristoratore, E.; Pascale, C. Active and Passive Remote Sensing for Supporting the Evaluation of the Urban Seismic Vulnerability. *Ital. J. Remote Sens.* **2010**, *42*, 129–141. [CrossRef]
490. Silván-Cárdenas, J.L.; Wang, L. Retrieval of Subpixel Tamarix Canopy Cover from Landsat Data along the Forgotten River Using Linear and Nonlinear Spectral Mixture Models. *Remote Sens. Environ.* **2010**, *114*, 1777–1790. [CrossRef]
491. Liu, X.; Li, X.; Zhang, X. Determining Class Proportions within a Pixel Using a New Mixed-Label Analysis Method. *IEEE Trans. Geosci. Remote Sens.* **2009**, *48*, 1882–1891.
492. Gilichinskya, M.; Sandströma, P.; Reesea, H.; Kivinenb, S.; Moenb, J.; Nilsona, M. *Application of National Forest Inventory for Remote Sensing Classification of Ground Lichen in Nothern Sweden*; International Archives of the Photogrammetry, Remote Sensing and Spatial Information Sciences-ISPRS Archives: Haifa, Israel, 2010; pp. 146–152.
493. Sarapirome, S.; Kulrat, C. Comparison on urban classifications using landsattm and linear spectral mixture analysis extracted images: Nakhon Ratchasima municipal area, Thailand. *Suranaree J. Sci. Technol.* **2010**, *17*, 401–411.
494. Cavalli, R.M.; Pascucci, S.; Pignatti, S. Optimal Spectral Domain Selection for Maximizing Archaeological Signatures: Italy Case Studies. *Sensors* **2009**, *9*, 1754–1767. [CrossRef] [PubMed]
495. Alves Aguiar, D.; Adami, M.; Fernando Silva, W.; Friedrich Theodor Rudorff, B.; Pupin Mello, M.; dos Santos Vila da Silva, J. Modis Time Series to Assess Pasture Land. In Proceedings of the 2010 IEEE International Geoscience and Remote Sensing Symposium, Honolulu, HI, USA, 20–24 July 2010; pp. 2123–2126.
496. Eckmann, T.C.; Still, C.J.; Roberts, D.A.; Michaelsen, J.C. Variations in Subpixel Fire Properties with Season and Land Cover in Southern Africa. *Earth Interact.* **2010**, *14*, 1–29. [CrossRef]
497. Meusburger, K.; Bänninger, D.; Alewell, C. Estimating Vegetation Parameter for Soil Erosion Assessment in an Alpine Catchment by Means of QuickBird Imagery. *Int. J. Appl. Earth Obs. Geoinf.* **2010**, *12*, 201–207. [CrossRef]
498. Meusburger, K.; Konz, N.; Schaub, M.; Alewell, C. Soil Erosion Modelled with USLE and PESERA Using QuickBird Derived Vegetation Parameters in an Alpine Catchment. *Int. J. Appl. Earth Obs. Geoinf.* **2010**, *12*, 208–215. [CrossRef]
499. Schmidt, M. Monitoring Aquatic Weeds in a River System Using SPOT 5 Satellite Imagery. *J. Appl. Remote Sens* **2010**, *4*, 043528. [CrossRef]
500. Soenen, S.A.; Peddle, D.R.; Hall, R.J.; Coburn, C.A.; Hall, F.G. Estimating Aboveground Forest Biomass from Canopy Reflectance Model Inversion in Mountainous Terrain. *Remote Sens. Environ.* **2010**, *114*, 1325–1337. [CrossRef]
501. Ustin, S.L.; Hart, Q.J.; Duan, L.; Scheer, G. Vegetation Mapping on Hardwood Rangelands in California. *Int. J. Remote Sens.* **1996**, *17*, 3015–3036. [CrossRef]
502. Hunt, E.R., Jr.; Barlow, M.M.; Mahelona, C.L.; Laycock, W.A.; Heising, S.J.; Smith, R.P.; Foreman, J. Progress of the Wyoming Hyperspectral Imagery Pilot Project: Analysis of AVIRIS Data for Rangeland Assessment. In *Hyperspectral Remote Sensing and Applications*; Shen, S.S., Ed.; SPIE: Denver, CO, USA, 1996; pp. 291–297.

503. Bowers, T.L.; Rowan, L.C. Remote Mineralogic and Lithologic Mapping of the Ice River Alkaline Complex, British Columbia, Canada, Using AVIRIS Data. *Photogramm. Eng. Remote Sens.* **1996**, *62*, 1379–1386.
504. Meer, F.V.D. Metamorphic Facies Zonation in the Ronda Peridotites: Spectroscopic Results from Field and GER Imaging Spectrometer Data. *Int. J. Remote Sens.* **1996**, *17*, 1633–1657. [CrossRef]
505. Rosenthal, W. Estimating Alpine Snow Cover with Unsupervised Spectral Unmixing. In Proceedings of the IGARSS'96. 1996 International Geoscience and Remote Sensing Symposium, Lincoln, NE, USA, 31 May 1996; Volume 4, pp. 2252–2254.
506. Van Der Meer, F. Spectral Mixture Modelling and Spectral Stratigraphy in Carbonate Lithofacies Mapping. *ISPRS J. Photogramm. Remote Sens.* **1996**, *51*, 150–162. [CrossRef]
507. Ben-Dor, E.; Kruse, F.A.; Dietz, J.B.; Braun, A.W.; Banin, A. Spatial Distortion and Quantitative Geological Mapping of Makhtesh Ramon, NEGEV, ISRAEL, by Using the GER 63 Channel Scanner Data. *Can. J. Remote Sens.* **1996**, *22*, 258–268. [CrossRef]
508. Kerdiles, H.; Grondona, M.O. NOAA-AVHRR NDVI Decomposition and Subpixel Classification Using Linear Mixing in the Argentinean Pampa. *Int. J. Remote Sens.* **1995**, *16*, 1303–1325. [CrossRef]
509. Dwyer, J.L.; Kruse, F.A.; Lefkoff, A.B. Effects of Empirical versus Model-Based Reflectance Calibration on Automated Analysis of Imaging Spectrometer Data: A Case Study from the Drum Mountains, Utah. *Photogramm. Eng. Remote Sens.* **1995**, *61*, 1247–1254.
510. Lacaze, B.; Hill, J.; Mehl, W. Evaluation of Green Vegetation Fractional Cover in Mediterranean Ecosystems from Spectral Unmixing of Landsat TM and AVIRIS Data. In *Multispectral and Microwave Sensing of Forestry, Hydrology, and Natural Resources*; Mougin, E., Ranson, K.J., Smith, J.A., Eds.; SPIE: Rome, Italy, 1995; pp. 339–346.
511. Rowan, L.C.; Bowers, T.L.; Crowley, J.K.; Anton-Pacheco, C.; Gumiel, P.; Kingston, M.J. Analysis of Airborne Visible-Infrared Imaging Spectrometer (AVIRIS) Data of the Iron Hill, Colorado, Carbonatite-Alkalic Igneous Complex. *Econ. Geol.* **1995**, *90*, 1966–1982. [CrossRef]
512. Lavreau, J. Models of Spectral Unmixing: Simplex versus Least Squares Method of Resolution. In Proceedings of the Multispectral and Microwave Sensing of Forestry, Hydrology, and Natural Resources, SPIE, Rome, Italy, 30 September 1995; Volume 2314, pp. 397–407.
513. Van Der Meer, F. Spectral Unmixing of Landsat Thematic Mapper Data. *Int. J. Remote Sens.* **1995**, *16*, 3189–3194. [CrossRef]
514. Bianchi, R.; Cavalli, R.M.; Marino, C.M.; Pignatti, S.; Poscolieri, M. Use of Airborne Hyperspectral Images to Assess the Spatial Distribution of Oil Spilled during the Trecate Blow-out (Northern Italy). In Proceedings of the Remote Sensing for Agriculture, Forestry, and Natural Resources; International Society for Optics and Photonics, Paris, France, 26–28 September 1995; Volume 2585, pp. 352–362.
515. Hall, F.G.; Peddle, D.R.; LeDrew, E.F. Remote Sensing of Biophysical Variables in Boreal Stands of Picea Mariana. In Proceedings of the 1995 International Geoscience and Remote Sensing Symposium, IGARSS'95—Quantitative Remote Sensing for Science and Applications, Firenze, Italy, 10–14 July 1995; Volume 2, pp. 976–977.
516. Cracknell, A.P. Review Article Synergy in Remote Sensing-What's in a Pixel? *Int. J. Remote Sens.* **1998**, *19*, 2025–2047. [CrossRef]
517. Shahid, K.T.; Schizas, I.D. Spatial-Aware Hyperspectral Nonlinear Unmixing Autoencoder with Endmember Number Estimation. *IEEE J. Sel. Top. Appl. Earth Obs. Remote Sens.* **2022**, *15*, 20–41. [CrossRef]
518. Foody, G.M. Status of Land Cover Classification Accuracy Assessment. *Remote Sens. Environ.* **2002**, *80*, 185–201. [CrossRef]
519. Stehman, S.V.; Foody, G.M. Key Issues in Rigorous Accuracy Assessment of Land Cover Products. *Remote Sens. Environ.* **2019**, *231*, 111199. [CrossRef]
520. Milella, M. *Saperi Della Cultura e Agire Formativo*; Morlacchi Editore: Perugia, Italy, 2003.
521. Congalton, R.G.; Green, K. *Assessing the Accuracy of Remotely Sensed Data: Principles and Practices*; CRC Press: Boca Raton, FL, USA, 2019.
522. Cavalli, R. Retrieval of Sea Surface Temperature from MODIS Data in Coastal Waters. *Sustainability* **2017**, *9*, 2032. [CrossRef]
523. Gupta, H.V.; Kling, H.; Yilmaz, K.K.; Martinez, G.F. Decomposition of the Mean Squared Error and NSE Performance Criteria: Implications for Improving Hydrological Modelling. *J. Hydrol.* **2009**, *377*, 80–91. [CrossRef]
524. Cavalli, R.; Betti, M.; Campanelli, A.; Cicco, A.; Guglietta, D.; Penna, P.; Piermattei, V. A Methodology to Assess the Accuracy with Which Remote Data Characterize a Specific Surface, as a Function of Full Width at Half Maximum (FWHM): Application to Three Italian Coastal Waters. *Sensors* **2014**, *14*, 1155–1183. [CrossRef] [PubMed]
525. Bradley, A.P. The Use of the Area under the ROC Curve in the Evaluation of Machine Learning Algorithms. *Pattern Recognit.* **1997**, *30*, 1145–1159. [CrossRef]
526. Cavalli, R.M.; Colosi, F.; Palombo, A.; Pignatti, S.; Poscolieri, M. Remote Hyperspectral Imagery as a Support to Archaeological Prospection. *J. Cult. Herit.* **2007**, *8*, 272–283. [CrossRef]
527. Jia, S.; Qian, Y. Spectral and Spatial Complexity-Based Hyperspectral Unmixing. *IEEE Trans. Geosci. Remote Sens.* **2007**, *45*, 3867–3879.
528. Comber, A.; Fisher, P.; Brunsdon, C.; Khmag, A. Spatial Analysis of Remote Sensing Image Classification Accuracy. *Remote Sens. Environ.* **2012**, *127*, 237–246. [CrossRef]
529. Roberts, D.A.; Gardner, M.; Church, R.; Ustin, S.; Scheer, G.; Green, R.O. Mapping Chaparral in the Santa Monica Mountains Using Multiple Endmember Spectral Mixture Models. *Remote Sens. Environ.* **1998**, *65*, 267–279. [CrossRef]
530. Morales-Barquero, L.; Lyons, M.B.; Phinn, S.R.; Roelfsema, C.M. Trends in Remote Sensing Accuracy Assessment Approaches in the Context of Natural Resources. *Remote Sens.* **2019**, *11*, 2305. [CrossRef]

531. Cerra, D.; Agapiou, A.; Cavalli, R.; Sarris, A. An Objective Assessment of Hyperspectral Indicators for the Detection of Buried Archaeological Relics. *Remote Sens.* **2018**, *10*, 500. [CrossRef]
532. AVIRIS—JPL-NASA. Available online: https://aviris.jpl.nasa.gov/data/free_data.html (accessed on 31 January 2023).
533. Grupo de Inteligencia Computacional. Available online: https://www.ehu.eus/ccwintco/index.php/Hyperspectral_Remote_Sensing_Scenes (accessed on 31 January 2023).
534. MultiSpec. Available online: https://engineering.purdue.edu/~biehl/MultiSpec/hyperspectral.html (accessed on 31 January 2023).
535. Remote Sensing Laboratory. Available online: https://rslab.ut.ac.ir/data (accessed on 31 January 2023).
536. Cuprite Reference Map. Available online: https://www.usgs.gov/media/images/aviris-scene-flown-over-cuprite-nevada (accessed on 20 May 2023).
537. Cavalli, R.M. Capability of Remote Sensing Images to Distinguish the Urban Surface Materials: A Case Study of Venice City. *Remote Sens.* **2021**, *13*, 3959. [CrossRef]
538. Justice, C.; Belward, A.; Morisette, J.; Lewis, P.; Privette, J.; Baret, F. Developments in the'validation'of Satellite Sensor Products for the Study of the Land Surface. *Int. J. Remote Sens.* **2000**, *21*, 3383–3390. [CrossRef]
539. Congalton, R.G.; Gu, J.; Yadav, K.; Thenkabail, P.; Ozdogan, M. Global Land Cover Mapping: A Review and Uncertainty Analysis. *Remote Sens.* **2014**, *6*, 12070–12093. [CrossRef]
540. Baret, F.; Weiss, M.; Allard, D.; Garrigue, S.; Leroy, M.; Jeanjean, H.; Fernandes, R.; Myneni, R.; Privette, J.; Morisette, J.; et al. VALERI: A Network of Sites and a Methodology for the Validation of Medium Spatial Resolution Land Satellite Products; 2021; hal-03221068. Available online: https://hal.inrae.fr/hal-03221068 (accessed on 10 May 2023).
541. PRODES. Available online: http://www.obt.inpe.br/OBT/assuntos/programas/amazonia/prodes (accessed on 22 March 2023).
542. Toutin, T. Review Article: Geometric Processing of Remote Sensing Images: Models, Algorithms and Methods. *Int. J. Remote Sens.* **2004**, *25*, 1893–1924. [CrossRef]
543. Cheng, X.; Wang, Y.; Jia, J.; Wen, M.; Shu, R.; Wang, J. The Effects of Misregistration between Hyperspectral and Panchromatic Images on Linear Spectral Unmixing. *Int. J. Remote Sens.* **2020**, *41*, 8862–8889. [CrossRef]
544. Strahler, A.H.; Boschetti, L.; Foody, G.M.; Friedl, M.A.; Hansen, M.C.; Herold, M.; Mayaux, P.; Morisette, J.T.; Stehman, S.V.; Woodcock, C.E. Global Land Cover Validation: Recommendations for Evaluation and Accuracy Assessment of Global Land Cover Maps. *Eur. Communities Luxemb.* **2006**, *51*, 1–60.
545. Gharbi, W.; Chaari, L.; Benazza-Benyahia, A. Joint Bayesian Hyperspectral Unmixing for Change Detection. In Proceedings of the 2020 Mediterranean and Middle-East Geoscience and Remote Sensing Symposium (M2GARSS), Tunis, Tunisia, 9–11 March 2020; pp. 37–40.
546. Park, J.-J.; Oh, S.; Park, K.-A.; Kim, T.-S.; Lee, M. Applying Hyperspectral Remote Sensing Methods to Ship Detection Based on Airborne and Ground Experiments. *Int. J. Remote Sens.* **2020**, *41*, 5928–5952. [CrossRef]

remote sensing

MDPI

Review

Review of Automatic Processing of Topography and Surface Feature Identification LiDAR Data Using Machine Learning Techniques

Zahra Gharineiat, Fayez Tarsha Kurdi * and Glenn Campbell

School of Surveying and Built Environment, Faculty of Health, Engineering and Sciences, University of Southern Queensland, Springfield Campus, Springfield, QLD 4300, Australia
* Correspondence: fayez.tarshakurdi@usq.edu.au

Abstract: Machine Learning (ML) applications on Light Detection And Ranging (LiDAR) data have provided promising results and thus this topic has been widely addressed in the literature during the last few years. This paper reviews the essential and the more recent completed studies in the topography and surface feature identification domain. Four areas, with respect to the suggested approaches, have been analyzed and discussed: the input data, the concepts of point cloud structure for applying ML, the ML techniques used, and the applications of ML on LiDAR data. Then, an overview is provided to underline the advantages and the disadvantages of this research axis. Despite the training data labelling problem, the calculation cost, and the undesirable shortcutting due to data downsampling, most of the proposed methods use supervised ML concepts to classify the downsampled LiDAR data. Furthermore, despite the occasional highly accurate results, in most cases the results still require filtering. In fact, a considerable number of adopted approaches use the same data structure concepts employed in image processing to profit from available informatics tools. Knowing that the LiDAR point clouds represent rich 3D data, more effort is needed to develop specialized processing tools.

Keywords: LiDAR; Machine Learning (ML); classification; modelling; point cloud

Citation: Gharineiat, Z.; Tarsha Kurdi, F.; Campbell, G. Review of Automatic Processing of Topography and Surface Feature Identification LiDAR Data Using Machine Learning Techniques. *Remote Sens.* **2022**, *14*, 4685. https://doi.org/10.3390/rs14194685

Academic Editor: Luis A. Ruiz

Received: 10 August 2022
Accepted: 17 September 2022
Published: 20 September 2022

1. Introduction

A Light Detection And Ranging (LiDAR) point cloud (airborne, terrestrial, static or mobile) is a list of 3D points covering the surface of a scanned scene. Topographical data obtained this way are rich in geometric features and lend themselves to the possibilities of automatic processing [1]. There are two major forms of automatic processing operations: automatic classification and automatic modelling [2]. Generally, one scanned scene will consist of classes that have different geometric natures or characteristics, e.g., an urban point cloud represents several classes such as terrain, buildings, vegetation, powerlines, roads, railways, and other artificial objects [3]. As each class in the scanned area will require a different modelling strategy depending on its specific geometric nature, e.g., the vegetation class modelling algorithm will need to be different from the building class modelling algorithm, it is necessary to classify the point cloud before starting the modelling stage.

In the first two decades since LiDAR technology's appearance, most of the suggested automatic processing algorithms belonged to the rule-based family [4]. In truth, a single rule-based algorithm actually consists of a list of procedures connected through a proposed workflow and depends on the physical structure of the point cloud [4]. Recently though, in the domain of topographical LiDAR data processing, the general trend has been to employ Machine Learning (ML) algorithms instead of rule-based ones, and the use of ML techniques has become a popular research topic [5].

Supervised ML algorithms assign observations to data classes previously generated, either manually or automatically, from the use of training data that could sometimes be generated automatically [6]. Alternatively, unsupervised ML algorithms do not need training data and can be classified into four families: classification tree methods such as the Random Forest (RF) algorithms, grouping and separability methods such as Support Vector Machines (SVM), k-Nearest Neighbors (KNN), and rule application methods such as Convolutional Neural Networks (CNN) [7].

This paper reviews the state-of-the-art ML algorithms developed for topographical LiDAR data processing. The novelty of this paper is the classification and analysis of the ML algorithms according to four different dimensions. First, the methods of point cloud generation for input into ML approaches are analyzed and discussed. Second, the different concepts of point cloud structure that are commonly used are studied and compared. Third, the suggested approaches are classified according to the most employed ML techniques, and then the main ML techniques are summarized. Finally, the most current applications of ML techniques are classified and cited.

2. Input Data

Notwithstanding the quality of the employed laser scanning technology, airborne, terrestrial, static, or mobile, all methods allow the creation of a 3D point cloud that covers the scanned area. A LiDAR point cloud consists of a point list of co-ordinates X, Y, and Z defined in 3D Euclidean space. For each point, in addition to the three coordinates, laser intensity, waveform, and Red Green Blue (RGB) colors can be provided [8]. Furthermore, for the same scanned scene, additional data such as multispectral images, maps, and orthophotos can often be provided. As a result, in the literature, the suggested ML approaches for LiDAR data processing are not just limited to the LiDAR point cloud alone. The following subsections explain the different point cloud generation methods for input into ML algorithms.

2.1. LiDAR Point Clouds

The 3D point cloud is the primary output of a laser scanning operation (Figure 1). This subsection deals with approaches that use only the point cloud, whereas the approaches that use other additionally acquired data will be discussed in the following subsections. The obvious advantage of approaches that use only the LiDAR point cloud is that they are always available for use in all scanning projects. The point cloud does not just represent a simple list of 3D points in the Euclidian space, it may be used as the input data to create a Digital Surface Model (DSM) [1]. Furthermore, for each point, a list of neighboring points can be defined in 3D space [9–11], where all points included inside a sphere surrounding the focus point are considered, or in 2D space where all points included inside a cylinder surrounding the focus point are considered [5]. After this stage is completed, each point and its neighboring points allow for fitting a mean line or plane to analyze their relative topologic positions through several indicators such as standard deviation, mean square error, eigenvector, and eigenvalues [12]. Additionally, the eigenvector permits the calculation of a list of useful geometric features such as linearity, planarity, sphericity and change of curvature [13,14]. In this context, other approaches are to superimpose the point cloud on an empty 2D grid to allow for the analysis of the topological relationships between neighboring points [15], or assuming that they represent one object, using one LiDAR point and its neighborhood to allow calculation of a list of static moments that help to study some of their geometric characteristics [16]. While this has its uses, it is important to note that the employment of just the point cloud as input data does not produce promising results in the general case, e.g., when identifying roofs in an airborne LiDAR point cloud in an urban area, the range of roof point coordinates may be incorrectly allocated to an incorrect building because of underlying topography of the scanned area. That is why the application of the ML techniques in this case uses the point features instead of the point coordinates as input data [9]. Consequently, a long list of geometric features that can be calculated from the point cloud is needed to create a suitable environment to apply ML. The ML techniques that have been applied to airborne and terrestrial LiDAR point clouds are shown in the next two subsections, with the use of laser intensity observations discussed in the subsection that follows those.

Figure 1. (**a**) Aerial image of scanned scene; (**b–d**) 3D LiDAR point cloud visualization (**b**) using RGB colors; (**c**) using laser intensity values; (**d**) using Z coordinate values.

2.1.1. Airborne LiDAR Point Cloud

Airborne LiDAR point clouds provide two obstacles to the applications of ML techniques: variation in point density within the scanned scene [11] and the large number of LiDAR points [17]. Point density plays a vital role in selecting the neighboring points for the calculation of point features [9]. Point density can vary markedly within the same point cloud with the location within the scanning strip, the terrain topography and the geometry, and the orientation of the scanned object with regard to the scan line all having an affect [8]. For a large area, the data volumes can be excessive, meaning the training step will place heavy demands on the computer capacity and processing time [17,18]. Lin et al. [19] and Mao et al. [20] developed approaches to mitigate this problem and classify an urban point cloud into nine classes: powerlines, low vegetation, impervious surfaces, cars, fences, roofs, façades, shrubs, and trees. In this context, Mao et al. [20] developed a Receptive Field Fusion-and-Stratification Network (RFFS-Net). An innovative Dilated Graph Convolution (DGConv) and its extension, the Annular Dilated Convolution (ADConv), are fundamental components of elementary building blocks. The receptive field fusion procedure was applied with the Dilated and Annular Graph Fusion (DAGFusion) component. Thus, the detection of dilated and annular graphs with numerous receptive zones allows the acquisition of developed multi-receptive field feature implementation to improve classification accuracy. To efficiently extract only one class from the urban point cloud, Ao et al. [21] advised using a presence and background learning algorithm like a backpropagation neural network.

2.1.2. Terrestrial LiDAR Point Cloud

This subsection focuses only on the ML approaches that use a static or mobile terrestrial LiDAR point cloud as input data either indoors or outdoors. An indoor cloud may focus on certain scanned objects such as tables, chairs, decorative statues, and mechanical equipment [22,23] or it may carry out a panoramic scan [24] and use the LiDAR point cloud to then extract the individual objects. An urban outdoor LiDAR point cloud will most likely emphasize artificial or natural objects such as building facades and terrain [25] while a rural scene, like Zou et al. [26] examined, may use a terrestrial LiDAR point cloud of forestry areas to classify the tree species. In fact, most of the suggested approaches that use ML techniques to process terrestrial LiDAR data do not use additional data with the point cloud [17,27–32]. Point density variation has less influence in terrestrial when compared to airborne data. Nevertheless, some authors do use additional data as input, e.g., Xiu et al. [23] suggested a ML algorithm to process indoor point cloud represented by 9 dimensions: X, Y, Z, R, G, B, and normalized location. He et al. [25] developed a SectorGSnet framework for a ground segmentation of terrestrial outdoor LiDAR point clouds. This framework consisted of an encoder in addition to segmentation modules. It introduced a bird's-eye-view segmentation strategy that discretizes the point cloud into segments of different areas. The points within each partition are then fed into a multimodal Point-Net encoder to extract the required features. Li et al. [33] suggested a Rotation Invariant neural Network (RINet) which associated semantic and geometric features to improve the descriptive capacity of scanned objects and classify the terrestrial data into twelve classes.

Terrestrial laser scanning plays a major role in autonomous driving vehicles with Silva et al. [34] developing a Deep Feature Transformation Network (DFT-Net) involving a cascading mixture of edge convolutions and feature transformation layers to capture the local geometric features by conserving topological relationships between points. Alternatively, self-learning algorithms appear as a practical solution to understand the correspondence between adjacent LiDAR scan scenes [35]. Nunes et al. [36] used a momentum encoder network and a feature bank in a self-learning approach [37,38] that aimed to learn the structural context of the scanned scene. This approach applies the contrastive loss over the extracted segments to distinguish between similar and dissimilar objects. Finally, Huang et al. [39] used an unsupervised domain adaptation ML to classify terrestrial LiDAR data and suggested using Generative Adversarial Network (GAN) to calculate synthetic data from the source domain, so the output will be close to the target domain.

2.1.3. Point Cloud and Laser Intensity

In practice, LiDAR systems measure and provide the laser pulse return intensity (Figure 1c). The intensity of emitted laser pulse is greater than the intensity of the reflected laser pulse and with the difference being dependent on the double distance trajectory in addition to the nature of the reflecting surface off which the pulse has returned [40]. Unlike the RGB-measured values of the point cloud, the intensity could be detected regardless of the illumination and can be provided in both airborne and terrestrial LiDAR. Some authors have used the intensity and the 3D point cloud together as input data into their ML algorithms.

In this regard, Wen et al. [41] proposed a Directionally constrained fully Convolutional Neural network (D-FCN) where the input data were the original 3D point cloud in addition to the LiDAR intensity. Since road line markings have a higher reflectance, and hence higher intensity value than the surrounding ground, Fang et al. [42] considered the 3D LiDAR point cloud and the laser intensity as input data to their ML algorithm. Wang et al. [29] employed the intensity component in semantic outdoor 3D terrestrial dataset to achieve the cloud segmentation using Graph Attention Convolution (GAC) and Murray et al. [43] calculated a 2D image from the intensity component of LiDAR data. This image was used as input data for the CNN algorithm and then for the SVM.

2.2. Point Cloud and Imagery

In the image processing domain, many algorithms for feature extraction from images have been implemented where the image's spatial and textural features were extracted using mathematical descriptors, such as histograms of oriented gradients and SVMs [44]. The combination of LiDAR data with high-resolution images can provide highly relevant data for the analysis of scanned scene characteristics [45]. Indeed, numerous authors develop classification ML networks using LiDAR point clouds in addition to digital images as input data. Nahhas et al. [46] employed orthophotos in addition to airborne LiDAR point clouds to recognize the building class by using an autoencoder-based dimensionality reduction to convert low-level features into compressed features. Similarly, Vayghan et al. [3] used aerial images and LiDAR data to extract building and tree footprints in urban areas while Zhang et al. [47] fused the LiDAR data and a point cloud calculated from the aerial images to improve the accuracy of a ML building extraction algorithm. Shi et al. [48] suggested the use of an enhanced lightweight deep neural network with knowledge refinement to detect local features from LiDAR data and imagery while preserving solid robustness for day-night visual localization.

2.3. Multispectral LiDAR Data

Multispectral images have layers that represent the reflectance in a few wide and disconnected spectral bands within given specified spectral intervals [49]. In the case of airborne LIDAR data, some authors have used multispectral images in addition to the LiDAR point cloud as input data for ML algorithms, because most objects on the Earth's surface have indicative absorption features in certain discrete spectral bands which can help to create an accurate classification of the scanned scene [49]. Though the multispectral data are not always available, where they are, they can be an asset for processing efficacity. In this context, Marrs and Ni-Meister, [50] used LiDAR, hyperspectral, and thermal images on experimental forests and found that the combination of these two data can help improve the classification of tree species. Yu et al. [51] used multispectral LiDAR data for individual tree extraction and tree species recognition. Zhao et al. [52] used a FR-GCNet network to increase the classification accuracy of multispectral LiDAR point clouds, whereas Zhou et al. [53] applied an RF algorithm on a combination of hyperspectral images and LiDAR data for monitoring insects. Peng et al. [54] suggested a MultiView Hierarchical Network (MVHN) could be used to segment hyper spectral images and LiDAR point cloud together. For this purpose, the hyper spectral images were divided into multiple groups with the same number of bands to extract spectral features. Thereafter, ResNet framework was implanted to detect the spectral-spatial information of the merged features.

2.4. Full-Waveform Representation and Point Cloud

Some airborne laser systems, called full waveform, can record the complete power spectrum of the returned pulse. The different surface characteristics can influence the reflected signal, so analysis of the laser pulse full waveform has been used to improve the extraction of surface features [55] especially in forested areas. Five parameters are calculated from the waveform of the return pulse, e.g., the amplitude of the highest peak, the total energy, the full-width half-maximum return width, and the length of the sequence. In this context, Guan et al. [56] constructed a geometric tree model based on the full-waveform representation. Afterward, in order to classify the tree species, they applied a deep learning algorithm to the last model to extract the high-level features. Blomley et al. [57] classified the tree species using the RF algorithm based on the geometric features calculated from the full-waveform analysis.

Similarly, by means of an integrated system that acquired hyperspectral images, LiDAR waveforms, and point clouds, Yang et al. [58] classified tree species after systematic pixel-wised investigation of different features. For this purpose, the Canopy Height Model (CHM) was extracted from the LiDAR data, and multiple features from the hyperspectral images, including Gabor textural features. Shinohara et al. [59] suggested a semantic classification algorithm named Full-Waveform Network (FWNet) based on PointNet-based architecture [27], which extracted the local and global features of the input waveform data. The classifier in this case consisted of 1D convolutional operational layers. Due to the sensitivity of border points to the multi return difference value, to achieve the cloud segmentation, Shin et al. [60] used multiple returns in addition to the point cloud as training data using the PointNet++ network [61].

2.5. Different Other Data

Sometimes other data, not mentioned previously, may be used in addition to the LiDAR point cloud. For example, Zhang et al. [62] used the interaction of the high-resolution L-band repeat-pass Polarimetric Synthetic Aperture Radar Interferometry (PolInSAR) and low-resolution large-footprint full-waveform LiDAR data to estimate forest height. Park and Guldmann [63] utilized a city LiDAR point cloud in addition to building footprint data to extract building class before applying an RF algorithm and Feng and Guo [64] suggested a segment-based parameter learning approach that fuses a 2D land map and 3D point cloud together.

For detecting individual trees, Schmohl et al. [65] used an orthophoto to colorize the point cloud for additional spectral features along with laser intensity and the number of returns utilized as additional input. Kogut et al. [66] improved the classification accuracy of seabed laser scanning (bathymetry data) by using the Synthetic Minority Oversampling Technique (SMOTE) algorithm to evaluate the input data. Then, a Multi-Layer Perceptron (MLP) neural workflow was applied to classify the point cloud. Barbarella et al. [67] applied a ML network that trained a model able to classify a particular gravity-driven coastal hillslope geomorphic model (slope-over-wall) including most of the soft rocks. However, they used only geometric data which are morphometric feature maps computed from a Digital Terrain Model (DTM) calculated from the LiDAR point cloud.

Finally, Duran et al. [68] compared nine ML methods: logistic regression, linear discriminant analysis, K-NN, decision tree, Gaussian Naïve Bayes, MLP, adaboost, RF, and SVM to classify LiDAR and colored photogrammetric point clouds into four classes: buildings, ground, low and high vegetation with the highest accuracy being attained with MPL. For more details about these ML techniques, please see Mohammed et al. [69] and Kim, [70].

3. Concepts of Point Cloud Structure for Applying ML Algorithms

The 3D point cloud consists of a large number of 3D points covering the scanned area. These points are normally distributed in an irregular way depending on the scanning system quality and the scanned area geometric characteristics. In any event, to process, classify, and model the LiDAR data using ML techniques, most of the suggested approaches try to define a mathematical model that allows for the management, reduction, pooling, and convolution of these data [71]. Consequently, most ML approaches consist of two main steps, firstly preprocessing and then ML algorithm application. In this paper, the mathematical model in addition to all operations realized on it before applying the ML technique is named the data adaptation step (Figure 2). The data adaptation procedures may play several roles. Some ML informatics tools for imagery data processing or other data kinds, require the transformation of point cloud into novel data forms such as 2D and 3D matrices before they can be used. As informatics tools for processing LiDAR data require high time processing cost, two solutions are employed: either designing new ML tools that correspond to the LiDAR data concept or, more commonly, reducing the LiDAR data. At this stage, it is important to refer that the interpolation or reduction of LiDAR data is not always a preferable solution from the geomatics industrial viewpoint.

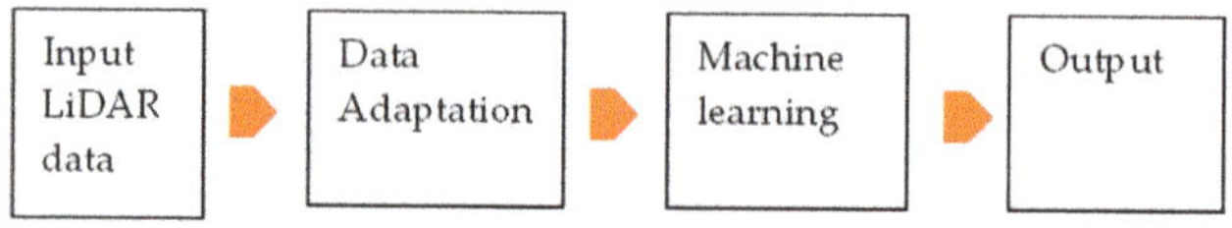

Figure 2. Structure of ML algorithm of LiDAR data processing.

In the next subsections, the main concepts of LiDAR data adaptation will be revealed and discussed.

3.1. Voxelization

Voxelization, a 3D matrixial representation, may sometimes solve the issue of the irregular distribution of the 3D point cloud [56]. In practice, the LiDAR points are distributed on the scanned surfaces which leads to a considerable number of empty voxels which cause additional calculation costs. Moreover, course spatial resolution (large voxel size) may cause the loss of information which will reduce the accuracy of data processing. Conversely, if the spatial resolution is too small, that may increase the calculation cost, and the memory usage [17].

In the literature, many authors suggest voxelizing the LiDAR point clouds. In this context, Maturana and Scherer [72] developed the VoxNet network using the occupancy grid algorithm. They divided the point cloud into many 3D grids and then normalized each grid unit to enter the volume build layers and maximum pooling layers. Gargoum et al. [73] suggested a voxel-based approach to classify the light poles of roads while Zou et al. [26] proposed a voxel-based deep learning method to identify tree species in a three-dimensional map. They extracted individual trees through point cloud density and used voxel rasterization to obtain features. Guan et al. [56] used a voxel-based upward growth algorithm to remove the ground point cloud and then segment a single tree species by European clustering and a voxel-based normalization algorithm. Shuang et al. [74] developed an Adaptive Feature Enhanced Convolutional Neural Network (AFERCNN) for 3D object detection. This algorithm is a point-voxel integrated network, where voxel features are extracted through the 3D voxel convolutional neural network. These features are projected to the 2D bird's eye view and the relationship between the features in both spatial dimension and channel dimension is learned. Wijaya et al. [75] applied a voxel-based 3D object detection deep neural network on terrestrial LiDAR data where they minimized the features from a 3D into a 2D bird-eye view map before generating object proposals to save processing time.

However, voxelization tries to conserve the LiDAR point cloud 3D structure by defining a spatial matrixial form that enables improved management of the point cloud. Hence, the form will be limited by the available, the used memory, and the requested processing time may represent the main limitations.

3.2. Graphic Structure

Using graphic structure to transform the 3D point cloud into a 2D regular grid has the main advantage of transforming the point cloud classification question into the general image processing one. Simonovsky and Komodakis [76] used edge labels to calculate Edge Conditional Convolution (ECC) in the neighborhood of regular grids. Then, an asymmetric edge operation was used to calculate the relationship between neighboring points. Wang et al. [77] developed a SpecGCN network where the maximum pooling was replaced with a recursive clustering. The nearest neighbor was applied to calculate a graph regular grid. Thereafter, they combined a spectral graph convolution using a local graph, with a pooling strategy. Nahhas et al. [46] suggested a deep learning approach based on using an interpolated LiDAR point cloud and orthophotos simultaneously. This approach employed object-based analysis to create objects, a feature-level fusion. Li et al. [78] developed a deep learning network named Attentive Graph Geometric Moments Convolution (AGGM Convolution) network to classify the LiDAR point cloud into four classes: trees, grass, roads, and buildings. The Dynamic Graph Convolution Neurol Network (DGCNN), suggested by Wang et al. [28], built the directed graph in both the Euclidean space and the feature space, and dynamically updated the feature layers. A similar approach suggested by Wang et al. [29] employed the attention mechanism in the graph-based methods. The extended approach is named Graph Attention Convolution Network (GACNet) for semantic point cloud segmentation.

In the same context, Wen et al. [79] presented a global-local Graph Attention Convolution Neural Network (GACNN) that could be directly applied to airborne LiDAR data. The graph attention convolution module includes two types of attention mechanisms: a local attention module that combines edge attention and density attention, and a global attention module. The local edge attention module is designed to dynamically learn convolution weights using the spatial relationships of neighboring points; thus, the receptive field of the convolution kernel can dynamically adjust to the structure of the point cloud. Zhao et al. [52] used a Feature Reasoning-based Graph Convolution Network (FR-GCNet) to increase the classification accuracy of airborne multispectral LiDAR data. Jing et al. [80] proposed a Graph-based neural Network with an Attention pooling strategy (AGNet) where the local features were extracted through the point topological structure. Chen et al. [81] improved the descriptiveness in the network ChebyNet [82] by increasing the width of input to avert the above drawbacks. The suggested network, named WGNet, is inspired by the image processing dilated convolution. This network is based on two modules, the local dilated connecting and context information awareness. Wan et al. [83] developed a Dilated Graph Attention-based Network (DGANet) for local feature extraction on 3D point clouds. It was based on the dilated graph attention modules which allow the network to learn the neighborhood representation by using the long-range dependencies given by the calculated dilated graph-like region for each point.

To conclude, the use of graphic structure facilitates the point cloud processing duty tasks by using image processing functions, but unfortunately at the cost of minimizing the 3D structure advantages.

3.3. Kernel-Based Convolution

The geometric structure of a point cloud can be defined through the Kernel correlation layer [41]. The kernel size value can be suggested according to a different number of neighboring points in the convolution layer. Points within the kernel can contribute to their center point [84]. At this stage, Klokov et al. [85] proposed a K-NN algorithm that uses the Euclidean metric to return the closest points inside the kernel. The kernel is defined by two parameters: the inner and the outer radius to ensure that the closest and unique points will be detected in each ring kernel. In the context of ML applications, Song et al. [86] employed the kernel correlation learning block approach to recognize the local and global features at different layers thus enhancing the network perception capacity. Zhang et al. [31] suggested a Local k-NNs Pattern in Omni-Direction Graph Convolution Neural Network named LKPO-GNN to capture both the global and local point cloud spatial layout. This approach converts the point cloud into an ordered 1D sequence, to feed the input data into a neural network and reduce the processing cost.

In fact, this approach allows applying all operations directly on the point cloud, but it still requires an optimized neighborhood searching procedure.

3.4. Reducing of Point Cloud Density (Downsampling)

Most ML approaches applied to LiDAR data try to reduce data density and keep the processing time within accepted limits. The successful use of the convolutional technique within the image processing field has encouraged authors to use the same approach in reducing LiDAR data and thus to solve the processing time issue. Although the most used point cloud structures apply the idea of point cloud reduction, the suggested approaches in this subsection conserve the point cloud structure and reduce the point density. However, the application of ML techniques is still in its infancy, and a lot of advancement is expected in future research.

In the context of point cloud reduction, Wen et al. [41] developed a D-FCN network architecture that included both downsampling and upsampling paths to enable multiscale point feature learning. Several authors, Hu et al. [30], Wei et al. [17], and Du et al. [22] used random downsampling to reduce the point cloud in the context of applying the ML algorithm such as developing consecutively feed-forward MLPsRandLA-Net and encoder–decoder structure (BushNet and ResDLPS-Net). Mao et al. [20] suggested three downsampling layers to classify the LiDAR data.

Though the downsampling reduces the data volume, it loses an important information quantity that may be useful to object recognition and modeling.

4. Employed ML Techniques

Currently, the advancement of digital technologies and data acquisition techniques in different disciplines can lead to the generation of excessively large data sets. To manage and process the oversized data sets, the questions of data classification and object recognition have become ones of crucial importance. In this context, ML techniques occupy an enviable position because they allow for automatic and efficient solutions. The ML techniques can be classified into four categories according to the required input data (see Mohammed et al. [69]): supervised learning, where labelled data are needed for training, unsupervised learning, where labelled data are not needed, semi-supervised learning that uses a mixture of classified and unclassified data, and reinforcement learning where no data are available. Of these, the supervised and unsupervised techniques may be considered the main two categories. In each one of these two groups, several algorithms are employed, e.g., supervised ML uses algorithms such as decision trees, rule-based classifiers, Naïve Bayesian classification, k-nearest neighbors' classifiers, RF, Neural Networks (NN), linear discriminant analysis, and SVM, whereas unsupervised ML uses k-means clustering, Gaussian mixture model, hidden Markov model, and principal component analysis.

In the LiDAR data-processing domain, the application of ML algorithms represents an emerging research area. Despite the great number of papers published in this area, very few new ML algorithms are employed. In the next subsections, more focused ML algorithms will be introduced and discussed.

4.1. Random Forest (RF) and Support Vector Machine (SVM)

Tarsha Kurdi et al. [87] summarized the applications of RF classifiers for automatic vegetation detection and modelling using LiDAR point clouds. Many authors used RF exclusively on LiDAR data [88], whereas other authors used additional data [89,90]. Yu et al. [91], and Yu et al. [51] estimated tree characteristics such as diameter, height, and stem volume using an RF classifier and Levick et al. [92] connected the DSM and field-measured wood volume using an RF algorithm. Chen et al. [88] used the feature selection method and an RF algorithm for landslide detection under forest canopy, where the DTM and the slope model were constructed for the scanned area, and the features were calculated at the pixel level. The same principle was used by Guan et al. [93] to identify the city classes in urban areas and Ba et al. [94] employed RF for detecting the tree species.

Man et al. [90] applied an RF classifier to calculate a two-dimensional distribution map of urban vegetation. In this study, individual tree segmentation was conducted on a CHM and point cloud data separately to obtain three-dimensional characteristics of urban trees. The results show that both the RF classification and object-based classification could extract urban vegetation accurately, with accuracies above 99%, and the individual tree segmentation based on point cloud data could delineate individual trees in three-dimensional space better than CHM segmentation. Arumäe et al. [95] calculated a model for predicting necessity thinning using the RF technique to retrieve the two indicative parameters for requiring thinning, height percentage and the canopy cover. Park and

Guldmann, [63] used an RF algorithm to classify building point clouds into four classes: rooftop, wall, ground, and high outlier. To overcome the complexity of building geometry of the Ming and Qing Dynasties' Official Architecture style (MQDOAs), Dong et al. [96] employed semantic roof segmentation. This method was composed of two stages. Some geometric features such as the normalized symmetrical distance, relative height, and local height difference are extracted and then the RF algorithm is applied to classify the roof point cloud. Feng and Guo [64] suggested a segment-based parameter learning approach in which a 2D land cover map is chosen to generate labelled samples, and a formalized operation is then implemented to train the RF classifier. Liao et al. [97] fed in point cloud super voxels and their convex connected patches into an RF algorithm. For this purpose, they consider three types of features: point-based, eigen-based, and grid-based.

The SVM algorithm tries to find a hyperplane in high dimensional feature space to classify some linearly correlative point distributions. While there could be many hyperplanes that separate the target classes, the hyperplane that optimizes the boundary between the classes is identified. Aside from just linear classification, SVM can carry out nonlinear classification using the kernel trick by indirectly drawing their inputs into high-dimensional feature spaces [69].

Though the SVM classifier is efficient for data classification when using rather small data, it is also used by Ba et al. [94] to recognize tree species. Murray et al. [43] trained an SVM on the passing and ongoing results of a CNN algorithm through pixel classification and the interpolation result of the intensity vector as input data. Hoang et al. [98] introduced a hybrid approach of a CNN and an SVM for 3D shape recognition, where eight layers of the CNN are utilized for geometric feature extraction and afterward an SVM is applied to classify them. Zhang et al. [99] suggested an object-based approach to classify an urban airborne LiDAR point cloud. First, different point features such as geometry, radiometry, topology, and echo characteristics are extracted and then the SVM classifier algorithm was applied to detect five classes: terrain, vegetation, building, powerlines, and vehicles. To detect powerlines, Shokri et al. [100] eliminated the undesirable points and then apply the SVM after calculating the point geometric features.

In conclusion, RF and SVM are less used in recent years, and both are more basic classification models. Therefore, most modern approaches focus on deep learning techniques.

4.2. Neural Network and Deep Learning

Deep learning represents a sort of ML, and it can be defined as a ML technique that employs a deep neural network such as the MLP neural network that contains two or more hidden layers [70]. A Perceptron Neural network consists of single neurons that have multiple inputs and generate a single output using an activation function. Figure 3 illustrates a deep learning algorithm functionality where the available data consist of two sections: labelled and unlabeled data. The labelled data will be used in training the suggested MLP neural network to correct the assumed weight values which will then be used in the same neural network to label the unlabeled data. For more information about deep learning techniques, please see Kim [70].

In the LiDAR data processing area, deep learning algorithms are widely applied especially for data classification. Zou et al. [26] used a low-level feature representation through voxel-based structure, and then classified tree species by using a deep learning model. In regard to Generative Adversarial Networks (GAN), Goodfellow et al. [101] have achieved a notable performance on pan-sharpening in the image processing domain. Zhang et al. [62] developed a PolGAN deep learning network to determine the forest tree heights. When applying a deep learning classification algorithm, Lin et al. [19] improved the labelling stage to produce training data because the data labelling procedure for generating training data consumes considerable time and effort. In this context, they suggested using weak labelling that needs little annotation effort. The pseudo labels are then considered as the input of a classification network [102]. Thereafter, an overlap region loss and an elevation attention unit are introduced for the classification network to obtain more accurate pseudo labels.

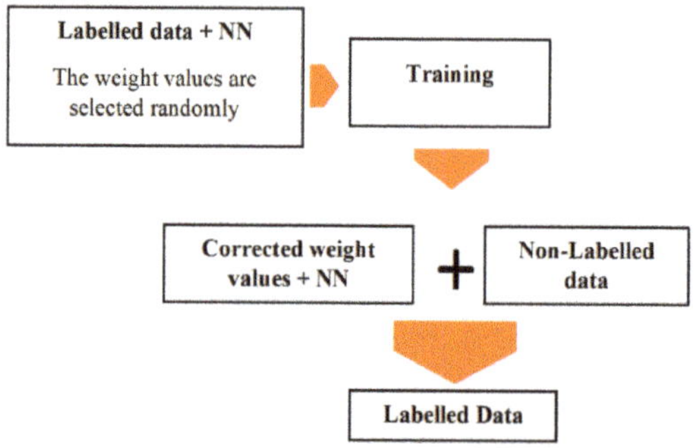

Figure 3. Deep learning functionality; NN is a Neural network.

Zhao et al. [52] used a Feature Reasoning-based Graph Convolution Network (FR-GCNet) to increase the classification accuracy of urban point clouds. Semantic labels were assigned to pixels using global and local features. Based on the graph convolution network, a global reasoning unit is embedded to find the global contextual features, while a local reasoning unit is added to learn edge features with attention weights in each local graph. Li et al. [103] compared three deep learning algorithms for classifying LiDAR point clouds, these algorithms are PointNet++ [61], SparseCNN [104] and KPConv [105]. They found that SparseCNN carries out a better classification accuracy than the other two approaches.

Where there are variations of point cloud density, Théodose et al. [106] suggested adapting an object detection deep learning approach. For this purpose, some data layers are randomly dismissed during the training step to grow the variability of the processed data. Sheikh et al. [32] proposed a Deep Feature Transformation Network (DFT-Net) to classify terrestrial LiDAR data. The suggested algorithm is based on graph analysis in which the edges are dynamically extracted for each layer. Hoang et al. [107] extracted and associated both global and regional features through Gaussian SuperVector and enhancing region illustration deep learning Network (GSV-NET) for 3D point cloud classification. Chen et al. [108] developed a Dynamic Point Feature Aggregation deep learning Network (DPFA-Net) by selectively performing the neighborhood feature aggregation, dynamic pooling, and an attention mechanism. In this semantic classification of the LiDAR point cloud framework, the features of the dynamic point neighborhood are aggregated via a self-attention mechanism. Finally, Song et al. [109] developed, in the context of automatic LiDAR data classification, a 2D and 3D Hough Network (2D&3DHNet) by linking 3D global Hough features and 2D local Hough features with a classification deep learning network.

4.3. Encoder–Decoder Structure

In the encoder–decoder structure, the network consists mainly of two subnetworks: the encoder sub-network and the decoder sub-network [110]. In the encoder part, consecutive downsampling procedures increase the receptivity of the extracted features but unfortunately, that reduces the point cloud resolution. In the decoder part, upsampling and convolution operations are employed for resolution recapture and feature combination.

In laser scanning, several authors developed an encoder–decoder algorithms to classify LiDAR data. Wen et al. [79] created an end-to-end encoder–decoder network named GACNN that is based on the graph attention convolution module and used it for detecting multiscale features of the LiDAR data and achieving point cloud classification. Wei et al. [17] proposed a network point cloud segmentation named BushNet which is the classic encoder–decoder structure. In this context, a minimum probability random sampling module is used for reducing the processing time and improving the convergence speed. Thereafter, the local multi-dimensional feature fusion module is applied to make the network more sensitive to bush point cloud features. Thus, the employed multi-channel attention module may improve the training efficiency.

Medina and Paffenroth [111] applied an encoder–decoder classifier for reduced LiDAR data by feeding the network with features calculated from the point neighborhood, which showed high efficiency in distinguishing the non-linear features. Mao et al. [20] developed an encoder–decoder architecture for point cloud classification named Receptive Field Fusion and Stratification Network (RFFS-Net) that is based on the PointConv network suggested by Wu et al. [112]. It consists of two steps: hierarchical graph generation and encoder–decoders feature extraction and aggregation. The input is provided by a hierarchical graph generation model and point features after which the point features are aggregated. Ibrahim et al. [113] used CNN architectures to semantically classify the terrestrial LiDAR data. They divided the point cloud into angle-wise slices that are transformed in the next step into enhanced pseudo images using the intensity and reflectivity values. Then, these images are employed to feed an encoder–decoder CNN model.

Finally, despite the promising results obtained by deep learning as well as encoder–decoder structure, more focus is needed on unsupervised learning techniques which may cancel the request for training data.

Having presented the main ML algorithms used to process LiDAR data, the next section will discuss current applications of ML technique on LiDAR point cloud.

5. Applications of ML on LiDAR Data

The use of laser scanning technology is widespread. It has been applied in urban, rural, and forested areas to target natural as well as artificial objects such as buildings (inside and outside), roads, railways, bridges, tunnels, and pipelines. Almost inevitably, a point cloud of a scanned area will consist of several object classes such as terrain, vegetation, buildings, standing water, noise, and artificial objects. As each class has a different modelling concept, it is essential to classify the point cloud into its main classes before starting the modelling step [5]. Once the point cloud of the scanned area is classified, the obtained classes can be analyzed and modelled according to the project goal. In this context, a large list of class modelling operations could be described. From the creation of laser scanning technology, most of the suggested approaches in the literature have been rule-based. Within the last five years, ML techniques have become an important approach for LiDAR data processing [2,8]. Unfortunately, ML techniques have hitherto only been used to a limited number of procedures, e.g., according to Hamedianfar et al. [114], the main applications of deep learning algorithms in forest areas are biomass estimation and tree species classification. In the next subsections, the main applications of ML techniques on LiDAR data are detailed.

5.1. Building Detection

The ML classifiers are sometimes focused on the building class in urban areas, with the aim of classifying the scanned scene into two classes: buildings and non-buildings. Nahhas et al. [46] suggested a deep learning approach based on the feature-level. CNN was used to transform compressed features into high-level features, which were used in building detection. Zhang et al. [47] used the U-NET model [115] to detect building polygons from orthophotos. Hence, to increase the point cloud density, the LiDAR, and photogrammetric point clouds are merged and employed for each polygon for feature extraction goals. Ojogbane et al. [116] improved a deep learning network suggested by Seydi et al. [117] to detect the building class. The suggested framework fuses the features obtained from interpolated airborne LiDAR data into DSM, in addition to a very high-resolution aerial imagery. Shin et al. [60] applied PointNet++ [61] for building extraction using multiple returns data.

While ML algorithms are employed by several authors for building recognition, in fact, the urban scene cannot just be simplified into building and non-building classes. Hence, the next section will go further through applying ML to achieve full classification.

5.2. Scene Segmentation

The classification question is widely discussed in this research area. One scanned scene consists of several classes, and the question that arises is: can the classification algorithm be used to extract the desired class list? Or can one algorithm only recognize certain classes? For this reason, we have chosen to identify the classification algorithms according to detected classes. With respect to airborne data, not all authors agree about the ideal number of classes. Wen et al. [41] developed a deep learning network that classified the airborne LiDAR data into nine classes: powerlines, low vegetation, cars, fences, roofs, facades, shrubs, and trees. Despite Wang and Gu [118] using the same number of classes, their class list is different: earth bar, grass, roads, buildings, trees, water, powerlines, cars, and ships. Li et al. [78] suggested a deep learning pixel-based analysis network to distinguish four classes in airborne data: trees, grass, roads, and buildings. Another class list is suggested by Ekhtari et al. [119] classified their scene into six classes: buildings, soil, grass, trees, asphalt, and concrete. An example of the final data set is shown in Figure 4. Zhao et al. [52] made small modifications to these classes as follows: roads, buildings, grass, trees, powerlines, and soil. Another modification to these classes is suggested by Shinohara et al. [59]: roads, buildings, transmission towers, trees, powerlines, and ground. Liao et al. [97] classified the airborne point cloud into three main classes: terrain, buildings, and vegetation using the RF algorithm. Zhao et al. [120] suggested a Point Expanded Multi-Scale Convolutional Network (PEMCNet) to classify the airborne LiDAR data containing point cloud, intensity, and return number, into five classes: ground, high vegetation, building, water, and raised road. To calculate the point features, it created point expanded grouping units that combined the extracted features at diverse scales. It is fair to say that the classes chosen in each study are a product of the study area and study aim rather than a desire to develop a universal class set.

Figure 4. An example of a 3D point cloud classified into six classes (buildings, soil, grass, trees, asphalt, and concrete) by Ekhtari et al. [119].

In the case of terrestrial data, a huge diversity of suggested class lists reflects the diversity of scanned scenes. Wang et al. [28], Qi et al. [27], Wang et al. [29], Hu et al. [30], Wei et al. [17], Zhang et al. [31], Xiu et al. [23], and Jing et al. [80] classified the terrestrial LiDAR data into several classes according to the scanned objects. To classify the terrestrial LiDAR data, Wen et al. [121] converted the LiDAR point cloud into a pseudo image and applied a semantic segmentation algorithm named Hybrid CNN-LSTM that has a neural network framework. Hence, the pseudo image is considered within Long Short-Term

Memory (LSTM) network that combines the different channel features generated by a convolutional neural network. Shuang et al. [122] proposed for terrestrial LiDAR point cloud classification, a Multi-Spatial Information and Dual Adaptive (MSIDA) network, which consists of encoding and dual adaptive sub-networks. To encode the point coordinates, each point and its neighborhood are transferred into a cylindrical and spherical coordinate system. The DA sub-network comprises a Coordinate System Attention Pooling Fusion (CSAPF) block in addition to a Local Aggregated Feature Attention (LAFA) one.

5.3. Vegetation Detection

Some classification algorithms are developed especially for forest areas, that focus on the vegetation class. In this case, they classify the scanned scene into two classes: vegetation and non-vegetation. Luo et al. [24] developed a semantic segmentation deep network to extract vegetation points from the LiDAR point cloud, where the tree points are grouped into a set of tree clusters using Euclidean distance clustering. A Pointwise Direction Embedding deep network (PDE-net) is employed to calculate the direction vectors of tree centers. Chen et al. [123] compared four ML algorithms: RF, Cubist, XGBoost, and CatBoost with rule-based algorithms to improve the estimation performance of forest biomass. The ML algorithms outperformed parametric stepwise regression, with the CatBoost network being superior, followed by XGBoost, RF, Cubist, and stepwise regression.

In the context of individual tree detection, Schmohl et al. [65] exploited the 3D LiDAR point cloud by using a 3D NN to detect individual trees. A sparse convolutional network was applied for feature calculation and feeding of the semantic segmentation output. Furthermore, they defined five semantic classes obtained from the dataset: terrain, buildings, low points, bridges, and vegetation. Luo et al. [124] proposed a tree detection algorithm through a deep learning framework based on a multi-channel information complementarity illustration. An adapted graph convolution network with local topological information was developed to extract the ground class thus avoiding parameters selection that did not consider different ground topographies. Then, a multichannel representation in addition to Multi-Branch Network (MBNet) was used through fusing multi-channel features. Corte et al. [125] used uncrewed aerial vehicle LiDAR point cloud to test four different ML approaches to detect individual trees and estimate their metrics such as diameter at breast height, total height, and timber volume. The tested methods were SVM, RF, NN, and Extreme Gradient Boosting. Windrim and Bryson [126] isolated individual trees, determine stem points, and further built a segmented model of the main tree stem that encompasses tree height, and diameter. This approach used deep learning models passing through multiple stages starting by ground characterization and removal, delineation of individual trees, and segmentation of tree points into stem and foliage. An example of output of their algorithm is shown in Figure 5. For extracting grasses and individual trees, Man et al. [90] extracted the two-dimensional distribution map of urban vegetation using the object-based RF classification method. Chen et al. [127] employed a PointNet network [27] for segmenting the individual tree crowns using the voxelization strategy.

Finally, Vayghan et al. [3] extracted high-elevation objects from the LiDAR data using the developed scan labelling method, and then the classification methods of a NN. Adaptive Neuro-Fuzzy Inference System (ANFIS), and Genetic Based K-Means algorithm (GBKMs) were used to separate buildings and trees with the purpose of evaluating their performance.

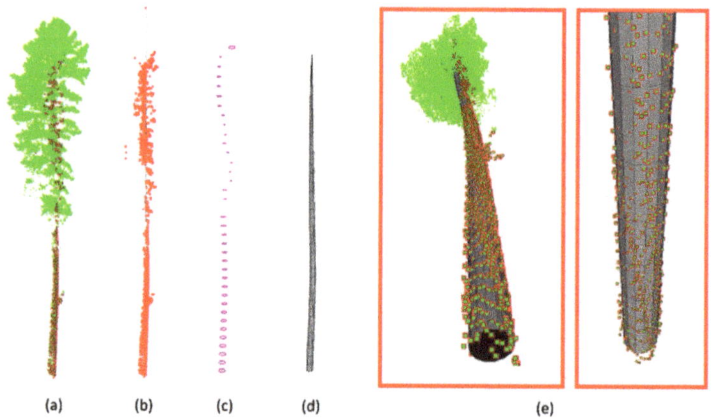

Figure 5. An example of the output of Windrim and Bryson's [126] (Figure 8) deep learning model where (**a**) is the segmented point cloud, (**b**) isolated stem points, (**c**) RANdom SAmple Consensus (RANSAC) algorithm circles attach stem section and (**d**) refined stem sections estimate based on robust least-squares fitting process. Panel (**e**) shows examples of the final fitted stem model.

5.4. Classification of Tree Species

Zou et al. [26] suggested a voxel-based deep learning method to classify terrestrial LiDAR point clouds of a forested area into species. They used three consecutive steps. After the extraction of individual trees using the density of the point clouds, a low-level feature voxel-based representation was constructed and then the classification of tree species was achieved by using a deep learning model.

Marrs and Ni-Meister [50] compared NNs, k-nearest neighbors, and RF approaches for recognizing tree species. The used variable reduction techniques and showed mixed results depending on the exact set of inputs to each machine learner. Dimensionality reduction based on classification tree nodes is a technique worth trying on multisource datasets. Mizoguchi et al. [128] classified individual tree species using terrestrial LiDAR based on CNN. The key component was the initial step of a depth image creation which well described the characteristics of each species from a point cloud.

Ba et al. [93] employed SVM and RF algorithms to test the discrimination level between tree genera. In this context, tree crowns were isolated and global morphology and internal structure features were computed. Yu et al. [51], Budei et al. [129], and Blomley et al. [57] estimated tree species based on an RF using tree features as predictors and tree species as a response for correctly extracted trees. Figure 6 shows an example of a successful detection phase. Yang et al. [58], and Nguyen et al. [130] both identified the tree species from LiDAR data in addition to other airborne measurements such as hyperspectral images using an SVM classifier. Hell et al. [131] tested the capacity of two deep learning networks PointCNN [132], and 3DmFV-Net [133] for the classification of four different tree species, both living and dead, using LiDAR data. It was shown that 3DmFV-Net is adequate for the geometry of the single trees, whereas PointCNN permits the incorporation of other features.

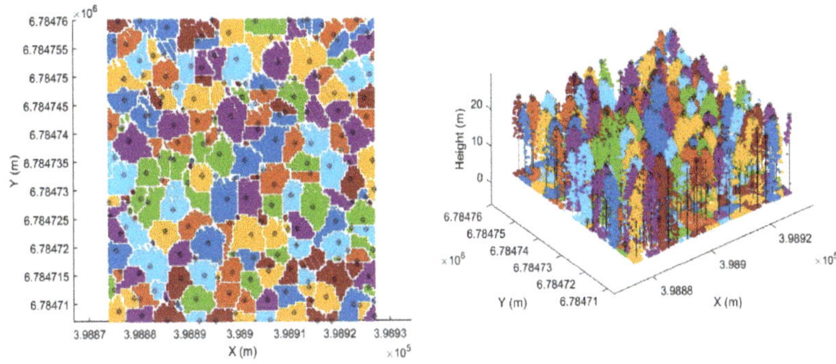

(a) (b)

Figure 6. An example of the results of Yu et al. [51] tree detection stage from (**a**) a plan view and (**b**) a 3D view.

5.5. Road Marking Classification

The high retro-reflective materials of road markings cause a high laser intensity with respect to the surrounding areas. On one hand, this feature allows easy identification of the road markings but unfortunately, the road markings are not only incomplete but also contain discontinues. That is why the road marking classification represents a challenging task [42]. In this context, Wen et al. [134] used a modified U-net model to segment road marking pixels to overcome the intensity variation, low contrast, and other obstacles. (Ma et al. [135] developed a capsule-based deep learning framework for road marking extraction and classification that consists of three modules. This approach starts with the segmentation of road surfaces. Thereafter, an Inverse Distance Weighting (IDW) interpolation is applied. Based on the convolutional and deconvolutional capsule operations, a U-shaped capsule-based network is created, and a hybrid network is developed using a revised dynamic routing algorithm and Softmax loss function. Fang et al. [42] proposed a graph attention network named GAT_SCNet to simultaneously group the road markings into 11 categories from LiDAR point clouds. The GAT_SCNet model builds serial computable subgraphs and uses a multi-head attention mechanism to encode the geometric and topological links between the node and neighbors to calculate the different descriptors of road marking.

5.6. Other Applications

In addition to the main applications presented previously, several important attempts to employ the ML for achieving other automatic operations on LiDAR data are documented in the literature. Ma et al. [136] proposed a workflow for the automatic extraction of road footprints from urban airborne LiDAR point clouds using deep learning PointNet++ [61]. In addition to the point cloud and laser intensity, the co-registered images and generated geometric features are employed to describe a strip-like road. In this context, graph-cut and constrained Triangulation Irregular Networks (TIN) are considered. Shajahan et al. [137] suggested a view-based method called a MultiView Convolutional Neural Network with Self-Attention (MVCNN-SA), which recognizes the roof geometric forms by considering multiple roof point cloud views.

In self-driving cars, several applications such as object recognition, automatic classification, and feature extraction are carried out using ML techniques [34,36,138–142]. The importance of data filtering before starting the modelling operation has been established and Gao et al. [143] proposed a filtering algorithm that uses deep learning and a multi-position sensor comparison approach to eliminate reflection noise. Nurunnabi et al. [144] introduced a local feature-based non-end-to-end deep learning approach that generated a binary classifier for terrain class filtering from which the feature relevance in addition to the models of different feature combinations were analyzed. Cao and Scaioni [145] applied a deep learning algorithm for semantic segmentation of terrestrial building point clouds. To reduce the number of labels, they suggested a label-efficient deep learning network (3DLEB-Net) that obtained per-point semantic labels of building point clouds with limited supervision. Shokri et al. [100] proposed an SVM approach for automated detection of powerlines from a LiDAR point cloud. In roadside laser scanning system applications, Zhang et al. [146] addressed the goal of a joint detection and tracking scheme by applying PV-RCNN [147] to automatic vehicle and pedestrian detection from the measured moving point cloud. Yin et al. [148] established a squeeze-excite mechanism in local aggregation procedures and employed deep residual learning through a suggested deep learning network that classified complicated piping elements. Amakhchan et al. [149] applied an MLP to filter the LiDAR building point cloud by eliminating the non-roof points. Mammoliti et al. [150] applied the semi-supervised clustering which combined semi-supervised learning and cluster analysis, to evaluate the rock mass discontinuities, orientation and spacing.

6. Conclusions

This paper has summarized and reviewed the state-of-the-art ML approaches applied to topographical LiDAR data. Four aspects were considered to analyze the studied methods. First, although all suggested approaches use an airborne or terrestrial LiDAR point cloud of the scanned scene as input data, some of them use, sometimes simultaneously, additional data such as real images, multispectral images, and waveforms to improve their efficiency. Of course, prima facie, using supplementary data may improve the conditions for obtaining the target result, but it is worth considering the contribution of the additional data to the final result. How critical the additional data are to the success of the target task needs to be verified.

Second, in literature, a long list of supervised and unsupervised ML techniques is available. As the unsupervised methods do not need labelled data, the use of these methods can solve the training data labelling problem. Unfortunately, most of the suggested approaches focus only on three supervised ML techniques: NN, RF, and SVM. More research is necessary to investigate the possible application, on LiDAR data, of other ML techniques, especially the unsupervised variety. These may provide opportunities for more efficient and lower cost solutions.

The third aspect is the concept of the LiDAR point cloud structure used within ML algorithms. Many of the proposed algorithms try to transform the question of 3D LiDAR data processing into 2D imagery processing so as to exploit the availability of the image processing informatics tools. These transformations lead to loss of information partly because of dimension reduction. Furthermore, the data reduction through downsampling techniques is similar to the pooling operation employed in image processing algorithms. This procedure is undesirable because it leads to the loss of information which may be beneficial to classify the data successfully. In this context, more research is needed to design a new methodology that simultaneously conserves the LiDAR data and saves the processing time.

Fourth, in regard to the new tools or trends for large-scale mapping and 3D modelling, ML techniques can mainly be employed to achieve five operations on topographical LiDAR data which are: buildings class detection, data classification, point cloud segmentation into vegetation and non-vegetation classes, separation of different tree species, and road marking classification. Some other applications of ML appear rarely in the literature. In fact, most feature-detection operations from topographical LiDAR data can be carried out with the help of classification procedures such as the detection of lines, planes, vertices, surfaces, breaklines, and borders. Filtering operations and modelling also represent an investigation area to apply ML techniques. Clearly, more effort and investigation are needed to improve and to apply ML algorithms on topographical LiDAR data.

Author Contributions: Conceptualization: F.T.K., Z.G. and G.C.; Methodology: Z.G. and F.T.K.; Investigation: F.T.K.; Resources: F.T.K. and Z.G.; Writing—original draft preparation: F.T.K., Z.G. and G.C.; Writing—review and editing: F.T.K., Z.G. and G.C.; Visualization: F.T.K. and Z.G.; Supervision: F.T.K., Z.G. and G.C. All authors have read and agreed to the published version of the manuscript.

Funding: This research received no external funding.

Data Availability Statement: Figure 4 was adopted from Ekhtari et al. [119], Figure 5 was adopted from Windrim and Bryson's [126], and Figure 6 was adopted from Yu et al. [51].

Acknowledgments: Thanks to Paul Reed Managing Director of East Coast Surveys (Aust) Pty Ltd. and CloudXPlus company to provide the dataset of Figure 1 which was measured in Queensland, Australia, http://www.eastcoastsurveys.com.au (accessed on 9 August 2022).

Conflicts of Interest: The authors declare no conflict of interest.

References

1. Tarsha Kurdi, F.; Gharineiat, Z.; Campbell, G.; Dey, E.K.; Awrangjeb, M. Full series algorithm of automatic building extraction and modelling from LiDAR data. In Proceedings of the 2021 Digital Image Computing: Techniques and Applications (DICTA), Gold Coast, Australia, 29 November–1 December 2021; pp. 1–8. [CrossRef]
2. Shan, J.; Toth, C.K. *Topographic Laser Ranging and Scanning Principles and Processing*, 2nd ed.; Taylor & Francis Group, LLC: Boca Raton, FL, USA, 2018; 630p; ISBN 13: 978-1-4987-7227-3.
3. Vayghan, S.S.; Salmani, M.; Ghasemkhani, N.; Pradhan, B.; Alamri, A. Artificial intelligence techniques in extracting building and tree footprints using aerial imagery and LiDAR data. *Geocarto Int.* **2022**, *37*, 2967–2995. [CrossRef]
4. Ayazi, S.M.; SaadatSeresht, M. Comparison of traditional and machine learning base methods for ground point cloud labeling. *Int. Arch. Photogramm. Remote Sens. Spatial Inf. Sci.* **2019**, *42*, 141–145. [CrossRef]
5. Tarsha Kurdi, F.; Gharineiat, Z.; Campbell, G.; Awrangjeb, M.; Dey, E.K. Automatic filtering of LiDAR building point cloud in case of trees associated to building roof. *Remote Sens.* **2022**, *14*, 430. [CrossRef]
6. De Geyter, S.; Bassier, M.; Vergauwen, M. Automated training data creation for semantic segmentation of 3D point clouds. Arch. Photogramm. *Remote Sens. Spatial Inf. Sci.* **2022**, *46*, 59–67. [CrossRef]
7. Michałowska, M.; Rapiński, J. A Review of tree species classification based on airborne LiDAR data and applied classifiers. *Remote Sens.* **2021**, *13*, 353. [CrossRef]
8. Shan, J.; Toth, C.K. *Topographic Laser Ranging and Scanning Principles and Processing*; Taylor & Francis Group, LLC: Boca Raton, FL, USA, 2008; 593p, ISBN 13: 978-1-4200-5142-1.
9. Dey, E.K.; Tarsha Kurdi, F.; Awrangjeb, M.; Stantic, B. Effective selection of variable point neighbourhood for feature point extraction from aerial building point cloud data. *Remote Sens.* **2021**, *13*, 1520. [CrossRef]
10. Ben-Shabat, Y.; Lindenbaum, M.; Fischer, A. Nesti-net: Normal estimation for unstructured 3D point clouds using convolutional neural networks. In Proceedings of the IEEE Conference on Computer Vision and Pattern Recognition, Long Beach, CA, USA, 16–20 June 2019; pp. 10112–10120. [CrossRef]
11. Weinmann, M.; Jutzi, B.; Hinz, S.; Mallet, C. Semantic point cloud interpretation based on optimal neighborhoods, relevant features and efficient classifiers. *ISPRS J. Photogramm. Remote Sens.* **2015**, *105*, 286–304. [CrossRef]
12. Sanchez, J.; Denis, F.; Coeurjolly, D.; Dupont, F.; Trassoudaine, L.; Checchin, P. Robust normal vector estimation in 3D point clouds through iterative principal component analysis. *ISPRS J. Photogramm. Remote Sens.* **2020**, *163*, 18–35. [CrossRef]
13. Thomas, H.; Goulette, F.; Deschaud, J.; Marcotegui, B.; LeGall, Y. Semantic classification of 3D point clouds with multiscale spherical neighborhoods. In Proceedings of the International Conference on 3D Vision (3DV), Verona, Italy, 5–8 September 2018; pp. 390–398. [CrossRef]
14. Nurunnabi, A.; Teferle, F.N.; Laefer, D.F.; Lindenbergh, R.C.; Hunegnaw, A. A two-step feature extraction algorithm: Application to deep learning for point cloud classification. *Int. Arch. Photogramm. Remote Sens. Spat. Inf. Sci.* **2022**, *46*, 401–408. [CrossRef]
15. Tarsha Kurdi, F.; Landes, T.; Grussenmeyer, P. Joint combination of point cloud and DSM for 3D building reconstruction using airborne laser scanner data. In Proceedings of the 4th IEEE GRSS/WG III/2+5, VIII/1, VII/4 Joint Workshop on Remote Sensing & Data Fusion over Urban Areas and 6th International Symposium on Remote Sensing of Urban Areas, Télécom Paris, Paris, France, 11–13 April 2007; p. 7.
16. Tarsha Kurdi, F.; Landes, T.; Grussenmeyer, P.; Koehl, M. Model-driven and data-driven approaches using Lidar data: Analysis and comparison. In *ISPRS Workshop, Photogrammetric Image Analysis (PIA07)*; Institut für Photogrammetrie und Fernerkundung (IPF): Munich, Germany, 2007; Part 3 W49A; Volume XXXVI, pp. 87–92; ISSN 1682-1750.
17. Wei, H.; Xu, E.; Zhang, J.; Meng, Y.; Wei, J.; Dong, Z.; Li, Z. BushNet: Effective semantic segmentation of bush in large-scale point clouds. *Comput. Electron. Agric.* **2022**, *193*, 106653. [CrossRef]
18. Winiwarter, L.; Esmorís Pena, A.M.; Weiser, H.; Anders, K.; Martínez Sánchez, J.; Searle, M.; Höfle, B. Virtual laser scanning with HELIOS++: A novel take on ray tracing-based simulation of topographic full-waveform 3D laser scanning. *Remote Sens. Environ.* **2021**, *269*, 112772. [CrossRef]

19. Lin, Y.; Vosselman, G.; Yang, M.Y. Weakly supervised semantic segmentation of airborne laser scanning point clouds. *SPRS J. Photogramm. Remote Sens.* **2022**, *187*, 79–100. [CrossRef]
20. Mao, Y.; Chen, K.; Diao, W.; Sun, X.; Lu, X.; Fu, K.; Weinmann, M. Beyond single receptive field: A receptive field fusion-and-stratification network for airborne laser scanning point cloud classification. *ISPRS J. Photogramm. Remote Sens.* **2022**, *188*, 45–61. [CrossRef]
21. Ao, Z.; Su, Y.; Li, W.; Guo, Q.; Zhang, J. One-Class Classification of Airborne LiDAR Data in Urban Areas Using a Presence and Background Learning Algorithm. *Remote Sens.* **2017**, *9*, 1001. [CrossRef]
22. Du, J.; Cai, G.; Wang, Z.; Huang, S.; Su, J.; Marcato Junior, J.; Smit, J.; Li, J. ResDLPS-Net: Joint residual-dense optimization for large-scale point cloud semantic segmentation. *ISPRS J. Photogramm. Remote Sens.* **2021**, *182*, 37–51. [CrossRef]
23. Xiu, H.; Liu, X.; Wang, W.; Kim, K.S.; Shinohara, T.; Chang, Q.; Matsuoka, M. Enhancing local feature learning for 3D point cloud processing using unary-pairwise attention. In Proceedings of the 32nd British Machine Vision Conference, Online, 22–25 November 2021.
24. Luo, H.; Khoshelham, K.; Chen, C.; He, H. Individual tree extraction from urban mobile laser scanning point clouds using deep pointwise direction embedding. *ISPRS J. Photogramm. Remote Sens.* **2021**, *175*, 326–339. [CrossRef]
25. He, D.; Abid, F.; Kim, Y.-M.; Kim, J.-H. SectorGSnet: Sector learning for efficient ground segmentation of outdoor LiDAR point clouds. *IEEE Access* **2022**, *10*, 11938–11946. [CrossRef]
26. Zou, X.; Cheng, M.; Wang, C.; Xia, Y.; Li, J. Tree classification in complex forest point clouds based on deep learning. *IEEE Geosci. Remote Sens. Lett.* **2017**, *14*, 2360–2364. [CrossRef]
27. Qi, C.R.; Su, H.; Mo, K.C.; Guibas, L.J. PointNet: Deep learning on point sets for 3D classification and segmentation. In Proceedings of the 2017 IEEE Conference on Computer Vision and Pattern Recognition (CVPR), Honolulu, HI, USA, 21–26 July 2017; pp. 77–85. [CrossRef]
28. Wang, Y.; Sun, Y.; Liu, Z.; Sarma, S.E.; Bronstein, M.M.; Solomon, J.M. Dynamic graph CNN for learning on point clouds. *ACM Trans. Graph.* **2019**, *38*, 1–12. [CrossRef]
29. Wang, L.; Huang, Y.C.; Hou, Y.L.; Zhang, S.M.; Shan, J. Graph attention convolution for point cloud semantic segmentation. In Proceedings of the 2019 IEEE/CVF Conference on Computer Vision and Pattern Recognition (CVPR), Long Beach, CA, USA, 15–20 June 2019; pp. 10288–10297. [CrossRef]
30. Hu, Q.; Yang, B.; Xie, L.; Rosa, S.; Guo, Y.; Wang, Z.; Trigoni, N.; Markham, A. Randla-Net: Efficient semantic segmentation of large-scale point clouds. In Proceedings of the IEEE/CVF Conference on Computer Vision and Pattern Recognition, Seattle, WA, USA, 13–19 June 2020; pp. 11108–11117.
31. Zhang, W.; Su, S.; Wang, B.; Hong, Q.; Sun, L. Local K-NNs pattern in omni-direction graph convolution neural network for 3D point clouds. *Neurocomputing* **2020**, *413*, 487–498. [CrossRef]
32. Sheikh, M.; Asghar, M.A.; Bibi, R.; Malik, M.N.; Shorfuzzaman, M.; Mehmood, R.M.; Kim, S.-H. DFT-Net: Deep feature transformation based network for object categorization and part segmentation in 3-dimensional point clouds. *Sensors* **2022**, *22*, 2512. [CrossRef]
33. Li, L.; Kong, X.; Zhao, X.; Huang, T.; Li, W.; Wen, F.; Zhang, H.; Liu, Y. RINet: Efficient 3D Lidar-based place recognition using rotation invariant neural network. *IEEE Robot. Autom. Lett.* **2022**, *7*, 4321–4328. [CrossRef]
34. Silva, J.; Pereira, P.; Machado, R.; Névoa, R.; Melo-Pinto, P.; Fernandes, D. Customizable FPGA-based hardware accelerator for standard convolution processes empowered with quantization applied to LiDAR data. *Sensors* **2022**, *22*, 2184. [CrossRef]
35. Xu, Y.; Lin, J.; Shi, J.; Zhang, G.; Wang, X.; Li, H. Robust self-supervised LiDAR odometry via representative structure discovery and 3D inherent error modeling. *IEEE Robot. Autom. Lett.* **2022**, *7*, 1651–1658. [CrossRef]
36. Nunes, L.; Marcuzzi, R.; Chen, X.; Behley, J.; Stachniss, C. SegContrast: 3D point cloud feature representation learning through self-supervised segment discrimination. *IEEE Robot. Autom. Lett.* **2022**, *7*, 2116–2123. [CrossRef]
37. He, K.; Fan, H.; Wu, Y.; Xie, S.; Girshick, R. Momentum contrast for unsupervised visual representation learning. In Proceedings of the 2020 IEEE/CVF Conference on Computer Vision and Pattern Recognition (CVPR), Seattle, WA, USA, 14–19 June 2020; pp. 9726–9735. [CrossRef]
38. Wang, B.; Ma, G.; Zhu, M. Fast momentum contrast learning for unsupervised person re-identification. *IEEE Signal Process. Lett.* **2021**, *28*, 2073–2077. [CrossRef]
39. Huang, J.; Yuan, J.; Qiao, C. Generation for unsupervised domain adaptation: A Gan-based approach for object classification with 3D point cloud data. In Proceedings of the ICASSP 2022–2022 IEEE International Conference on Acoustics, Speech and Signal Processing (ICASSP), Singapore, 22–27 May 2022; pp. 3753–3757. [CrossRef]
40. Toth, C.K. The strip adjustment and registration. In *Topographic Laser Ranging and Scanning Principles and Processing*; Shan, J., Toth, C.K., Eds.; CRC Press: Boca Raton, FL, USA, 2008; pp. 254–2287.
41. Wen, C.; Yang, L.; Li, X.; Peng, L.; Chi, T. Directionally constrained fully convolutional neural network for airborne LiDAR point cloud classification. *ISPRS J. Photogramm. Remote Sens.* **2020**, *162*, 50–62. [CrossRef]
42. Fang, L.; Sun, T.; Wang, S.; Fan, H.; Li, J. A graph attention network for road marking classification from mobile LiDAR point clouds. *Int. J. Appl. Earth Obs. Geoinf.* **2022**, *108*, 102735. [CrossRef]
43. Murray, X.; Apan, A.; Deo, R.; Maraseni, T. Rapid assessment of mine rehabilitation areas with airborne LiDAR and deep learning: Bauxite strip mining in Queensland, Australia. *Geocarto Int.* **2022**, 1–24. [CrossRef]

44. Cao, D.; Xing, H.; Wong, M.S.; Kwan, M.-P.; Xing, H.; Meng, Y.A. Stacking ensemble deep learning model for building extraction from remote sensing images. *Remote Sens.* **2021**, *13*, 3898. [CrossRef]
45. Shirmard, H.; Farahbakhsh, E.; Müller, R.D.; Chandra, R. A review of machine learning in processing remote sensing data for mineral exploration. *Remote Sens. Environ.* **2022**, *268*, 112750. [CrossRef]
46. Nahhas, F.H.; Shafri, H.Z.M.; Sameen, M.I.; Pradhan, B.; Mansor, S. Deep learning approach for building detection using LiDAR–orthophoto fusion. *Hindawi J. Sens.* **2018**, *2018*, 7212307. [CrossRef]
47. Zhang, P.; He, H.; Wang, Y.; Liu, Y.; Lin, H.; Guo, L.; Yang, W. 3D urban buildings extraction based on airborne lidar and photogrammetric point cloud fusion according to U-Net deep learning model segmentation. *IEEE Access* **2022**, *10*, 20889–20897. [CrossRef]
48. Shi, C.; Li, J.; Gong, J.; Yang, B.; Zhang, G. An improved lightweight deep neural network with knowledge distillation for local feature extraction and visual localization using images and LiDAR point clouds. *ISPRS J. Photogramm. Remote Sens.* **2022**, *184*, 177–188. [CrossRef]
49. Pu, R. *Hyperspectral Remote Sensing: Fundamentals and Practices*; Taylor & Francis Group: Boca Raton, FL, USA, 2017; p. 466; ISBN 9781498731591.
50. Marrs, J.; Ni-Meister, W. Machine learning techniques for tree species classification using co-registered LiDAR and hyperspectral data. *Remote Sens.* **2019**, *11*, 819. [CrossRef]
51. Yu, X.; Hyyppä, J.; Litkey, P.; Kaartinen, H.; Vastaranta, M.; Holopainen, M. Single-sensor solution to tree species classification using multispectral airborne laser scanning. *Remote Sens.* **2017**, *9*, 108. [CrossRef]
52. Zhao, P.; Guan, H.; Li, D.; Yu, Y.; Wang, H.; Gao, K.; Junior, J.M.; Li, J. Airborne multispectral LiDAR point cloud classification with a feature Reasoning-based graph convolution network. *Int. J. Appl. Earth Obs. Geoinf.* **2021**, *105*, 102634. [CrossRef]
53. Zhou, Q.; Yu, L.; Zhang, X.; Liu, Y.; Zhan, Z.; Ren, L.; Luo, Y. Fusion of UAV Hyperspectral Imaging and LiDAR for the Early Detection of EAB Stress in Ash and a New EAB Detection Index—NDVI $_{(776,678)}$. *Remote Sens.* **2022**, *14*, 2428. [CrossRef]
54. Peng, Y.; Zhang, Y.; Tu, B.; Zhou, C.; Li, Q. Multiview Hierarchical Network for Hyperspectral and LiDAR Data Classification. *IEEE J. Sel. Top. Appl. Earth Obs. Remote Sens.* **2022**, *15*, 1454–1469. [CrossRef]
55. Stilla, U.; Jutzi, B. Waveform analysis for small-footprint pulsed Laser systems. In *Topographic Laser Ranging and Scanning Principles and Processing*; Shan, J., Toth, C.K., Eds.; CRC Press: Boca Raton, FL, USA, 2008; pp. 234–253.
56. Guan, H.; Yu, Y.; Ji, Z.; Li, J.; Zhang, Q. Deep learning-based tree classification using mobile Lidar data. *Remote Sens. Lett.* **2015**, *6*, 864–873. [CrossRef]
57. Blomley, R.; Hovi, A.; Weinmann, M.; Hinz, S.; Korpela, I.; Jutzi, B. Tree species classification using within crown localization of waveform LiDAR attributes. *ISPRS J. Photogramm. Remote Sens.* **2017**, *133*, 142–156. [CrossRef]
58. Yang, G.; Zhao, Y.; Li, B.; Ma, Y.; Li, R.; Jing, J.; Dian, Y. Tree species classification by employing multiple features acquired from integrated sensors. *J. Sens.* **2019**, *2019*, 1–12. [CrossRef]
59. Shinohara, T.; Xiu, H.; Matsuoka, M. FWNet: Semantic segmentation for full-waveform LiDAR data using deep learning. *Sensors* **2020**, *20*, 3568. [CrossRef]
60. Shin, Y.H.; Son, K.W.; Lee, D.C. Semantic segmentation and building extraction from airborne LiDAR data with multiple return using PointNet++. *Appl. Sci.* **2022**, *12*, 1975. [CrossRef]
61. Qi, C.R.; Yi, L.; Su, H.; Guibas, L.J. Pointnet++: Deep hierarchical feature learning on point sets in a metric space. *Proc. Adv. Neural Inf. Process. Syst.* **2017**, *30*, 5099–5108.
62. Zhang, Q.; Ge, L.; Hensley, S.; Metternicht, G.I.; Liu, C.; Zhang, R. PolGAN: A deep-learning-based unsupervised forest height estimation based on the synergy of PolInSAR and LiDAR data. *ISPRS J. Photogramm. Remote Sens.* **2022**, *186*, 123–139. [CrossRef]
63. Park, Y.; Guldmann, J.M. Creating 3D city models with building footprints and LiDAR point cloud classification: A machine learning approach. *Comput. Environ. Urban Syst.* **2019**, *75*, 76–89. [CrossRef]
64. Feng, C.C.; Guo, Z. Automating parameter learning for classifying terrestrial LiDAR point cloud using 2D land cover maps. *Remote Sens.* **2018**, *10*, 1192. [CrossRef]
65. Schmohl, S.; Narváez Vallejo, A.; Soergel, U. Individual tree detection in urban ALS point clouds with 3D convolutional networks. *Remote Sens.* **2022**, *14*, 1317. [CrossRef]
66. Kogut, T.; Tomczak, A.; Słowik, A.; Oberski, T. Seabed modelling by means of airborne laser bathymetry data and imbalanced learning for offshore mapping. *Sensors* **2022**, *22*, 3121. [CrossRef] [PubMed]
67. Barbarella, M.; Di Benedetto, A.; Fiani, M. Application of Supervised Machine Learning Technique on LiDAR Data for Monitoring Coastal Land Evolution. *Remote Sens.* **2021**, *13*, 4782. [CrossRef]
68. Duran, Z.; Ozcan, K.; Atik, M.E. Classification of photogrammetric and airborne LiDAR point clouds using machine learning algorithms. *Drones* **2021**, *5*, 104. [CrossRef]
69. Mohammed, M.; Badruddin Khan, M.; Bashier, E.B.M. *Machine Learning Algorithms and Applications*, 1st ed.; CRC Press: Boca Raton, FL, USA, 2016; p. 226. [CrossRef]
70. Kim, P. *MATLAB Deep Learning with Machine Learning, Neural Networks and Artificial Intelligence*; Apress: Berkeley, CA, USA, 2017; p. 151. [CrossRef]
71. Mirzaei, K.; Arashpour, M.; Asadi, E.; Masoumi, H.; Bai, Y.; Behnood, A. 3D point cloud data processing with machine learning for construction and infrastructure applications: A comprehensive review. *Adv. Eng. Inform.* **2022**, *51*, 101501; ISSN 1474-0346. [CrossRef]

72. Maturana, D.; Scherer, S. Voxnet: A 3D convolutional neural network for real-time object recognition. In Proceedings of the 2015 IEEE/RSJ International Conference on Intelligent Robots and Systems (IROS), Hamburg, Germany, 28 September–3 October 2015; pp. 922–928. [CrossRef]
73. Gargoum, S.A.; Koch, J.C.; El-Basyouny, K. A voxel-based method for automated detection and mapping of light poles on rural highways using lidar data. *Transp. Res. Rec.* **2018**, *2672*, 274–283. [CrossRef]
74. Shuang, F.; Huang, H.; Li, Y.; Qu, R.; Li, P. AFE-RCNN: Adaptive feature enhancement RCNN for 3D object detection. *Remote Sens.* **2022**, *14*, 1176. [CrossRef]
75. Wijaya, K.T.; Paek, D.; Kong, S.H. Multiview attention for 3D object detection in Lidar point cloud. In Proceedings of the 2022 International Conference on Artificial Intelligence in Information and Communication (ICAIIC), Jeju Island, Korea, 21–24 February 2022; pp. 210–215. [CrossRef]
76. Simonovsky, M.; Komodakis, N. Dynamic edge-conditioned filters in convolutional neural networks on graphs. In Proceedings of the 2017 IEEE Conference on Computer Vision and Pattern Recognition (CVPR), Honolulu, HI, USA, 21–26 July 2017; pp. 29–38. [CrossRef]
77. Wang, C.; Samari, B.; Siddiqi, K. Local spectral graph convolution for point set feature learning. In *Computer Vision—ECCV 2018. Lecture Notes in Computer Science*; Ferrari, V., Hebert, M., Sminchisescu, C., Weiss, Y., Eds.; Springer: Cham, Switzerland, 2018; p. 11208. [CrossRef]
78. Li, D.; Shen, X.; Guan, H.; Yu, Y.; Wang, H.; Zhang, G.; Li, J.; Li, D. AGFP-Net: Attentive geometric feature pyramid network for land cover classification using airborne multispectral LiDAR data. *Int. J. Appl. Earth Obs. Geoinf.* **2022**, *108*, 102723; ISSN 0303-2434. [CrossRef]
79. Wen, C.; Li, X.; Yao, X.; Peng, L.; Chi, T. Airborne LiDAR point cloud classification with global-local graph attention convolution neural network. *ISPRS J. Photogramm. Remote Sens.* **2021**, *173*, 181–194. [CrossRef]
80. Jing, W.; Zhang, W.; Li, L.; Di, D.; Chen, G.; Wang, J. AGNet: An attention-based graph network for point cloud classification and segmentation. *Remote Sens.* **2022**, *14*, 1036. [CrossRef]
81. Chen, Y.; Luo, Z.; Li, W.; Lin, H.; Nurunnabi, A.; Lin, Y.; Wang, C.; Zhang, X.-P.; Li, J. WGNet: Wider graph convolution networks for 3D point cloud classification with local dilated connecting and context-aware. *Int. J. Appl. Earth Obs. Geoinf.* **2022**, *110*, 102786. [CrossRef]
82. Defferrard, M.; Bresson, X.; Vandergheynst, P. Convolutional neural networks on graphs with fast localized spectral filtering. In Proceedings of the Advances in Neural Information Processing Systems, Barcelona, Spain, 5–10 December 2016; pp. 3844–3852. [CrossRef]
83. Wan, J.; Xie, Z.; Xu, Y.; Zeng, Z.; Yuan, D.; Qiu, Q. DGANet: A dilated graph attention-based network for local feature extraction on 3D point clouds. *Remote Sens.* **2021**, *13*, 3484. [CrossRef]
84. Hua, B.; Tran, M.; Yeung, S. Pointwise convolutional neural networks. In Proceedings of the 2018 IEEE/CVF Conference on Computer Vision and Pattern Recognition (CVPR), Salt Lake City, UT, USA, 18–23 June 2018; pp. 984–993. [CrossRef]
85. Klokov, R.; Lempitsky, V. Escape from cells: Deep KD-networks for the recognition of 3D point cloud models. In Proceedings of the 2017 IEEE International Conference on Computer Vision (ICCV), Venice, Italy, 22–29 October 2017; pp. 863–872. [CrossRef]
86. Song, Y.; He, F.; Duan, Y.; Liang, Y.; Yan, X. A kernel correlation-based approach to adaptively acquire local features for learning 3D point clouds. *Comput.-Aided Des.* **2022**, *146*, 103196, ISSN 0010-4485. [CrossRef]
87. Tarsha Kurdi, F.; Amakhchan, W.; Gharineiat, Z. Random Forest machine learning technique for automatic vegetation detection and modelling in LiDAR data. *Int. J. Environ. Sci. Nat. Resour.* **2021**, *28*, 556234. [CrossRef]
88. Chen, W.; Li, X.; Wang, Y.; Chen, G.; Liu, S. Forested landslide detection using LiDAR data and the random forest algorithm: A case study of the three Gorges, China. *Remote Sens. Environ.* **2014**, *152*, 291–301. [CrossRef]
89. Huang, R.; Zhu, J. Using Random Forest to integrate LiDAR data and hyperspectral imagery for land cover classification. In Proceedings of the 2013 IEEE International Geoscience and Remote Sensing Symposium—IGARSS, Melbourne, Australia, 21–26 July 2013; pp. 3978–3981. [CrossRef]
90. Man, Q.; Dong, P.; Yang, X.; Wu, Q.; Han, R. Automatic extraction of grasses and individual trees in urban areas based on airborne hyperspectral and LiDAR data. *Remote Sens.* **2020**, *12*, 2725. [CrossRef]
91. Yu, X.; Hyyppä, J.; Vastaranta, M.; Holopainen, M.; Viitala, R. Predicting individual tree attributes from airborne laser point clouds based on the random forests technique. *ISPRS J. Photogramm. Remote Sens.* **2011**, *66*, 28–37. [CrossRef]
92. Levick, S.R.; Hessenmöller, D.; Schulze, E.D. Scaling wood volume estimates from inventory plots to landscapes with airborne LiDAR in temperate deciduous forest. *Carbon Balance Manag.* **2016**, *11*, 7. [CrossRef]
93. Guan, H.; Yu, J.; Li, J.; Luo, L. Random forests-based feature selection for land-use classification using LiDAR data and orthoimagery. *Int. Arch. Photogramm. Remote Sens. Spatial Inf. Sci.* **2012**, *XXXIX-B7*, 203–208. [CrossRef]
94. Ba, A.; Laslier, M.; Dufour, S.; Hubert-Moy, L. Riparian trees genera identification based on leaf-on/leaf-off airborne laser scanner data and machine learning classifiers in northern France. *Int. J. Remote Sens.* **2020**, *41*, 1645–1667. [CrossRef]
95. Arumäe, T.; Lang, M.; Sims, A.; Laarmann, D. Planning of commercial thinnings using machine learning and airborne lidar data. *Forests* **2022**, *13*, 206. [CrossRef]
96. Dong, Y.; Li, Y.; Hou, M. The point cloud semantic segmentation method for the Ming and Qing dynasties' official-style architecture roof considering the construction regulations. *ISPRS Int. J. Geo-Inf.* **2022**, *11*, 214. [CrossRef]

97. Liao, L.; Tang, S.; Liao, J.; Li, X.; Wang, W.; Li, Y.; Guo, R. A Supervoxel-based random forest method for robust and effective airborne LiDAR point cloud classification. *Remote Sens.* **2022**, *14*, 1516. [CrossRef]
98. Hoang, L.; Lee, S.H.; Kwon, K.R. A 3D shape recognition method using hybrid deep learning network CNN–SVM. *Electronics* **2020**, *9*, 649. [CrossRef]
99. Zhang, J.; Lin, X.; Ning, X. SVM-based classification of segmented airborne LiDAR point clouds in urban areas. *Remote Sens.* **2013**, *5*, 3749–3775. [CrossRef]
100. Shokri, D.; Rastiveis, H.; Sheikholeslami, S.M.; Shah-hosseini, R.; Li, J. Fast extraction of power lines from mobile LiDAR point clouds based on SVM classification in non-urban area. *Earth Obs. Geomat. Eng.* **2022**, *5*, 63–73. [CrossRef]
101. Goodfellow, I.; Pouget-Abadie, J.; Mirza, M.; Xu, B.; Warde-Farley, D.; Ozair, S.; Courville, A.; Bengio, Y. Generative adversarial nets. Advances in neural information processing systems. In Proceedings of the International Conference on Neural Information Processing Systems (NIPS 2014), Montreal, QC, Canada, 8–13 December 2014; Volume 27, pp. 2672–2680.
102. Wei, J.; Lin, G.; Yap, K.H.; Hung, T.Y.; Xie, L. Multi-path region mining for weakly supervised 3D semantic segmentation on point clouds. In Proceedings of the 2020 IEEE/CVF Conference on Computer Vision and Pattern Recognition (CVPR), Seattle, WA, USA, 13–19 June 2020; pp. 4383–4392. [CrossRef]
103. Li, N.; Kähler, O.; Pfeifer, N. A comparison of deep learning methods for airborne Lidar point clouds classification. *IEEE J. Sel. Top. Appl. Earth Obs. Remote Sens.* **2021**, *14*, 6467–6486. [CrossRef]
104. Graham, B.; Engelcke, M.; Maaten, L.V.D. 3D semantic segmentation with submanifold sparse convolutional networks. In Proceedings of the 2018 IEEE/CVF Conference on Computer Vision and Pattern Recognition, Salt Lake City, UT, USA, 18–23 June 2018; pp. 9224–9232. [CrossRef]
105. Thomas, H.; Qi, C.R.; Deschaud, J.E.; Marcotegui, B.; Goulette, F.; Guibas, L.J. KPConv: Flexible and deformable convolution for point clouds. In Proceedings of the 2019 IEEE/CVF International Conference on Computer Vision (ICCV), Seoul, Korea, 27 October–2 November 2019; pp. 6411–6420. [CrossRef]
106. Théodose, R.; Denis, D.; Chateau, T.; Frémont, V.; Checchin, P. A deep learning approach for LiDAR resolution-agnostic object detection. *IEEE Trans. Intell. Transp. Syst.* **2022**, *23*, 14582–14593. [CrossRef]
107. Hoang, L.; Lee, S.H.; Lee, E.J.; Kwon, K.R. GSV-NET: A Multi-modal deep learning network for 3D point cloud classification. *Appl. Sci.* **2022**, *12*, 483. [CrossRef]
108. Chen, J.; Kakillioglu, B.; Velipasalar, S. Background-aware 3-D point cloud segmentation with dynamic point feature aggregation. *IEEE Trans. Geosci. Remote Sens.* **2022**, *60*, 1–12. [CrossRef]
109. Song, W.; Li, D.; Sun, S.; Zhang, L.; Xin, Y.; Sung, Y.; Choi, R. 2D&3DHNet for 3D object classification in LiDAR point cloud. *Remote Sens.* **2022**, *14*, 3146.
110. Badrinarayanan, V.; Kendall, A.; Cipolla, R. SegNet: A deep convolutional encoder-decoder architecture for image segmentation. *IEEE Trans. Pattern Anal. Mach. Intell.* **2017**, *39*, 2481–2495. [CrossRef] [PubMed]
111. Medina, F.P.; Paffenroth, R. Machine learning in LiDAR 3D point clouds. In *Advances in Data Science*; Association for Women in Mathematics Series, 26; Demir, I., Lou, Y., Wang, X., Welker, K., Eds.; Springer: Cham, Switzerland, 2021. [CrossRef]
112. Wu, W.; Qi, Z.; Fuxin, L. Pointconv: Deep convolutional networks on 3D point clouds. In Proceedings of the IEEE/CVF Conference on Computer Vision and Pattern Recognition (CVPR), Long Beach, CA, USA, 15–20 June 2019; pp. 9613–9622. [CrossRef]
113. Ibrahim, M.; Akhtar, N.; Ullah, K.; Mian, A. Exploiting Structured CNNs for Semantic Segmentation of Unstructured Point Clouds from LiDAR Sensor. *Remote Sens.* **2021**, *13*, 3621. [CrossRef]
114. Hamedianfar, A.; Mohamedou, C.; Kangas, A.; Vauhkonen, J. Deep learning for forest inventory and planning: Acritical review on the remote sensing approaches so far and prospects for further applications. *Forestry* **2022**, *95*, 451–465. [CrossRef]
115. Ronneberger, O.; Fischer, P.; Brox, T. U-Net: Convolutional networks for biomedical image segmentation. In Proceedings of the International Conference on Medical Image Computing and Computer-Assisted Intervention, Munich, Germany, 5–9 October 2015; pp. 234–241, LNCS. 9351. [CrossRef]
116. Ojogbane, S.S.; Mansor, S.; Kalantar, B.; Khuzaimah, Z.B.; Shafri, H.Z.M.; Ueda, N. Automated building detection from airborne LiDAR and very high-resolution aerial imagery with deep neural network. *Remote Sens.* **2021**, *13*, 4803. [CrossRef]
117. Seydi, S.T.; Hasanlou, M.; Amani, M. A new end-to-end multi-dimensional CNN framework for land cover/land use change detection in multi-source remote sensing datasets. *Remote Sens.* **2020**, *12*, 2010. [CrossRef]
118. Wang, Q.; Gu, Y. A Discriminative tensor representation model for feature extraction and classification of multispectral LiDAR data. *IEEE Trans. Geosci. Remote Sens.* **2020**, *58*, 1568–1586. [CrossRef]
119. Ekhtari, N.; Glennie, C.; Fernandez-Diaz, J.C. Classification of airborne multispectral lidar point clouds for land cover mapping. *IEEE J. Sel. Top. Appl. Earth Obs. Remote Sens.* **2018**, *11*, 2068–2078. [CrossRef]
120. Zhao, G.; Zhang, W.; Peng, Y.; Wu, H.; Wang, Z.; Cheng, L. PEMCNet: An efficient multi-scale point feature fusion network for 3D LiDAR point cloud classification. *Remote Sens.* **2021**, *13*, 4312. [CrossRef]
121. Wen, S.; Wang, T.; Tao, S. Hybrid CNN-LSTM architecture for LiDAR point clouds semantic segmentation. *IEEE Robot. Autom. Lett.* **2022**, *7*, 5811–5818. [CrossRef]
122. Shuang, F.; Li, P.; Li, Y.; Zhang, Z.; Li, X. MSIDA-Net: Point Cloud Semantic Segmentation via Multi-Spatial Information and Dual Adaptive Blocks. *Remote Sens.* **2022**, *14*, 2187. [CrossRef]

123. Chen, M.; Qiu, X.; Zeng, W.; Peng, D. Combining sample plot stratification and machine learning algorithms to improve forest aboveground carbon density estimation in northeast China using airborne LiDAR data. *Remote Sens.* **2022**, *14*, 1477. [CrossRef]
124. Luo, Z.; Zhang, Z.; Li, W.; Chen, Y.; Wang, C.; Nurunnabi, A.; Li, J. Detection of individual trees in UAV LiDAR point clouds using a deep learning framework based on multichannel representation. *IEEE Trans. Geosci. Remote Sens.* **2022**, *60*, 1–15. [CrossRef]
125. Corte, A.P.D.; Souza, D.V.; Rex, F.E.; Sanquetta, C.R.; Mohan, M.; Silva, C.A.; Zambrano, A.M.A.; Prata, G.; Alves de Almeida, D.R.; Trautenmüller, J.W.; et al. Forest inventory with high-density UAV-Lidar: Machine learning approaches for predicting individual tree attributes. *Comput. Electron. Agric.* **2020**, *179*, 105815. [CrossRef]
126. Windrim, L.; Bryson, M. Detection, segmentation, and model fitting of individual tree stems from airborne laser scanning of forests using deep learning. *Remote Sens.* **2020**, *12*, 1469. [CrossRef]
127. Chen, X.; Jiang, K.; Zhu, Y.; Wang, X.; Yun, T. Individual tree crown segmentation directly from UAV-borne LiDAR data using the PointNet of deep learning. *Forests* **2021**, *12*, 131. [CrossRef]
128. Mizoguchi, T.; Ishii, A.; Nakamura, H.; Inoue, T.; Takamatsu, H. *Lidar-Based Individual Tree Species Classification Using Convolutional Neural Network*; Proc. SPIE 10332, Videometrics, Range Imaging, and Applications XIV; SPIE Optical Metrology: Munich, Germany, 2017; p. 103320O. [CrossRef]
129. Budei, B.; St-Onge, B.; Hopkinson, C.; Audet, F.A. Identifying the genus or species of individual trees using a three-wavelength airborne Lidar system. *Remote Sens. Environ.* **2018**, *204*, 632–647. [CrossRef]
130. Nguyen, H.; Demir, B.; Dalponte, M. Weighted support vector machines for tree species classification using Lidar data. In Proceedings of the IGARSS 2019–2019 IEEE International Geoscience and Remote Sensing Symposium, Yokohama, Japan, 28 July–2 August 2019; pp. 6740–6743. [CrossRef]
131. Hell, M.; Brandmeier, M.; Briechle, S.; Krzystek, P. Classification of tree species and standing dead trees with Lidar point clouds using two deep neural networks: PointCNN and 3DmFV-Net. *PFG* **2022**, *90*, 103–121. [CrossRef]
132. Li, Y.; Bu, R.; Sun, M.; Wu, W.; Di, X.; Chen, B. PointCNN: Convolution on x-transformed points. In Proceedings of the Advances in Neural Information processing systems 31 (NIPS 2018), Montreal, QC, Canada, 3–8 December 2018; pp. 820–830. [CrossRef]
133. Ben-Shabat, Y.; Lindenbaum, M.; Fischer, A. 3DmFV: Three-dimensional point cloud classification in real-time using convolutional neural networks. *IEEE Robot Autom. Lett.* **2018**, *3*, 3145–3152. [CrossRef]
134. Wen, C.; Sun, X.; Li, J.; Wang, C.; Guo, Y.; Habib, A. A deep learning framework for road marking extraction, classification and completion from mobile laser scanning point clouds. *ISPRS J. Photogramm. Remote Sens.* **2019**, *147*, 178–192. [CrossRef]
135. Ma, L.; Li, Y.; Li, J.; Yu, Y.; Junior, J.M.; Goncalves, W.N.; Chapman, M.A. Capsule-based networks for road marking extraction and classification from mobile LiDAR point clouds. *IEEE Trans. Intell. Transp. Syst.* **2021**, *22*, 1981–1995. [CrossRef]
136. Ma, H.; Ma, H.; Zhang, L.; Liu, K.; Luo, W. Extracting urban road footprints from airborne LiDAR point clouds with PointNet++ and two-step post-processing. *Remote Sens.* **2022**, *14*, 789. [CrossRef]
137. Shajahan, D.A.; Nayel, V.; Muthuganapathy, R. Roof classification from 3-D LiDAR point clouds using multiview CNN with self-attention. *IEEE Geosci. Remote Sens. Lett.* **2020**, *17*, 1465–1469. [CrossRef]
138. Silva, A.; Fernandes, D.; Névoa, R.; Monteiro, J.; Novais, P.; Girão, P.; Afonso, T.; Melo-Pinto, P. Resource-constrained onboard inference of 3D object detection and localisation in point clouds targeting self-driving applications. *Sensors* **2021**, *21*, 7933. [CrossRef]
139. Lee, Y.; Park, S. A Deep Learning-Based Perception Algorithm Using 3D LiDAR for Autonomous Driving: Simultaneous Segmentation and Detection Network (SSADNet). *Appl. Sci.* **2020**, *10*, 4486. [CrossRef]
140. Kang, D.; Wong, A.; Lee, B.; Kim, J. Real-time semantic segmentation of 3D point cloud for autonomous driving. *Electronics* **2021**, *10*, 1960. [CrossRef]
141. Sun, Y.; Zuo, W.; Huang, H.; Cai, P.; Liu, M. PointMoSeg: Sparse Tensor-Based End-to-End Moving-Obstacle Segmentation in 3-D Lidar Point Clouds for Autonomous Driving. *IEEE Robot. Autom. Lett.* **2022**, *6*, 510–517. [CrossRef]
142. Peng, K.; Fei, J.; Yang, K.; Roitberg, A.; Zhang, J.; Bieder, F.; Heidenreich, P.; Stiller, C.; Stiefelhagen, R. MASS: Multi-Attentional Semantic Segmentation of LiDAR Data for Dense Top-View Understanding. *IEEE Trans. Intell. Transp. Syst.* **2022**, *23*, 15824–15840. [CrossRef]
143. Gao, R.; Li, M.; Yang, S.J.; Cho, K. Reflective noise filtering of large-scale point cloud using transformer. *Remote Sens.* **2022**, *14*, 577. [CrossRef]
144. Nurunnabi, A.; Teferle, F.N.; Li, J.; Lindenbergh, R.C.; Hunegnaw, A. An efficient deep learning approach for ground point filtering in aerial laser scanning point clouds. *Int. Arch. Photogramm. Remote Sens. Spatial Inf. Sci.* **2021**, *XLIII-B1-2021*, 31–38. [CrossRef]
145. Cao, Y.; Scaioni, M. 3DLEB-Net: Label-efficient deep learning-based semantic segmentation of building point clouds at LoD3 level. *Appl. Sci.* **2021**, *11*, 8996. [CrossRef]
146. Zhang, J.; Xiao, W.; Mills, J.P. Optimizing moving object trajectories from roadside Lidar data by joint detection and tracking. *Remote Sens.* **2022**, *14*, 2124. [CrossRef]
147. Shi, S.; Guo, C.; Jiang, L.; Wang, Z.; Shi, J.; Wang, X.; Li, H. PV-RCNN: Point-voxel feature set abstraction for 3D object detection. In Proceedings of the IEEE/CVF Conference on Computer Vision and Pattern Recognition (CVPR), Seattle, WA, USA, 13–19 June 2020; pp. 10526–10535. [CrossRef]
148. Yin, C.; Cheng, J.C.P.; Wang, B.; Gan, V.J.L. Automated classification of piping components from 3D LiDAR point clouds using SE-PseudoGrid. *Autom. Constr.* **2022**, *139*, 104300. [CrossRef]

149. Amakhchan, W.; Tarsha Kurdi, F.; Gharineiat, Z.; Boulaassal, H.; El Kharki, O. Automatic filtering of LiDAR building point cloud using multilayer perceptron Neuron Network. In Proceedings of the 3rd International Conference on Big Data and Machine Learning (BML'22), Istanbul, Turkey, 22–23 December 2022. Available online: https://bml.maasi.org/ (accessed on 9 August 2022).
150. Mammoliti, E.; Di Stefano, F.; Fronzi, D.; Mancini, A.; Malinverni, E.S.; Tazioli, A.A. Machine learning approach to extract rock mass discontinuity orientation and spacing, from laser scanner point clouds. *Remote Sens.* **2022**, *14*, 2365. [CrossRef]

MDPI
St. Alban-Anlage 66
4052 Basel
Switzerland
www.mdpi.com

Remote Sensing Editorial Office
E-mail: remotesensing@mdpi.com
www.mdpi.com/journal/remotesensing

www.ingramcontent.com/pod-product-compliance
Lightning Source LLC
Chambersburg PA
CBHW061848210926
51850CB00006B/1006